Quantum Mechanics

Bruce Cameron Reed

Quantum Mechanics

An Enhanced Primer

Second Edition

 Springer

Bruce Cameron Reed
Department of Physics (Emeritus)
Alma College
Alma, MI, USA

ISBN 978-3-031-14019-8   ISBN 978-3-031-14020-4 (eBook)
https://doi.org/10.1007/978-3-031-14020-4

This Springer imprint is published by the registered company Springer Nature Switzerland AG
The registered company address is: Gewerbestrasse 11, 6330 Cham, Switzerland

Preface

This book is a revised edition of my *Quantum Mechanics*, which was originally published by Jones & Bartlett in 2008. This new version gestated for longer than I had intended. Following the Jones & Bartlett edition, my professional interests turned to other directions, and the usual responsibilities and commitments of a mid/senior-level academic position kept accumulating. But I always wanted to get back to this topic, which has fascinated me for decades. Not only did I want to refresh myself, but I also wanted to take the opportunity to fix errors and confusing statements that had been pointed out to me by my own students and other diligent readers, add some new problems and sections, improve some of the figures, and incorporate new references. I hope that readers—particularly the students for whom it is intended—enjoy reading and learning from this revised edition as much as I enjoyed preparing it.

The rationale for this book remains the same as it was in 2008: A text directed at the needs of physics and chemistry students at smaller colleges and universities who are encountering their first serious course in quantum mechanics (QM), typically at about the junior level. The curricula of such institutions often leave room for only one semester of QM as opposed to the two or more that are common at larger schools, so such students face the task of trying to absorb in a short time both the underlying physics and mathematical formalisms of quantum theory. My goal is to give students a sense of "where it all leads" before they encounter that full mathematical formalism in, say, a senior-level or first graduate course by providing a treatment of the basic theories and results of wave mechanics that falls intermediate between sophomore-level "modern physics" texts and those typical of more advanced treatments. Students should think of this book as being a primer to help prepare them for more advanced work.

Background preparation assumed here includes advanced calculus (partial differentiation and multiple integration), some exposure to basic concepts of differential equations (particularly the series solution method), vector calculus in Cartesian, spherical, and cylindrical coordinates, and the standard menu of physics courses: mechanics, electricity and magnetism, and modern physics. Some previous exposure to the historical context from which quantum mechanics developed is helpful but not strictly necessary; some of this is reviewed in Chap. 1.

I am firmly of the belief that it is only by working through the derivations of a number of classical problems and by doing numerous homework exercises can students come to develop a feeling for the conceptual content of quantum mechanics and the necessary analytic skills involved in applying it. Present-day physics represents the cumulative knowledge of a chain of reasoning and experiment that stretches back over centuries, and I believe that it is important for students to have a sense of how we came to be where we are. Consequently, this book emphasizes the development of exact or approximate analytic solutions of Schrödinger's equation, that is, solutions that can be expressed in algebraic form. To this end, the layout of this book follows a fairly conventional ordering. Chapter 1 reviews some of the developments of early twentieth-century physics that indicated the need for a radical re-thinking of fundamental physical laws on the atomic scale: Planck's quantum hypothesis, the Rutherford–Bohr atomic model, and de Broglie's matter-wave concept. Chapter 2 develops Schrödinger's equation, the fundamental "law" of quantum mechanics. Chapter 3 explores applications of Schrödinger's equation to one-dimensional problems, working up to a treatment of alpha-decay as a barrier-penetration phenomenon. Chapter 4 is a sort of intermission wherein some of the mathematical formalisms of quantum mechanics such as operators, expectation values, the uncertainty principle, Ehrenfest's theorem, the orthogonality theorem, the superposition principle, the virial theorem, and Dirac notation are introduced.

Chapter 5 returns to specific solutions of Schrödinger's equation with an analysis of the harmonic oscillator potential. Extension of Schrödinger's equation to three dimensions appears in Chap. 6, along with a treatment of separation of variables and angular momentum operators for central potentials. This material sets the stage for a rigorous analysis of the Coulomb potential in Chap. 7. Chapter 8 deals with some more advanced aspects of angular momentum and can be considered optional. Chapter 9 explores techniques for undertaking approximate analytic solutions of Schrödinger's equation in situations where full analytic solutions are difficult or impossible to achieve: the WKB method, perturbation theory, and the variational method. In view of the central role of computers in almost all contemporary research, Chap. 10 explores a simple algorithm for numerically integrating Schrödinger's equation with a Microsoft Excel spreadsheet. This comes with an important caveat, however: no physicist, either beginning student or seasoned researcher, should ever let playing with the computer become a substitute for first exploring a problem conceptually and analytically.

The emphasis in this book is on the time-independent Schrödinger equation, but some aspects of time dependence are taken up briefly in Chap. 11. A few particularly lengthy mathematical derivations of limited physical content are gathered together in Appendix A. Answers to a number of the end-of-chapter problems appear in Appendix B. Appendices C and D list a number of useful integrals and physical constants.

The new material in this edition is spread among several chapters. Chapter 1 now includes a section on the classical electromagnetic-radiation atomic collapse problem. A new subsection in Sect. 3.3 describes how the finite rectangular well can be analyzed with matrix algebra, and a short new subsection in Sect. 3.10 gives a more general analysis of resonant scattering from potential wells. Section 4.8 now includes a discussion of the time evolution of the spreading of a Gaussian wavepacket, and Sect. 4.10 explores the concept of momentum-space wavefunctions. In Sect. 5.3, the discussion of the termination condition for Hermite polynomials in the harmonic oscillator problem has been clarified, and a new Appendix, A.2, gives a detailed derivation of the normalization of Hermite polynomials. Several new end-of-chapter problems have been added.

In each chapter, I try to get directly to the essential physics and illustrate it with examples that contain actual numbers and which can serve as vehicles to introduce, where practicable, powerful ancillary techniques such as dimensional analysis and numerical integration. Each chapter contains a number of problems; it would not be unreasonable for students to attempt most of them throughout the course of a semester.

For students, I have three pieces of advice. First, quantum mechanics is unique among the subdisciplines of physics in that so many of its essential concepts seem contrary to experience and intuition. There is no one formula or description via which you can comprehend it quickly. It will take time and thought: mull things over in your mind, discuss them with your classmates and professors, and, when you come to understand something, *write it down in your own words*. In this spirit, I have tried to keep the tone of this work informal while preserving a sensible level of rigor.

Second, a number of problems appear at the end of each chapter. Only by doing them for yourself can you become familiar with the tools of the trade. Problems are classified as elementary (E), intermediate (I), or advanced (A), although these are somewhat arbitrary designations. For handy reference, brief summaries of important concepts and formulae appear before each problem set.

Third, there are likely many quantum mechanics texts in your school's library. Consult them. What may seem opaque as expressed by one author may be clearer in the words of another. One source I have found particularly valuable over the years is *Introduction to Quantum Mechanics with Applications to Chemistry* by Linus Pauling and E. B. Wilson (New York: Dover Publications, 1985). Originally published in 1935 when quantum physics was still new, this venerable work contains detailed analyses of many classic problems. Another work I find appealing for its lucid explanations is *An Introduction to Quantum Physics* by A. P. French and E. F. Taylor (New York: W. W. Norton, 1978). Any serious student of physics should also always have at hand a good reference to the many special functions that crop up in mathematical physics; Hans Weber and George Arfken's *Essential Mathematical Methods for Physicists* (Amsterdam: Elsevier, 2004) can be strongly recommended and is referenced frequently in the present work. Also, a good reference for techniques of numerical analysis is indispensable; a standard work in this area is *Numerical Recipes: The Art of Scientific Computing* by William H. Press, Brian P. Flannery, Saul A. Tuekolsky, and William T. Vetterling (Cambridge, UK: Cambridge University Press, 1986).

Quantum mechanics and its applications are a vibrant, central part of much present-day research. To give students a flavor of the diversity of current work, references to semi-popular articles appearing in Physics Today (abbreviated Phys.

Today), a monthly publication of the American Institute of Physics, appear occasionally throughout the text. These articles are sometimes brief "update" pieces and sometimes feature-length works, but all should be largely accessible to undergraduates and contain references to the original research literature. Students are strongly encouraged to explore them.

Some sections in Chaps. 3, 6, and 7 have subsections. Citations are designated with square brackets, beginning anew with [1] in each chapter.

My interest in quantum mechanics was stimulated in my own student days by a number of excellent teachers at both the University of Waterloo and Queen's University, and has only grown over the intervening years. A number of instructors, students, special friends, colleagues, and family members from those years and on up to the present have supported and encouraged me. It gives me pleasure to especially acknowledge John Altholz, Karen Ball, Dick Bowker, Peter Burns, David Cassidy, John Coster-Mullen, Peter Dawson, Michael DeRobertis, Eugene Deci, Carleen Dewit, Patrick Furlong, John Gibson, Dick Groves, Bob Hayward, Miriam Hiebert, Lorraine Hill, Art Hobson, Lisa Jylänne, Patricia Kinnee, Tim Koeth, Vern Koslowsky, Gilles Labrie, Harry Lustig, Lorne Nelson, John Palimaka, Klaus Rohe, John Schreiner, Ray Smith, Frank Settle, Ute Stargardt, Roger Stuewer, George Wagner, and Pete Zimmerman. Unfortunately, some of these wonderful people are no longer among us, but are fondly remembered: *Requiesce in pace.* Also, I continue to be grateful to individuals who gave generously of their time and expertise in reviewing first-edition chapters of this work prior to publication: Xi Chen (Central College), James Clemens (Miami University), Jospeh Ganem (Loyola College), Noah Graham (Middlebury College), Rick McDaniel (Henderson State University), Soma Mukherjee (University of Texas, Brownsville), David Olsgaard (Simpson College), Vasilis Pagonis (McDaniel College), Harvey Picker (Trinity College), Darrell Schroeter (Occidental College), Blair Tuttle (Pennsylvania State University, Erie), and Ann Wright (Hendrix College). I claim exclusive ownership of any errors that remain.

Thanks would not be complete without a tip of my hat to my editor at Springer, Angela Lahee, who made this book a published reality.

Finally, this work is dedicated to my wife Laurie, who has once again borne with my distraction as "one last tome" (I promise!) occupied my time and thoughts. Our various cats over the years, Fred, Leo, Stella, Cassie, Nyx, and Newton, have amply proven Aldous Huxley's adage that "If you want to write, keep cats."

Bedford, Nova Scotia, Canada
2022

Bruce Cameron Reed

Contents

About the Author

Bruce Cameron Reed is the Charles A. Dana Professor of Physics (Emeritus) at Alma College in Alma, Michigan. During a 35-year career in both Canada and the USA, he taught courses ranging from first-year algebra-based mechanics to senior-level quantum mechanics. He earned a Ph.D. in Physics at the University of Waterloo (Canada). In addition to the present volume, he has authored seven other texts, including five on the World War II Manhattan Project, and has published some 200 journal articles, review papers, book reviews, and semi-popular articles in the areas of astronomy, data analysis, quantum mechanics, nuclear physics, the history of physics, and pedagogical aspects of physics and astronomy. In 2009, he was elected to Fellowship in the American Physical Society (APS) for his work on the history of the Manhattan Project. From 2009 to 2013, he edited the APS's *Physics and Society* newsletter, and subsequently served for six years as Secretary-Treasurer of the Society's Forum on History of Physics. Now formally retired, he lives in Nova Scotia with his wife Laurie and a variable number of cats. He remains busy with teaching classes at local universities, writing papers, consulting, and serving in journal editing and reviewing positions.

Chapter 1
Foundations

Summary This chapter surveys some of the historical issues in physics which prompted the development of quantum mechanics: Planck's blackbody radiation theory, Balmer's empirical expression for the wavelengths of the visible spectral lines of hydrogen, the Rutherford-Bohr planetary atomic model, de Broglie's matter-wave hypothesis, and the "classical" prediction that atoms should collapse over extremely short time scales. By the mid-1920s it was clear that classical concepts of mechanics and electromagnetism would need to be modified to understand atomic structure, particularly the existence of discrete energy states and the wavelike behavior of particles on atomic scales. This work set the stage for the development of quantum mechanics proper in 1925–26.

The tremendous success of Newton's laws of mechanics on scales from projectiles to stars is staggering. These laws also possess a strong intuitive appeal; the relative ease with which even beginning students can assimilate and apply them is due in no small part to the fact that one can often literally see what the mathematics is describing. Arcing baseballs, spinning and colliding billiard balls, rolling tires, flowing water, oscillating pendula and to some extent even orbiting satellites are the stuff of everyday life and common experience; it easy to formulate a picture in your mind's eye around which a quantitative analysis can be developed. Conversely, quantum mechanics deals with the nature of matter at the atomic level, far removed from the domain of practical experience. While it now constitutes the foundation of our understanding of everything from the structure of atoms and molecules to the life cycles of stars, it is also often counterintuitive. Its formulation, use, and interpretation are so vastly different from Newtonian mechanics that questions regarding familiar concepts such as position and velocity largely lose their meaning.

The counterintuitiveness of quantum mechanics renders it practically impossible to develop the subject with the smoothness and logical flow of Newtonian mechanics. To some extent one must simply plunge in and trust that the customs of quantum culture will become familiar as one's experience with it grows. Despite this, however, quantum mechanics is not without its points of contact to classical physics, and these can be exploited to ease the transition to a quantum mode of thinking. The danger in this is that familiar ideas and unquestioned assumptions can be difficult to release;

© The Author(s), under exclusive license to Springer Nature Switzerland AG 2022
B. C. Reed, *Quantum Mechanics*, https://doi.org/10.1007/978-3-031-14020-4_1

there is an art to sensing how far analogies can be extended. Scientific theories, however revolutionary, do not arise in isolation, and classical mechanics formed the background against which the architects of quantum mechanics constructed their edifice. Qualitatively, quantum mechanics can be viewed as a theory which correctly describes the behavior of microscopic material particles but which merges into the predictions of Newtonian mechanics on macroscopic scales.

The purpose of this chapter is to review some of the facts which led physicists in the early part of the twentieth century to the realization that Newtonian mechanics is incomplete in the realm of atomic-scale phenomena. You may well have encountered some of this material previously: the experiments of Michael Faraday and J. J. Thomson (Sect. 1.1), the nature of atomic spectra (Sect. 1.2), the Rutherford-Bohr model of the atom (Sect. 1.3), and wave-particle duality (Sect. 1.4). The intention here is not in-depth explanation but rather the context of early twentieth-century physics and the discrepancies between experiment and theory that forced new developments.

1.1 Faraday, Thomson, and Electrons

The roots of modern scientific speculation on atomic structure trace their origins to the early nineteenth century. From the chemical experiments of Antoine Lavoisier and John Dalton, the concept of an atom as the smallest unit of matter representative of a given element was firmly established by about 1810. Michael Faraday's electrolysis experiments of the early 1830's suggested that electrical current involved the transport of "ions", what we now recognize as individual atoms or molecules bearing net positive or negative electrical charges. Faraday thus perceived that nature *quantizes* certain parameters: ions could apparently not possess any arbitrary amount of electrical charge, but rather only integral multiples of a fundamental unit of charge. In 1881, Irish physicist George Stoney first estimated the fundamental unit of negative charge to be about -10^{-20} Coulomb [1]. Later, in 1891, he coined the term "electron" to describe the carriers of the fundamental unit of negative charge.

Knowledge of Avogadro's number N_A along with knowledge of the density and atomic weight of an element makes it possible to estimate the effective sizes of atoms constituting that element. If one imagines a substance to be composed of tightly-packed spherical atoms of radius R, the distance between atomic centers will be $2R$. Each atom will effectively occupy a volume of space equivalent to a cube of side length $2R$, namely $8R^3$. If the atomic weight of the substance is A grams per mole, the mass of an individual atom will be A/N_A and the density will be $\rho = A/8N_A R^3$, which we can turn into an expression for the atomic radius:

$$R = \left(\frac{A}{8N_A \rho} \right)^{1/3} . \tag{1.1}$$

By applying this expression to light and heavy elements we can get an idea of the range of atomic sizes. At one extreme is lithium, the lightest naturally solid element. With $A = 7$ g/mole (1.1×10^{-26} kg/atom) and $\rho = 0.534$ g/cm^3, we find $R \sim 1.40 \times 10^{-10}$ m, or 1.40 Å (1 Å = 10^{-10} m). At the other end of the periodic table is uranium, the heaviest naturally-occurring element, with $A = 238$ g/mole (4×10^{-25} kg/atom) and $\rho = 18.95$ g/cm^3, which give $R \sim 1.48$ Å. Despite a difference in mass of a factor of 80, uranium and lithium atoms effectively act as if they are about the same size! This is a simplistic analysis, but it does demonstrate the chemical wisdom that essentially all atoms act as if they are on the order of 1 Å in radius. The Ångstrom is a convenient unit of length for atomic-scale problems and will be used extensively throughout this text.

In 1897, J. J. Thomson undertook his famous cathode-ray deflection experiments ("cathode ray" was an early term for electron), from which he determined the charge-to-mass ratio (e/m) of the electron [2]. In modern units, he arrived at a value of about -2×10^{11} Coulomb/kg; in combination with Stoney's charge estimate of $\sim -10^{-20}$ Coulomb, it became clear that the mass of the electron must be on the order of 10^{-31} kg, some four orders of magnitude smaller than atomic masses. Moreover, Thomson found the same result for a variety of different cathode materials, demonstrating that electrons are universal constituents of all atoms.

At the time of Thomson's e/m experiments, Robert Millikan's oil drop experiments to precisely determine the value of e lay well in the future (1909). However, let us break with historical sequence to make a point about the size of the electron itself. First, modern values for the mass and charge of the electron:

$$m_e = 9.10954 \times 10^{-31} \text{ kg},$$

and

$$e = -1.60219 \times 10^{-19} \text{ C}.$$

Now, a common derivation in elementary electricity and magnetism texts is that of an expression for the self-potential-energy V of a solid sphere of radius r bearing an electrical charge e distributed uniformly throughout. The result is

$$V = \frac{3e^2}{20\pi\varepsilon_o r}. \tag{1.2}$$

If we assume that this self-energy is equivalent to the sphere's Einsteinian rest energy $E = mc^2$ (another break with historical sequence: Einstein did not develop his special theory of relativity until 1905), we can solve for the radius of the sphere:

$$r = \frac{3e^2}{20\pi\varepsilon_o mc^2}. \tag{1.3}$$

For an electron the numbers give

$$r_e = \frac{3(-1.602 \times 10^{-19}\,\mathrm{C})^2}{20\pi(8.85 \times 10^{-12}\,\mathrm{C^2/Nm^2})(9.11 \times 10^{-31}\,\mathrm{kg})(2.998 \times 10^8\,\mathrm{m/s})^2}$$

$$= 1.69 \times 10^{-15}\,\mathrm{m} \sim 10^{-5}\ \text{atomic radii.}$$

Electrons are minute in comparison to their host atoms; essentially, they are point masses. Current experimental evidence is that electrons *are* in fact point masses, and that protons have radii of about 0.8×10^{-15} m. The above analysis predicts a radius of $\sim 10^{-18}$ m for the proton; we can interpret the discrepancy as indicating that the origin of mass cannot purely be energy. This is not to imply that $E = mc^2$ is somehow in error, only that this approach cannot be an adequate model for the origin of mass. Modern versions of quantum field theory hold that mass is built up of particles known as quarks and gluons and fields that link them.

Considerations along the above lines stimulated development of models of atoms wherein pointlike negatively-charged electrons were held in stable configurations by their mutual repulsion within a larger positively charged cloud; Thomson himself was one of the prime movers in this effort. (Atomic model-building had a venerable history well before Thomson; an engaging account can be found in Pais [3].) However, these efforts met with little success, for they failed to explain atomic and molecular spectra. In addition, Thomson's 1898 atomic model proved inadequate for explaining the continuous spectrum of blackbody radiation. These phenomena played pivotal roles in the development of quantum mechanics.

1.2 Spectra, Radiation, and Planck

In 1814, German scientist Joseph Fraunhofer passed sunlight through a prism and examined the resulting continuous spectrum of colors with a magnifying glass. To his surprise, he observed hundreds of dark lines crossing the spectrum at specific wavelengths. Later, in the 1850's, Gustav Kirchoff observed the same effect on passing light from artificial sources through samples of gas. Work along these lines culminated in Kirchoff's three law of spectroscopy:

(i) Light from a hot solid object, on passing through a prism, yields a continuous spectrum with no lines.

(ii) The same light, when passed through a cool gas, yields the same continuous spectrum but with certain wavelengths of light removed, a so-called absorption spectrum.

(iii) If the light emitted by a hot gas alone is viewed through a prism, one observes a series of bright lines at certain wavelengths superimposed on an otherwise dark background, a so-called emission spectrum.

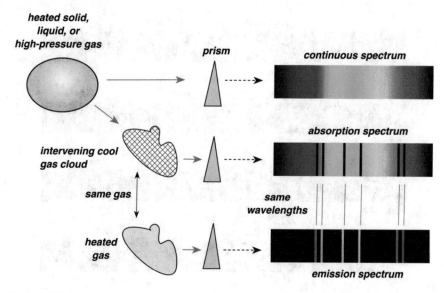

Fig. 1.1 Kirchoff's laws

The key observation was that if the same gas is used in (ii) and (iii), the absorption and emission lines appear at the *same wavelengths* in both cases. These phenomena are illustrated schematically in Fig. 1.1. It soon became clear that every chemical element or compound produces a characteristic pattern of spectral lines; spectroscopy serves as a diagnostic test for the presence or absence of an element or compound in a given sample. Spectra of a few elements are shown in Fig. 1.2. Hydrogen displays the simplest pattern of any element, a series of lines separated by ever-decreasing gaps until a "continuum" of emission is reached at a wavelength of about 3646 Å.

Explaining line spectra posed a tremendous challenge to builders of atomic models. Because Maxwell's equations revealed that an accelerating electrical charge radiates energy in the form of light, it was logical to suggest that spectral lines arose from the motion of Thomson's "corpuscular" electrons in the vicinity of whatever constitutes the massive, positively-charged part of the atom. Thomson proposed an atomic model consisting of pointlike electrons embedded within a much larger spherical cloud of positive charge, the so-called "raisin bread" or "plum pudding" model. In such an arrangement, the electrons will oscillate back-and-forth through the cloud in simple harmonic motion as would a billiard ball dropped into a frictionless hole bored through the Earth. While these oscillations lead to frequencies on the order of that of visible light, Thomson's model suffered from two serious difficulties. The first is that only one frequency results from an atom of a given size; how are we to account for whole series of spectral lines? The second difficulty was even more catastrophic: as the electron radiates away energy, the amplitude of its motion should steadily decrease until it comes to rest in the center of the atom. In short, matter should

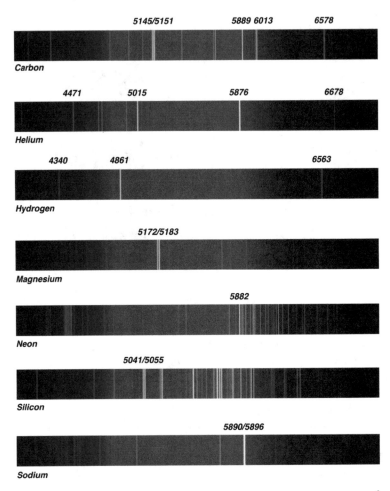

Fig. 1.2 Spectra of selected elements. Wavelengths of prominent lines are indicated in Å. From Ref. [4]

collapse! An explicit calculation of this effect in the context of the Bohr model of the atom is explored in Sect. 1.5, which can be considered optional.

How, then, to explain discrete-line spectra at all? Thomson worked with immense dedication over many years on a variety of models of this general sort, but with no real success. Understanding the stability of atoms and the origin of line spectra had to await the augmentation of classical models by new quantum postulates.

There was, however, one piece of evidence which gave early atomic theorists hope that nature might yield her secrets to mathematical analysis. In 1885, Johann Balmer discovered that the spectral lines of hydrogen followed a simple mathematical regularity, namely, that they could be computed from the formula [5]

$$\lambda_n = 3646 \left(\frac{n^2}{n^2 - 4} \right) \text{Å}, \quad n = 3, 4, 5, \ldots . \tag{1.4}$$

Note that as $n \to \infty$, $\lambda_n \to 3646\,\text{Å}$, the wavelength at which the hydrogen spectral lines "pile up" into a continuum.

Balmer's formula was purely empirical; he had no model for an underlying physical mechanism. But this does not diminish its value: when a phenomenon can be fitted into a simple arithmetic scheme, one is justified in having hope that a deeper understanding may not be far off. Indeed, Balmer's formula was a key clue for Niels Bohr's development of the first really successful theory of atomic structure.

We remark in passing that spectroscopists often refer to spectral lines by their reciprocal wavelengths. Balmer's formula can be rearranged to reflect this:

$$\frac{1}{\lambda_n} = R_H \left(\frac{1}{4} - \frac{1}{n^2} \right) \text{Å}^{-1}, \quad n = 3, 4, 5, \ldots , \tag{1.5}$$

where R_H is known as the Rydberg constant for hydrogen; its experimental value is $109{,}678\,\text{cm}^{-1}$. This is named after Swedish physicist Johannes Rydberg, who suggested that the spectral lines of any element could be described by a generalized version of this expression:

$$\frac{1}{\lambda_{n,m}} = K \left(\frac{1}{m^2} - \frac{1}{n^2} \right) \quad (n > m), \tag{1.6}$$

where K is a constant specific to the element involved. Different series of spectral lines are given by various values of n for a fixed value of m. We will return to this modified Balmer formula later in this chapter.

Continuous spectra also posed a challenge to classical physics. Experimentally, any solid body at a temperature T above absolute zero emits a continuous spectrum of electromagnetic radiation. By the end of the nineteenth century, experiments had also established that the spectrum of "thermal radiation" emitted by a body is independent of the nature of the body, depending only on T. These experiments were usually carried out with so-called "blackbody" cavities. This can be thought of as a heated, internally hollow lump of metal with a hole in it from which some of the thermal radiation can escape and so be sampled by a detector. Figure 1.3 shows spectra for the light emitted by such cavities at temperatures of 5000, 6000, and 7000 K over the wavelength range 2000–10,000 Å; what is plotted is the power emitted per square meter of surface area of the sampling hole per Ångstrom wavelength interval as a function of wavelength. (The range of human vision spans approximately 3500–7000 Å). As T increases, the wavelength of peak emission shifts to shorter (bluer) wavelengths. If you are familiar with stellar astronomy, you may know that blue stars have hotter surfaces than yellow stars, which are in turn hotter than red-dwarf stars.

Attempts to find a mathematical description of thermal radiation based on classical models failed disastrously. The generally-held concept was somewhat analogous to Thomson's oscillating electrons: each atom of a material body would be in constant

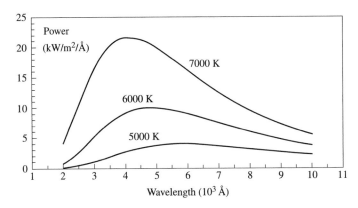

Fig. 1.3 Thermal radiation spectra

motion, jostling about with its neighbors. When a charged particle oscillates at frequency ν (in, say, simple harmonic motion as studied in freshman-level physics classes), it will emit electromagnetic radiation at that same frequency; for this reason the atoms were abstracted as "oscillators". (Recall that, in 1900, knowledge of the detailed internal structure of atoms was essentially nil.) The cavity will fill with electromagnetic radiation bouncing around at the speed of light, and an equilibrium ("thermal equilibrium") will soon be established as the oscillators emit and absorb radiation. Like standing waves on a vibrating string, however, the cavity can only "support" electromagnetic waves of certain frequencies depending on its dimensions. From Maxwell's theory, the number of possible "standing-wave" frequencies available between frequencies ν and $\nu + d\nu$, usually designated as $N(\nu)d\nu$, was known to be given by

$$N(\nu)d\nu = \frac{8\pi V}{c^3}\nu^2 d\nu, \tag{1.7}$$

where V is the volume of the cavity and c is the speed of light. An approximate derivation of this expression is given in Sect. 6.1. Now, what is desired is an expression for the energy within the cavity in the form of electromagnetic waves in some specified range of frequencies; this is what the measurements ultimately sample. In effect, we need to know how many oscillators are oscillating at each available frequency and what amount of energy is associated with each mode of oscillation. In classical physics, however, there is no connection between the energy of an oscillator and its frequency: you can set the amplitude and frequency of a vibrating string separately to be whatever values you please. To build a connection to energy, the approach adopted was to appeal to the success of Maxwell and Boltzmann's work on statistical mechanics. Their research showed that if a particle is in an environment at absolute (Kelvin) temperature T, then the probability of its having energy between E and $E + dE$ is proportional to an exponential function of T:

$$p(E)dE = Ae^{-E/kT}dE, \qquad (1.8)$$

where A is a constant of integration and k is Boltzmann's constant. To determine A we insist that if we add up probabilities over all possible energies, they must sum to unity. This is equivalent to saying that the particle must have *some* energy between zero and infinity, that is, we demand

$$\int_0^\infty p(E)\,dE = A \int_0^\infty e^{-E/kT}\,dE = 1.$$

This integral evaluates to kT, giving $A = 1/kT$. The average energy of a particle in such a situation is then given by the integral of probability-weighted energy:

$$\langle E \rangle = \int_0^\infty E p(E)\,dE = \frac{1}{kT}\int_0^\infty E e^{-E/kT}dE = kT. \qquad (1.9)$$

(For a fuller elaboration of this averaging technique, see Sect. 4.2.) Now, if the number of oscillators wiggling about with frequencies between v and $v + dv$ is given by (1.7), then the amount of radiant energy in the cavity between these frequencies will be given by this number times the average energy of an oscillator at temperature T, kT:

$$E(v)dv = \frac{8\pi V}{c^3}kTv^2 dv. \qquad (1.10)$$

This relationship proved to be in good agreement with experimental data at low frequencies. But at high frequencies one has a disaster: the amount of energy in the cavity becomes infinite, a situation known as the ultraviolet catastrophe. Something was apparently wrong with either classical electromagnetism or statistical mechanics, or perhaps even both. It is probably fair to say that at the dawn of the twentieth century, thermal radiation was *the* outstanding problem of theoretical physics.

A way around this difficulty was found by Max Planck of the University of Berlin late in the year 1900 in a paper usually credited with marking the birth of quantum mechanics proper [6]. However, his solution came at a substantial price: abandoning the classical notion that the energy and frequency of an oscillator are independent. Planck proposed that the energy of an oscillator is restricted to only certain multiples of its frequency, namely $E_n = nhv$, where n is an integer-valued "quantum number" ($n = 0, 1, 2, 3, \ldots$) and h is a constant of nature.

Maintaining the assumption of a Maxwellian exponential probability distribution, the probability of an oscillator having energy E is now given by

$$p(E) = Ae^{-nhv/kT}, \qquad (1.11)$$

where A is again a normalization constant. Note that no "dE" appears in this expression: the idea now is that an oscillator has some probability of having a particular energy as opposed to any arbitrary energy between E and $E + dE$. Forcing the probabilities to again add up to unity means that we must have

$$A \sum_{n=0}^{\infty} e^{-nh\nu/kT} = 1,$$

or

$$A \sum_{n=0}^{\infty} (e^{-h\nu/kT})^n = 1. \tag{1.12}$$

We are now summing over discrete energy states as opposed to integrating over energy as a continuous variable. The sum appearing in (1.12) is of the form Σx^n, where $x = e^{-h\nu/kT}$, that is, we have a geometric series. Such a sum evaluates to $(1 - x)^{-1}$, hence

$$A = (1 - e^{-h\nu/kT}). \tag{1.13}$$

With this, the probability of an oscillator having energy E can be written as

$$p(E) = e^{-nh\nu/kT}(1 - e^{-h\nu/kT}). \tag{1.14}$$

With this new probability recipe, the average energy of oscillators of frequency ν becomes

$$\langle E \rangle = \sum_{n=0}^{\infty} E p(E) = \sum_{n=0}^{\infty} (nh\nu) p(E) = h\nu(1 - e^{-h\nu/kT}) \sum_{n=0}^{\infty} n e^{-nh\nu/kT} \tag{1.15}$$

$$= h\nu(1 - e^{-h\nu/kT}) \frac{e^{-h\nu/kT}}{(1 - e^{-h\nu/kT})^2} = \frac{h\nu}{(e^{h\nu/kT} - 1)}.$$

In working (1.15), we have used another result from series analysis, namely

$$\sum_{n=0}^{\infty} n x^n = \frac{x}{(1 - x)^2} \quad (|x| < 1),$$

where we again have $x = e^{-h\nu/kT}$. Again assuming that the number of oscillators with frequencies between ν and $\nu + d\nu$ is given by (1.7), we find a new expression for the energy in the cavity in this frequency range:

$$E_\nu d\nu = \frac{8\pi h V}{c^3} \frac{\nu^3}{(e^{h\nu/kT} - 1)} d\nu. \tag{1.16}$$

This is Planck's famous *blackbody radiation formula*. Note that frequency is still a continuous variable; the energy that an oscillator can have is discretized by $E = nh\nu$ once its frequency is specified. Planck found this expression to be a perfect match to the experimental data, provided that the constant h is chosen suitably. This is now known as Planck's constant and has the value

$$h = 6.626 \times 10^{-34} \text{ J-s.}$$

Planck's assumption is now practically an element of popular culture; hindsight tends to blind us to perceiving just how radical it was to his contemporaries. That electrical charges come in discrete units is one thing, but to suggest that the most venerable quantity in all of physics, energy itself, could be exchanged only in discrete lumps, that is, that it is *quantized*, was to disavow two and a half centuries of accumulated wisdom. As with all fundamental postulates, the difficulty in understanding it and its consequences stem from just that: like $F = ma$, it is a postulate, not derivable from anything more fundamental. The justification for Planck's hypothesis is that it has never been found to lead to conclusions in conflict with experiment.

The development of blackbody theory actually spanned the better part of two decades, with Planck modifying his assumptions and calculations many times; the derivation given here represents only a very cursory survey of the issue. Readers interested in exploring more deeply this fundamental revolution in physics are directed to Thomas Kuhn's comprehensive *Black-Body Theory and the Quantum Discontinuity, 1894–1912* [7].

Planck's hypothesis lay essentially dormant until 1905, when Einstein adapted it to explain the photoelectric effect, the emission of electrons from a metal surface being bombarded by light [8]. This phenomenon had proven impossible to understand when light was considered as a wave (as in the Maxwellian tradition), but proved straightforward to comprehend if the light were regarded instead as a stream of particles (*light-quanta* or *photons*, although the latter term was not coined until the 1920's) of energy $E = h\nu$, where ν is the frequency of the wave one would normally associate with the light. It was as if the incident light were colliding billiard-ball style with electrons, knocking them out of the metal. While Einstein's work resolved the photoelectric effect, it thrust into the foreground an apparent paradox: is light a wave or a stream of particles? On one hand it behaves optically as any sensible wave, that is, it is refractible, diffractible, focusible, and so forth. On the other hand it can behave dynamically like billiard balls. Which picture do we adopt? The answer is that we have to live with both. While in some circumstances one can conveniently analyze an experiment with an explicitly wave picture in mind (double-slit diffraction in classical optics, say), in others a particle view provides for a convincing analysis (the photoelectric effect). The inescapable fact is that photons *simultaneously* possess both wave and particle properties. Material particles such as electrons and entire atoms and molecules also prove to do the same; see Sect. 1.4. Indeed, it has been amply verified experimentally that photons interfere with themselves even as they pass one-by-one through an optical system! For readers interested in learning about the details of such experiments and many other aspects of the fundamentals of quantum mechanics,

Greenstein and Zajonc's *The Quantum Challenge* can be highly recommended [9]. We shall have more to say about this wave-particle duality in coming chapters.

The idea of light carrying momentum was not new. Indeed, Maxwell had predicted that electromagnetic radiation of energy E would possess momentum in the amount $p = E/c$. (If you are familiar with special relativity, recall the Einstein energy-momentum-rest mass relation $E^2 = p^2 c^2 + m_o^2 c^4$, and set the rest mass m_o to be zero for a photon). This can be expressed in terms of the frequency and wavelength of the light as $p = E/\lambda \nu$ since $c = \lambda \nu$. If light is also treated as consisting of particles characterized by the Planck relation $E = h\nu$, then $p = h/\lambda$ or

$$\lambda = \frac{h}{p}. \tag{1.17}$$

Equation (1.17) applies only to photons, even though it was derived from a mixture of both classical and quantum concepts. Nevertheless, in the 1920's, Arthur Compton verified that photons and electrons indeed interact billiard-ball style, precisely as this predicts. This deceptively simple equation will figure in subsequent discussions.

The substitutions leading to (1.17) are simple. Yet, it seems remarkable that the connection between Planck's quantized oscillator energies and the discrete wavelengths (= discrete energies) of light involved in spectral lines was not made for nearly another decade. This breakthrough had to await further experimental probing of the internal structure of atoms.

1.3   The Rutherford-Bohr Atom

In early 1911, Ernest Rutherford, in collaboration with Hans Geiger and Ernest Marsden, proposed a model of atomic structure radically different from Thomson's plum-pudding scenario [10]. Based on an analysis of the distribution of alpha-particles (helium nuclei) scattered through thin metal foils, they put forth a *nuclear* model of the atom wherein the positive charge and with it most of the atom's mass is concentrated in a tiny volume of space at the center of the atom, around which negative electrons orbit like planets about the Sun. Their experiments indicated a size on the order of 10^{-14} m for the positively charged nucleus: larger than an electron, but some four orders of magnitude smaller than the atom itself. Evidently, atoms are mostly empty space.

The stage was now set for a leap of logic of the sort alluded to at the end of the previous section. Planck had established that atoms constituting the source of thermal radiation from material bodies could exist only in certain discrete energy states. If Rutherford's planetary electrons were restricted to orbits of certain special energies, could electrons "transiting" from orbits of higher energy (E_2, say) to orbits of lower energy (E_1, say) be the source of spectra and thermal radiation, with the energy difference appearing as a photon of frequency ν_{21} given by Planck's hypothesis written in the form $E_2 - E_1 = h\nu_{21}$? These were among the assumptions made

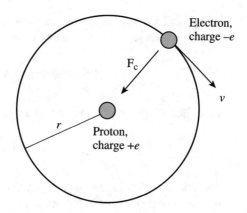

Fig. 1.4 Bohr model for hydrogen

by a young Danish physicist, Niels Bohr, in a trilogy of papers published in the *Philosophical Magazine* in 1913 that marked the first truly successful theory of the inner workings of atoms [11–13].

Let us explore the consequences of Bohr's postulates for a simple model of the hydrogen atom. The situation is illustrated in Fig. 1.4, where an electron of mass m_e and charge $-e$ is in a circular orbit of radius r and speed v around a central proton of charge $+e$.

Bohr began by assuming that the energetics of the electron's orbit are dictated by the Newtonian dynamics of circular orbits, namely that if the electron is in a circular orbit then it must experience a centripetal force of magnitude $m_e v^2 / r$ toward the proton. Identifying the source of this centripetal force to be the Coulomb attraction between electron and proton gives

$$\text{magnitude of Coulomb force} = \frac{m_e v^2}{r} = \frac{e^2}{4\pi \varepsilon_o r^2}. \tag{1.18}$$

A more refined version of this calculation accounts for the motion of the proton and the electron about their mutual center of mass. This can be effected by substituting the reduced mass of the system, $m_e m_p / (m_e + m_p)$, in place of the electron mass m_e wherever it appears; m_p is the mass of the proton. Also, the theory can be extended to include hydrogen-like ions, that is, systems wherein a single electron orbits a nucleus of charge $+Ze$, by replacing e^2 by Ze^2.

Since the physically relevant quantities are wavelengths and energies, we start out by deriving an expression for the total energy (kinetic + potential), of the electron in its orbit. From (1.18),

$$\text{Kinetic energy} = \frac{1}{2} m_e v^2 = \frac{e^2}{8\pi \varepsilon_o r}. \tag{1.19}$$

The potential energy is just the electrostatic potential of an electron (charge $-e$) and a proton (charge $+e$) separated by distance r:

$$\text{Potential energy} = -\frac{e^2}{4\pi \varepsilon_o r}. \tag{1.20}$$

From Eqs. (1.19) and (1.20) we find the total energy E_{total}:

$$E_{total} = \text{KE} + \text{PE} = -\frac{e^2}{8\pi \varepsilon_o r}. \tag{1.21}$$

The interpretation of the negative sign is that positive energy must be supplied from some external source to dissociate the electron from the atom. We say that the electron is in a *bound energy state*. Curiously, the total energy is equal to one-half of the potential energy; this result is actually a specific case of a very general result known as the *virial theorem*, which is treated in Sect. 4.9.

Imagine the electron "falling" from infinity [$E = 0$ with $r = \infty$ in (1.21)] down to an orbit of radius r about the proton. In doing so it would lose energy, which is assumed to appear as a photon of frequency v. At this point, Bohr made two assumptions: (i) that this frequency is equal to one-half of the electron's final orbital frequency about the proton, and (ii) consistent with Planck's hypothesis, the energy of the photon is given by nhv, where n is an integer and h is Planck's constant. For an electron in a circular orbit of speed v and radius r, the time for one orbit is $2\pi r/v$; the orbital frequency is then $v/2\pi r$, and Bohr's hypotheses give

$$E_{photon} = |E_{total}| = nhv = \frac{nh}{2}\left(\frac{v}{2\pi r}\right) = \frac{e^2}{8\pi \varepsilon_o r}. \tag{1.22}$$

Solving the last two members of this expression for v gives

$$v = \frac{e^2}{2\varepsilon_o nh}. \tag{1.23}$$

Now solving (1.19) for the orbital radius and substituting this result for v gives

$$r_n = \left(\frac{\varepsilon_o h^2}{\pi m_e e^2}\right) n^2 = a_o n^2, \quad (n = 1, 2, 3, ...). \tag{1.24}$$

The interpretation of (1.24) is that the electron is restricted to orbits of radius $a_o n^2$ where a_o, now known as the *Bohr radius*, represents the radius of the smallest permissible orbit; we cannot have $n = 0$, for this would imply $E = -\infty$ in (1.21). n is called the *principal quantum number*.

Using the following modern values for the constants:

$$\varepsilon_o = 8.8542 \times 10^{-12} \ \text{C}^2/\text{Nm}^2,$$

$$h = 6.6261 \times 10^{-34} \, \text{J-s},$$

$$m_e = 9.1094 \times 10^{-31} \, \text{kg},$$

$$e = 1.6022 \times 10^{-19} \, \text{C},$$

one finds

$$a_o = 5.2918 \times 10^{-11} \, \text{m} = 0.52918 \, \text{Å}. \tag{1.25}$$

This is a remarkable result, exactly of the order of atomic dimensions as deduced in Sect. 1.1. Bohr's assumptions result in a theoretical explanation of the sizes of atoms, an otherwise empirical detail.

Example 1.1 The lone proton in a hydrogen atom has an effective radius of about 10^{-15} m. Suppose that the entire atom is magnified so that the proton has a radius of 1 mm, about the size of a pinhead. If the orbiting electron is in the $n = 1$ state in (1.24), what would be the radius of its magnified orbit?

As a multiple of the proton's effective radius, the distance of the $n = 1$ Bohr orbit is $a_o/10^{-15}$ m $\sim (5.29 \times 10^{-11}$ m$)/10^{-15}$ m $\sim 52,900$. If the proton is expanded to a radius of 1 mm, then the electron will be orbiting at about $(1 \, \text{mm}) (52,900) \sim 52.9$ m. If the expanded proton were placed at the center of an American football field, the electron's orbit will reach to just beyond the goal lines!

More importantly, Eq. (1.24) leads to an expression for the wavelengths of photons emitted when electrons transit between possible orbits, and also to a theoretically predicted value of the Rydberg constant. To arrive at the wavelength expression, imagine that the electron falls from an initial orbital radius r_i to some final orbital radius r_f, with $r_f < r_i$. According to (1.21), $E_f < E_i$ (both E_f and E_i are negative); if the energy lost by the electron is presumed to appear as a photon in accordance with Planck's hypothesis, then

$$E_{photon} = E_i - E_f = \frac{e^2}{8\pi \varepsilon_o} \left(\frac{1}{r_f} - \frac{1}{r_i} \right) = h\nu_{i \to f}. \tag{1.26}$$

This would correspond to a photon of wavelength $\lambda = c/\nu$, that is,

$$\frac{1}{\lambda_{i \to f}} = \frac{e^2}{8\pi \varepsilon_o hc} \left(\frac{1}{r_f} - \frac{1}{r_i} \right). \tag{1.27}$$

Substituting (1.24) for the orbital radii gives

$$\frac{1}{\lambda_{n_i \to n_f}} = \frac{m_e e^4}{8\varepsilon_o^2 h^3 c}\left(\frac{1}{n_f^2} - \frac{1}{n_i^2}\right) \qquad (n_f < n_i). \qquad (1.28)$$

Now recall the Balmer-Rydberg expression for the visible-region hydrogen-line wavelengths, (1.5):

$$\frac{1}{\lambda_n} = R_H\left(\frac{1}{4} - \frac{1}{n^2}\right). \qquad (1.29)$$

Equations (1.28) and (1.29) are remarkably similar in form; indeed, they correspond precisely if we take $n_f = 2$ and if the Rydberg constant is given by

$$R_H = \frac{m_e e^4}{8\,\varepsilon_o^2 h^3 c}. \qquad (1.30)$$

Substitution of the appropriate numerical values yields $R_H = 109{,}737\,\text{cm}^{-1}$, in close agreement with the experimental value of $109{,}678\,\text{cm}^{-1}$. The slight discrepancy (0.05%) can be accounted for by the motion of the electron and proton about their mutual center of mass (Problem 1.11). With this agreement, it is clear that (1.28) will produce the observed wavelengths of the Balmer lines when $n_f = 2$. According to the Bohr model, the Balmer series of hydrogen lines corresponds to electrons transiting to the $n = 2$ orbit from "higher" orbits.

Equation (1.28) is quite general, and can be used to calculate the wavelengths of whole series of spectral lines for hydrogen. Inverting it gives the wavelengths directly:

$$\lambda_{n_i \to n_f} = \frac{8\,\varepsilon_o^2 h^3 c}{m_e e^4}\left(\frac{1}{n_f^2} - \frac{1}{n_i^2}\right)^{-1} = 911.75\left(\frac{1}{n_f^2} - \frac{1}{n_i^2}\right)^{-1}\ \text{Å}, \qquad (1.31)$$

where the numerical value of 911.75 Å accounts for electron/proton center-of-mass motion.

Example 1.2 Compute the wavelength of the $7 \to 4$ transition for a hydrogen atom. What is the energy of the photon emitted in such a transition, in eV?
From (1.31), and retaining Ångstroms,

$$\lambda_{7\to 2} = 911.75\left(\frac{1}{4^2} - \frac{1}{7^2}\right)^{-1} = 911.75\left(\frac{1}{16} - \frac{1}{49}\right)^{-1} = 911.75\left(\frac{33}{784}\right)^{-1}$$

$$= 21{,}661\ \text{Å}.$$

With this wavelength, 2.166×10^{-6} m, such a photon would lie in the infrared part of the electromagnetic spectrum, far to the red of human visual response. The energy is

$$E = \frac{hc}{\lambda} = \frac{(6.6261 \times 10^{-34}\,\text{J} \cdot \text{s})(2.9979 \times 10^{8}\,\text{m/s})}{2.1661 \times 10^{-6}\,\text{m}} = 9.171 \times 10^{-20}\,\text{J}.$$

With 1 eV = 1.602×10^{-19} eV, this corresponds to about 0.57 eV.

Table 1.1 lists hydrogen spectral wavelengths computed from this formula for a number of (n_i, n_f) pairs. As described earlier, all lines corresponding to the same n_f is known as a *series* of lines. Each such series is characterized by a *series limit* corresponding to $n_i = \infty$; the spacing of the lines decreases toward the series limits. The range of human vision runs from about 3500 to 7000 Å; only the Balmer series lies entirely within these limits. The series corresponding to $n_f = 3$ lies in the infrared region of the spectrum, and had been observed in 1908 by Paschen [14] to occur at exactly the wavelengths predicted by the Bohr formula. After publication of Bohr's theory, the series corresponding to $n_f = 1, 4$, and 5 were discovered by Lyman [15], Brackett [16], and Pfund [17]. These discoveries lent immense credibility to Bohr's theory.

The energy of an electron in quantum state n in the Bohr model is given by combining Eqs. (1.21) and (1.24):

$$E_n = -\frac{m_e e^4}{8\varepsilon_o^2 h^2 n^2} = -\frac{13.606}{n^2} \text{ eV} \quad (n = 1, 2, 3, \ldots), \quad (1.32)$$

where the factor of 13.606 eV, known as the Rydberg energy, accounts for the electron/proton center-of-mass motion; the various constants were carried out to five significant figures. This is one of the most famous results of early quantum mechanics. It tells us that to ionize a hydrogen atom from its *ground state* $(n = 1)$ requires an expenditure of 13.6 eV of energy, precisely the observed value.

Table 1.1 Hydrogen transition wavelengths (Å)

Initial state							
Final state	2	3	4	5	6	Infinity	Series
1	1215.7	1025.7	972.5	949.7	937.8	911.8	Lyman
2	–	6564.6	4862.7	4341.7	4102.9	3647.0	Balmer
3	–	–	18756.0	12821.5	10941.0	8205.8	Paschen
4	–	–	–	40522.2	26258.4	14588.2	Brackett
5	–	–	–	–	74597.7	22794.3	Pfund
6	–	–	–	–	–	32824.2	Humphreys

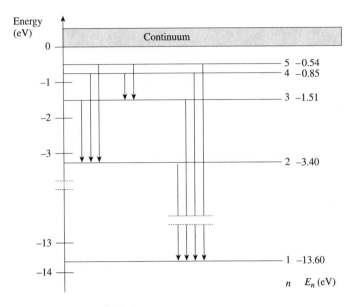

Fig. 1.5 Energy level diagram for hydrogen

Figure 1.5 shows an *energy-level diagram* for hydrogen based on the Bohr model. Energy is plotted increasing upwards on the vertical axis; the horizontal axis has no physical meaning and serves only to improve the readability of the figure. The arrows show a number of possible quantum jumps or transitions, their lengths are proportional to the frequencies of the corresponding emitted photons since $E = h\nu$. Electrons that are initially not bound to the atom (so-called free electrons with $E > 0$) can become so by releasing just enough energy to bring them into one of the stationary states; thus, there is a continuum of energy levels with $E > 0$.

Incidentally, we have been assuming that photons themselves possess no electrical charge. This appears to be a safe assumption: Current experimental limits on any photon charge indicate an upper limit of $10^{-32}e$; see [18]. Along a similar line, it is now possible to synthesize atoms of *antihydrogen*, a combination of an anti-proton and an anti-electron, and subject them to spectroscopic tests to check that the transition properties are identical to those of "normal" hydrogen; see [19, 20]. One can only wonder what Bohr would have nade of such advanced tests of his theory.

An intriguing aspect of Bohr's model concerns the orbital angular momentum of the electron, usually designated L. Classically, a mass m in a circular orbit of radius r and speed v has angular momentum $L = mvr$. Applying this to the electron in the Bohr model with the help of (1.23) and (1.24),

$$L = m_e vr = m_e \left(\frac{e^2}{2\varepsilon_o n h} \right) \left(\frac{\varepsilon_o h^2}{\pi m_e e^2} \right) n^2 = \left(\frac{h}{2\pi} \right) n, \quad (n = 1, 2, 3, \ldots).$$

$$(1.33)$$

The physical interpretation of this result is that the orbital angular momentum of the electron is quantized in units of $h/2\pi$. Planck's constant, which was originally introduced to explain thermal radiation, now appears as a fundamental unit of angular momentum. Many treatments of the Bohr model begin by assuming that angular momentum is quantized in this way. Whether one regards E or L as being the fundamentally quantized attribute is arbitrary; the relationship is symbiotic. The fundamental quantum of angular momentum, $h/2\pi$, is given its own symbol, $\hbar$, read as "h-bar":

$$\hbar = \frac{h}{2\pi} = 1.05459 \times 10^{-34} \, \text{Js.} \qquad (1.34)$$

We will see in Chap. 7 that angular momentum plays a central role in the wave mechanics of the hydrogen atom.

At this point, it is useful to look back at the illustration of Kirchoff's Laws, Fig. 1.1, and relate it to Bohr's model. A cool gas will have most of its atoms in the lowest energy (ground) state. When light is passed through such a gas, only those photons with energy equal to the various differences between the ground state and higher energy levels of the gas atoms will be absorbed out, causing certain wavelengths to be removed from the light beam. In the case of a hot gas, we will see photons emitted from de-excitations of the gas atoms, the reverse of the cool-gas excitation process. The continuous spectrum of a hot solid object (e.g., a light-bulb filament) is a manifestation of interactions between atoms.

Example 1.3 One of the results of Planck's blackbody radiation theory is that any environment at absolute (Kelvin) temperature T will teeming with photons of average energy $2.7kT$ where k is Boltzmann's constant. At room temperature (300 K), would this average energy be sufficient to ionize a hydrogen atom from the $n = 1$ state? If not, what temperature would result in ionization?

The average photon energy evaluates as

$$\langle E \rangle = 2.7(1.381 \times 10^{-23} \, \text{J/K}) \, T = 3.729 \times 10^{-23} \, T \, \text{J},$$

if T is in Kelvins. This is equivalent to $2.327 \times 10^{-4} T$ eV. At T = 300 K, $\langle E \rangle$ = 0.07 eV, far below the ionization energy of 13.6 eV. $\langle E \rangle = 13.6$ eV would require $T \sim 58{,}000$ K. In cosmology, the time after the Big bang at which the Universe had cooled to this temperature is known as the recombination time; after this time, neutral hydrogen could form.

Example 1.4 X-rays are photons with wavelengths from 0.06 to 125 Å. In 1913–14, Henry Moseley provided supporting evidence for the Bohr model by measuring the wavelengths of photons emitted from (among others) $2 \rightarrow 1$ electronic transitions for metals of a variety of atomic numbers Z. To what range in Z does the X-ray part of the spectrum correspond for $2 \rightarrow 1$ transitions in the Bohr model?

From (1.31) and the comment given regarding incorporating different Z values following (1.18), we can write the Bohr transition wavelengths as

$$\lambda_{n_i \rightarrow n_f} = \frac{912}{Z^2} \left(\frac{1}{n_f^2} - \frac{1}{n_i^2} \right)^{-1} \text{Å},$$

where we have rounded the Rydberg constant to the nearest Ångstrom. For $(n_i, n_f) = (2, 1)$ this becomes (again to the nearest Å)

$$\lambda_{2 \rightarrow 1} = \frac{1216}{Z^2} \text{Å}$$

For $\lambda = 125$ Å, $Z \sim 3$, and for $\lambda = 0.06$ Å, $Z \sim 142$. Thus, essentially the whole of the periodic table was available to Moseley.

In the years following its publication, Bohr's theory underwent a number of refinements. Inclusion of elliptical orbits and relativistic effects led to some understanding of the "fine structure" of spectral lines. Experimental work by Henry Moseley on X-rays and by James Franck and Gustav Hertz on collisions of electrons with atomic and molecular gases added compelling evidence for the existence of quantized energy levels [21–23]. But, despite its success and powerful intuitive appeal, Bohr's model was not without problems. First of all was the arbitrariness of the orbital frequencies assumption: It clearly works, but was there any way a more fundamental understanding might be sought? Despite years of effort by Bohr and his colleagues to extend the theory to more complex atoms, they had only very limited success [24]. Deeper understanding of these matters had to await a wave formulation of quantum mechanics in 1926.

1.4 de Broglie Matter Waves

In his 1924 Ph.D. thesis, French physicist Louis de Broglie ("de broy") proposed that if photons could behave like particles and transport momentum according as

$$p = \frac{h}{\lambda}, \tag{1.35}$$

then might material particles possessing momentum be accompanied by a *matter-wave* of wavelength given by inverting this relation: $\lambda = h/p = h/mv$?

The mathematical manipulation here is trivial, but the physical hypothesis seems absurd: We do not observe matter to be wavy; it is solid and localized. But just because we cannot see a particular phenomenon does not mean that it does not exist. The controlling factor in (1.35) is the minuteness of Planck's constant. A mass of 1 kg moving at 1 m/s would have an associated *de Broglie wavelength* of $\sim 10^{-34}$ m, some 20 orders of magnitude smaller than an atomic nucleus. We could scarcely hope to observe the "matter-wave" associated with such a macroscopic object. On the other hand, if an extremely tiny mass is involved, such as that of an electron, then it turns out that the associated wave can be comparable in size to its parent particle.

Table 1.2 shows a few de Broglie wavelengths computed via the relationship now named after him [25]:

$$\lambda = \frac{h}{mv} = \frac{h}{\sqrt{2Km}}, \quad (v \ll c), \tag{1.36}$$

where K is the kinetic energy of the mass involved.

If particles really do possess associated waves, then it is clear that we can hope to detect them only in atomic-scale phenomena. Leaving aside for the moment the question of what they mean, let us ask how such matter-waves might manifest themselves.

Waves can be diffracted to produce constructive and destructive interference patterns. According to de Broglie, electrons fired through a double-slit apparatus should somehow interfere with each other (and themselves!). The problem is that the slit separation must be on the order of the wavelength involved, far too small (at least in de Broglie's day) to be physically manufactured. However, as de Broglie himself pointed out, nature provides natural diffraction gratings with spacings on the order of Ångstroms: the regularly spaced rows of atoms in metallic crystals. Less than three years after de Broglie advanced his hypothesis, experimental verification came independently from American and British groups of researchers. Working at the Bell Telephone Laboratories in the United States, Clinton Davisson and Lester Germer observed, by accident, interference effects with electron scattering from crystals of nickel [26]. In Britain, George P. Thomson (J. J.'s son) verified the effect by scattering electrons through thin metallic foils [27]. In 1961, Claus Jönsson demonstrated electron diffraction with a true double-slit arrangement, utilizing an elaborate array of

Table 1.2 de Broglie wavelengths

Situation	λ (Å)	Comment
Electron, energy 1 eV	12.3	Molecule-size
Proton, energy 1 keV	0.009	Subatomic size
800 kg car @ 20 m/s	4.1×10^{-28}	$\ll$ Nuclear size

electrostatic lenses to magnify the image [28–30]. More recently, the wave properties of Carbon-60 "buckyball" molecules have been observed by similarly passing them through a series of slits, as has the diffraction of electrons by a standing light wave [31, 32]. All these experiments and many others have demonstrated that electrons, atoms, and molecules behave in full agreement with de Broglie's hypothesis.

The wave nature of electrons is put to stunning practical use in electron microscopy. The resolution of a microscope, that is, the smallest distance by which two objects can be separated and still viewed as distinct, is proportional to the wavelength of light used. Since electrons behave as waves of wavelengths much shorter than that of visible light, much greater detail is resolvable. Images of individual molecules and atoms can be obtained in this way.

The evidence for de Broglie's matter waves is incontrovertible. But what does it mean to say that a particle has a wave nature? How do we reconcile the classical point-mass view of matter with wave properties? As a schematic model, consider Fig. 1.6, which is supposed to represent an electron.

The victim in this game of wavy particles is the classical idea that the position of the particle can be specified with an arbitrary degree of precision. In effect (but not in reality), the electron acts as if it is smeared out into a so-called wave packet that pervades all space. The amplitude of this wave, however, is not constant: it proves to be strongly peaked in the vicinity of the electron over a length scale Δx on the order of the size of the electron, as the figure suggests. For practical purposes, we can no longer specify with precision where the electron is; we can only speak meaningfully of the probability of finding it at some position. We will explore these "probability waves" further in subsequent chapters. It is this wavy nature of matter that lies at the heart of the Heisenberg uncertainty principle, which we will examine in Chap. 4.

These considerations in no way invalidate the application of Newtonian dynamics to objects such as golf balls or planets. The de Broglie waves accompanying such masses are so infinitesimally small in comparison with their length-scales that we cannot hope to detect them. Only when the size of the system under study is comparable to $\lambda = h/mv$ will quantum effects become noticeable. As an example, refer to Table 1.2. A 1 keV proton incident on a hydrogen atom has a de Broglie wavelength much less than the size of the atom; for practical purposes, this will be a particle interaction. On the other hand, an electron of energy 1 eV has a de Broglie wavelength on the order of the size of that same hydrogen atom: to an imagined observer riding on the atom, the incoming electron would appear amorphous.

Fig. 1.6 A particle and its de Broglie matter-wave. This is a cartoon!

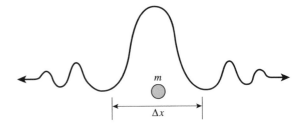

m

Δx

The "wave-particle duality" is perhaps the most counterintuitive aspect of quantum mechanics, and has, not surprisingly, inspired an extensive body of literature. A excellent discussion appears in the first two chapters of the third volume of Feynman's *Lectures on Physics* [33], and a very readable and humorous look at the situation by David Mermin can be found in [34].

Example 1.5 What speed must an electron have if it's de Broglie wavelength is 5000 Å? A photon of this wavelength would be in the visible part of the electromagnetic spectrum.

From (1.36), solve for v to give

$$v = \frac{h}{m\lambda} = \frac{(6.626 \times 10^{-34}\,\text{J-s})}{(9.109 \times 10^{-31}\,\text{kg})(5 \times 10^{-7}\,\text{m})} = 1455\,\text{m/s}.$$

See also Problem 1.20. If this speed were due to the electron's random thermal motion in an environment at absolute temperature T, the environment would be at $T \sim 0.04\,\text{K}$. Electrons of such speeds are termed "cold."

Example 1.6 de Broglie further hypothesized that the allowed obits for an electron in a hydrogen atom are given by the condition that an integral number "n" de Broglie wavelengths fit around the circumference of the orbit. Show that this condition leads to the Bohr quantization condition on angular momentum, Eq. (1.33).

Let the electron have speed v in an orbit of radius r. The circumference of the orbit is then $2\pi r$. de Broglie's hypothesis is then that an integral number of wavelengths fit into this circumference:

$$\frac{2\pi r}{\lambda} = n.$$

Setting $\lambda = h/m_e v$ gives

$$\frac{2\pi m_e vr}{h} = n.$$

However, $m_e vr$ is just the angular momentum of the electron in its orbit, so we have

$$L = \left(\frac{h}{2\pi}\right) n = \hbar n,$$

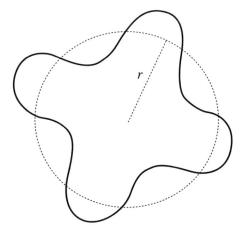

Fig. 1.7 de Broglie quantization condition

precisely the Bohr condition on L. If the condition on the electron's speed that
emerged from Bohr's orbital frequency assumption is used here (Eq. 1.23), the
result is the same restriction on orbital radii and energies that emerged from
his approach.

Figure 1.7 shows a sketch of what de Broglie had in mind: an integral number of
wavelengths (in the case shown, $n = 4$) fitting around the circumference of an orbit
like a standing wave. It should be emphasized that the amplitude of the wave shown
in the sketch is purely arbitrary; de Broglie's relation gives no information on this.

1.5   The Radiative Collapse Problem (Optional)

It was mentioned in Sect. 1.2 that one of the dilemmas of early atomic theories was
that they predicted that atoms should collapse almost instantaneously. This section,
which can be considered optional, develops an analysis of this issue. This is drawn
from the author's *The Bohr Atom: A Guide* [35].

One of the results of Maxwell's work was that an accelerated electrical charge
radiates away energy in the form of electromagnetic radiation. The rate of energy
loss depends on the charge and the magnitude of the acceleration a. For an electron,
the rate of energy loss is given by

$$\frac{\mathrm{d}E}{\mathrm{d}t} = -\frac{e^2 a^2}{6\pi\varepsilon_o c^3},\tag{1.37}$$

where c is the speed of light. This formula was developed by Irish physicist Joseph
Larmor in 1897.

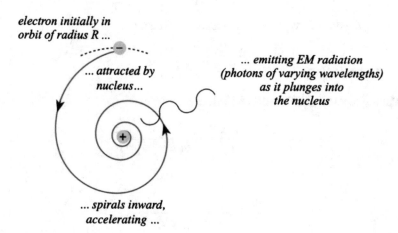

Fig. 1.8 Schematic illustration of electron spiral into nucleus

Figure 1.8 sketches the scenario considered here: An electron begins with orbital radius R around a nucleus of charge Ze, and then spirals inward due to energy loss, eventually merging with the nucleus in a catastrophic matter/anti-matter explosion. Refer to (1.21) for the total energy of such a system: If there is energy loss, E will become progressively more negative, which corresponds to the orbital radius becoming progressively smaller. The essential question is: If R has a value characteristic of atomic sizes, say $\sim 1\,\text{Å}$, how long will the spiral-in take?

If the spiral-in takes much longer than the characteristic time for one orbit, we can model the electron as being in a circular orbit at any time, with the acceleration being the corresponding centripetal acceleration. The validity of this assumption is investigated *a posteriori*.

As with the foregoing treatment of the Bohr atom, treat the nucleus as immobile. With (1.18), we can write the acceleration when the electron is at radius r as

$$a = \frac{v_e^2}{r} = \frac{Ze^2}{4\pi \varepsilon_o m_e r^2}. \tag{1.38}$$

The rate of energy loss is then

$$\frac{dE}{dt} = -\frac{e^2}{6\pi \varepsilon_o c^3}\left(\frac{Ze^2}{4\pi \varepsilon_o m_e r^2}\right)^2 = -\frac{Z^2 e^6}{96\pi^3 \varepsilon_o^3 m_e^2 c^3}\frac{1}{r^4}. \tag{1.39}$$

Now, at any moment, the total energy of the system will be, from (1.21),

$$E = -\frac{Ze^2}{8\pi \varepsilon_o r}. \tag{1.40}$$

Taking the derivative of this with respect to time gives

$$\frac{dE}{dt} = \frac{Ze^2}{8\pi \varepsilon_o r^2} \left(\frac{dr}{dt} \right).$$ (1.41)

Putting together (1.39) and (1.41) gives

$$\frac{Ze^2}{8\pi \varepsilon_o r^2} \left(\frac{dr}{dt} \right) = -\frac{Z^2 e^6}{96\pi^3 \varepsilon_o^3 m_e^2 c^3} \frac{1}{r^4},$$ (1.42)

or

$$\left(\frac{dr}{dt} \right) = -\frac{Ze^4}{12\pi^2 \varepsilon_o^2 m_e^2 c^3} \frac{1}{r^2}.$$ (1.43)

Separating variables and integrating from $r = R$ at $t = 0$ to $r = 0$ at $t = t_{collapse}$ gives

$$\int_R^0 r^2 dr = -\left(\frac{Ze^4}{12\pi^2 \varepsilon_o^2 m_e^2 c^3} \right) \int_0^{t_{collapse}} dt.$$ (1.44)

Integrating and substituting the appropriate numbers gives

$$t_{collapse} = \left(\frac{4\pi^2 \varepsilon_o^2 m_e^2 c^3}{Ze^4} \right) R^3 \sim \frac{\left(1.05 \times 10^{20} \text{ m}^{-3} \right)}{Z} R^3.$$ (1.45)

For $R = 1\,\text{Å}$ and $Z = 1$ (hydrogen), the collapse time is $\sim 10^{-10}$ s: about *one-tenth of a nanosecond*. Clearly, this model cannot be correct. It was the issue of this brief collapse time that prompted Bohr to arbitrarily assume that electrons had to be in some sort of non-radiating states wherein they did not emit electromagnetic radiation even as they orbited the nucleus; these came to be called (somewhat confusingly) "stationary" states.

Is the assumption that the spiral-in will occur over many orbits valid? We can check this by comparing the typical orbital period to the spiral-in period. For an orbit of radius R and speed v, the orbital period will be $2\pi R/v$. Using (1.18) for the speed gives

$$t_{orbit} = \frac{2\pi R}{v} = 2\pi R \sqrt{\frac{4\pi \varepsilon_o m_e R}{Ze^2}} = \frac{4\pi^{3/2} \varepsilon_o^{1/2} m_e^{1/2} R^{3/2}}{Z^{1/2} e}.$$ (1.46)

Hence

$$\frac{t_{collapse}}{t_{orbit}} = \left(\frac{4\pi^2\varepsilon_o^2 m_e^2 c^3}{Ze^4}\right) R^3 \left(\frac{Z^{1/2}e}{4\pi^{3/2}\varepsilon_o^{1/2}m_e^{1/2}}\right) \frac{1}{R^{3/2}}$$

$$= \left(\frac{\pi^{1/2}\varepsilon_o^{3/2}m_e^{3/2}c^3}{Z^{1/2}e^3}\right) R^{3/2}.$$

(1.47)

The term in brackets evaluates to 2.66×10^{20} m$^{-3/2}$; for $R = 1$Å, $t_{collapse}/t_{orbit} \sim$ 266,000. The spiral-in time is much longer than the orbital time, so treating the spiral-in as a succession of circular orbits is plausible.

Summary

The need to introduce the hypothesis that some physical quantities such as energy and angular momentum are quantized grew out of the inability of classical physics to provide adequate understanding of phenomena such as thermal radiation, atomic spectra, and atomic structure. The way was led by Max Planck, who in 1900 proposed that the mechanism of thermal radiation could be modeled by assuming that the source of blackbody radiation lay in "oscillators" restricted to possessing energies given by

$$E = nh\nu,$$

where n is an integer ($n = 1, 2, 3, \ldots$), h is Planck's constant, and ν is the frequency of vibration. Later, Niels Bohr used this assumption in conjunction with a semiclassical planetary electron model of the hydrogen atom. The essence of Bohr's model is that the orbiting electron is restricted to certain "stationary" energy states wherein it does not radiate. Atomic spectra are then presumed to be due to electrons executing transitions between different stationary states, with the difference in energy ΔE between the states appearing as a photon of energy $\Delta E = h\nu$ in accordance with Planck's hypothesis. The energy level of a Bohr atom in quantum state n ($n = 1, 2, 3, \ldots$) is given by

$$E_n = -\frac{m_e Z^2 e^4}{8\varepsilon_o^2 h^2 n^2} = -13.595\frac{Z^2}{n^2} \text{ eV} \quad (n = 1, 2, 3, \ldots),$$

where Z is the atomic number of the nucleus concerned. The wavelength of the photon emitted by an electron transiting from Bohr orbit n_i to orbit n_f ($n_i > n_f$) is given by

$$\lambda_{n_i \to n_f} = \frac{911.75}{Z^2}\left(\frac{1}{n_f^2} - \frac{1}{n_i^2}\right)^{-1} \text{ Å}.$$

A corollary of the Bohr model is that the orbital angular momentum of the electron is quantized in units of $h/2\pi$.

A central concept of the quantum aspect of nature is that of matter-waves, due to de Broglie. On the scale of atomic phenomena, we are forced to abandon the notion of point particles in favor of treating them as being effectively smeared out over a characteristic de Broglie wavelength given by

$$\lambda = \frac{h}{mv}.$$

The idea of matter-waves proves to be intimately related to the probability distribution interpretation of quantum mechanics.

Problems

1.1(E) Table 1.3 gives densities and atomic weights for a number of common elements. Use (1.1) to calculate approximate sizes for these species.

1.2(E) Consider a gas of "classical" Maxwell-Boltzmann atoms whose energies are distributed in accordance with (1.8). In an environment of temperature $T = 6000$ K K (the approximate temperature of the surface of the Sun), what is the ratio of the number of atoms with energies between $E + dE$ to those with energies between $2E$ and $2E + dE$ when $E = 5$ eV?

1.3(I) Equation (1.16) gives the amount of radiant energy $E(v)dv$ between frequencies v and $v + dv$ within a blackbody cavity. According to Einstein, the energy of an individual photon of frequency v is $E = hv$. If the radiant energy is presumed to be comprised of individual photons, then the number of photons with frequencies between v and $v + dv$ in the cavity would be given by $n(v)dv = E(v)dv/hv$. The total number of photons within the cavity will then be given by the integral of this expression for $v = 0$ to ∞. Given that

$$\int_0^\infty \frac{x^2 dx}{(e^x - 1)} \sim 2.404,$$

determine the density of photons in a blackbody cavity at $T = 300$ K, about room temperature. Intergalactic space has a temperature of about 2.7 K; what is the density of photons in that environment?

Table 1.3 Problem 1.1

Element	Density (g/cm^3)	A (g/mol)
Al	2.7	27
Fe	7.88	56
Cu	8.96	64
Pb	11.34	207
Pt	21.45	195

1.4(E) To establish the relationship between (1.16) and Fig. 1.3 requires some knowledge of thermodynamics, which shows that the rate of energy escape (power) between wavelengths λ and $\lambda + d\lambda$ through a hole of area A in a blackbody cavity of temperature T is given by

$$P(\lambda)d\lambda = A\left\{\frac{2\pi hc^2}{\lambda^5\left(e^{hc/\lambda kt} - 1\right)}\right\}d\lambda.$$

Consider a blackbody at $T = 7000$ K. What is the power emitted through a hole of area 1 m^2 between $\lambda = 4000$ and 4001 Å? Does your result accord with Fig. 1.3?

1.5(E) Derive an expression for the speed of an electron in Bohr orbit n in terms of the speed of light. Is it justifiable to neglect relativistic effects in the development of the Bohr model?

1.6(E) Repeat the derivation of the Bohr model for a lone electron orbiting a nucleus of Z protons. Use your results to compute ionization energies for a single electron orbiting nuclei of (a) helium, (b) carbon, and (c) uranium.

1.7(E) Derive an expression for the centripetal force experienced by the electron in Bohr orbit n as a function of n and physical constants. What is the value of the force when $n = 1$? Modern force-probe microscopes are capable of measuring nano-Newton scale forces, albeit with stationary objects.

1.8(E) An electron transits between two energy levels separated by energy E, emitting a photon of wavelength λ in accordance with the Planck/Bohr relation $E = hc/\lambda$. (a) Show that if λ is given in units of Ångstroms, then $E = 12,398/\lambda$ electron-volts. (b) As a consequence of some perturbing effect, the separation of the energy levels is altered by an amount dE electron-volts. Show that the corresponding change in the wavelength of the emitted photon is $d\lambda = -\lambda^2 dE/12,398$ Å. If a transition leading to a photon of wavelength 4000 Å is perturbed by $dE = 0.01$ eV, what is $d\lambda$?

1.9(I) Consider an electron and a proton separated by some distance r. They will attract each other by both electrical and gravitational forces. What is the ratio F_{elec}/F_{grav}? See also the following problem.

1.10(E) Both Newton's gravitational law and Coulomb's law are inverse-square laws: the force of attraction between the Sun (S) and the Earth (E) has magnitude GM_SM_E/r^2, whereas the force of attraction between an electron and a proton in a hydrogen atom is $e^2/4\pi\varepsilon_o r^2$. Derive an expression for the equivalent of the Bohr radius for the gravitational case. What is the value of the quantum number of Earth's orbit? Would distance differences between individual quantum states in the solar system be observable?

1.11(A) The text indicates how one can take into account the motions of the proton and electron around their mutual center of mass in the Bohr model. Using the condition that the center of mass of an isolated system must stay in the same place, derive the correction indicated in the text, and hence derive the factor of 911.75 Å in (1.31).

HINT: Do not forget the kinetic energy of the proton about the electron/proton center of mass.

1.12(I) It is remarked in the text that the effect of the motions of the electron and proton around their common center of mass in the Bohr model can be accounted for by substituting the reduced mass of the system, $m_e m_p / (m_e + m_p)$ in place of the electron mass m_e wherever the latter appears. An atom of deuterium consists of a lone electron orbiting a nucleus consisting of a proton and a neutron fused together; such a nucleus has mass close to $2m_p$. Show that the difference in wavelength between the Balmer $3 \to 2$ transition for deuterium versus ordinary hydrogen is given by

$$(\lambda_h - \lambda_d)_{3 \to 2} \sim \frac{144}{5} \frac{\varepsilon_o^2 h^3 c}{m_p e^4}.$$

Evaluate this result numerically. If you have available a spectrometer capable of resolving spectral lines that are separated by no less than $2\,\text{Å}$, would you see the deuterium and hydrogen lines as separate?

1.13(I) From the discussion in Sect. 1.3 and the result of Problem 1.12, it should be evident that when a lone electron orbiting a nucleus of charge Ze transits from Bohr orbit n_i to orbit n_f, the wavelength of the photon emitted is given by

$$\lambda_{n_i \to n_f} = \frac{8 \varepsilon_o^2 h^3 c}{\mu Z^2 e^4} \left(\frac{1}{n_f^2} - \frac{1}{n_i^2} \right)^{-1},$$

where μ is the reduced mass of the electron/nucleus system. From this, the $4 \to 2$ transition for hydrogen ($Z = 1$) should yield a photon of the same wavelength as for the $8 \to 4$ transition for helium ($Z = 2$), *if* the reduced masses are the same. However, the reduced masses are slightly different, which makes the wavelengths slightly different. Compute the difference (in Å) between the wavelengths of these hydrogen and (once-ionized) helium transitions. In terms of the atomic mass unit u, the masses of an electron, proton and helium nucleus are respectively $m_e = 0.0005485u$, $m_p = 1.0072765u$, and $m_{He} = 4.0015062u$.

1.14(E) Write a computer program or spreadsheet to compute the wavelengths of photons emitted by a single electron orbiting a hydrogen-like atom of nuclear charge $+Ze$ as it transits between any pair of states (n_i, n_f). Test your program by comparing with Table 1.1 for $Z = 1$. What are the wavelengths of the "Balmer series" for helium? For uranium?

1.15(I) Derive expressions for the lowest and highest-possible wavelength photons emitted during transitions corresponding to a given series of spectral lines. Hence show that only the Lyman and Balmer series in hydrogen are not overlapped by lines from other series.

1.16(E) The Ritz combination principle was an empirical result that predated the Bohr model. This principle states that the inverse of the wavelength of a spectral line of some atomic species can often be predicted by computing the difference in the inverse wavelengths of two other spectral lines of the same species. Show that in the context of the Bohr model, hydrogen $p \rightarrow m$ transitions can be predicted from the $p \rightarrow n$ and $m \rightarrow n$ transitions according as

$$\frac{1}{\lambda_{p \rightarrow m}} = \frac{1}{\lambda_{p \rightarrow n}} - \frac{1}{\lambda_{m \rightarrow n}}.$$

Verify numerically by computing the $5 \rightarrow 3$ transition on the basis of the values for the $5 \rightarrow 1$ and $3 \rightarrow 1$ wavelengths given in Table 1.1.

1.17(I) A diatomic molecule such as H_2 can be modeled as two point particles each of mass m joined by a rigid rod of length L. Assuming that the molecule rotates about an axis perpendicular to the rod through its midpoint with angular momentum given by the Bohr condition (1.33), derive an expression for the quantized values of the rotational energy.

1.18(I) As another (and quite unrealistic) model of an atom, consider a solid sphere of mass M and radius R rotating about an axis through its center in accordance with the Bohr quantization condition on angular momentum. Show that the possible rotational energies are given by $E_n = 5n^2h^2/16\pi^2MR^2$, where n is the quantum number. From this result derive an expression for the wavelength of a photon emitted as this system undergoes a transition from state $n + 1$ to state n. Evaluate your answer numerically for the case of M equal to the mass of a proton (essentially the mass of a hydrogen atom), $R = 1\,\text{Å}$, and $n = 2$. Approximately what value would n have to be to give wavelengths in the optical region of the spectrum, $\lambda \sim 5000$ Å?

1.19(E) Compute the de Broglie wavelength for an electron of energy 1 keV. If this electron were interacting with a hydrogen atom, would classical mechanics suffice?

1.20(E) Particles in motion have associated de Broglie wavelengths. As a rule, if one is within a few de Broglie wavelengths of a particle, it loses its identity as a particle and acts more like a wave, with the result that classical mechanics no longer suffices to describe collisions. Rutherford investigated the scattering of α-particles (helium nuclei) from stationary gold nuclei, using classical mechanics to analyze the collisions. This question investigates the issue from a wave-mechanics perspective. (a) Determine the de Broglie wavelength of an α-particle of kinetic energy 6 million electron-volts. (b) The electrical potential energy of charges Q_1 and Q_2 separated by distance r is given by $U = Q_1Q_2/4\pi\varepsilon_o r$. By equating the initial kinetic energy of the α-particle to the electrical potential energy, find the distance of closest approach of the α-particle to the gold nucleus, and so render a judgment as to how well-justified Rutherford was in his use of classical mechanics.

1.21(E) In kinetic theory, atoms are modeled as point masses of mass m, and the mean speed v of an atom in an environment at absolute temperature T is $mv^2 = 3kT$, where k is Boltzmann's constant. Derive an expression for the de Broglie wavelength for an atom of mass m at absolute temperature T in terms of h, k, T, and m. At room temperature, hydrogen atoms typically travel many thousands of Ångstroms between collisions; is it fair to treat them as acting as point masses under such circumstances?

1.22(E) In a paper published in the October 14, 1999 edition of *Nature* [*Nature* **401**, 680–682 (1999)], a group at Universität Wien in Austria reported experimental detection of de Broglie wave interference of Carbon-60 "buckyball" molecules. Such molecules consist of 60 carbon atoms arranged in a roughly spherical shape reminiscent of a soccer ball. In their experiment, the C-60 molecules had a most probable velocity of 220 m/s. Their data is well-fit by a de Broglie wavelength of 2.5 picometers. Verify this wavelength from the information given.

1.23(E) An atom of mass m is initially at rest with respect to an outside observer and has an electron in an excited state. The electron transits to a lower energy level, emitting a single photon of wavelength λ in the process. The emitted photon will have momentum $p = h/\lambda$. To conserve momentum, the atom must recoil with some speed v. Derive an expression giving the ratio of the photon's energy to the kinetic energy of the recoiling atom in terms of λ, m, c, and h. A classical analysis will suffice for the recoiling atom. Evaluate the ratio numerically for a hydrogen atom emitting a photon of wavelength 5000 Å. Based on your result, do you think it was reasonable to neglect the energy involved with the recoiling atom in the derivation of the Bohr model? Along these lines, the momentum transferred to an atom upon its absorbing a photon has actually been measured [36].

References

1. G.J. Stoney, Phil. Mag. **11**, 381 (1881)
2. J.J. Thomson, Phil. Mag. **44**, 293 (1897)
3. A. Pais, *Inward Bound* (Oxford University Press, Oxford, 1986)
4. https://commons.wikimedia.org/wiki/File:Carbon_Spectra.jpg https://commons.wikimedia.org/wiki/File:Helium_spectra.jpg https://commons.wikimedia.org/wiki/File:Hydrogen_Spectra.jpg https://commons.wikimedia.org/wiki/File:Magnesium_Spectra.jpg https://commons.wikimedia.org/wiki/File:Neon_spectra.jpg https://commons.wikimedia.org/wiki/File:Silicon_Spectra.jpg https://commons.wikimedia.org/wiki/File:Sodium_Spectra.jpg
5. J.J. Balmer, Ann. Phys. **261**, 80 (1885)
6. M. Planck, Ann. Phys. **306**, 69 (1900)
7. T. Kuhn, *Black-Body Theory and the Quantum Discontinuity, 1894–1912* (University of Chicago Press, Chicago, 1978)
8. A. Einstein, Ann. Phys. **322**, 132 (1905)
9. G. Greenstein, A. Zajonc, *The Quantum Challenge*, 2nd edn. (Jones & Bartlett, Sudbury, Massachusetts, 2006)
10. E. Rutherford, Phil. Mag. **21**, 669 (1911)
11. N. Bohr, Phil. Mag. **26**, 1 (1913)
12. N. Bohr, Phil. Mag. **26**, 476 (1913)
13. N. Bohr, Phil. Mag. **26**, 857 (1913)
14. F. Paschen, Ann. Phys. **332**, 537 (1908)
15. T. Lyman, Phys. Rev. **3**, 504 (1914)
16. F. Brackett, Nat. **109**, 209 (1922)
17. H.A. Pfund, J. Opt. Sci. Am. **9**, 193 (1924)

18. B. Altschul, Phys. Rev. Lett. **98**(26), 261801 (2007)
19. G.B. Andresen et al., Nat. **468**(7324), 673 (2010)
20. M. Ahmadi et al., Nat. **578**(7795), 375 (2020)
21. H. Moseley, Phil. Mag. **26**, 1024 (1913)
22. H. Moseley, Phil. Mag. **27**, 703 (1914)
23. G. Hertz, Z. Phys. **22**, 18 (1924)
24. A.P. French, P.J. Kennedy, *Niels Bohr: A Centenary Volume* (Harvard, New Haven, 1985)
25. A. P. French, E. F. Taylor, *An Introduction to Quantum Physics* (W. W. Norton, New York 1978), See Sect. 2–3 for a relativistic treatment
26. C. Davisson, L.H. Germer, Phys. Rev. **30**, 707 (1927)
27. G.P. Thomson, Proc. Roy. Soc. **A117**, 600 (1928)
28. C. Jönsson, Z. Phys. **161**, 454 (1961)
29. D. Brandt, S. Hirschi, Am. J. Phys. 42(1), 4 (1974). This is a translation of Jönsson's paper
30. A. Tonomura, J. Endo, T. Matsuda, T. Kawasaki, H. Ezawa, Am. J. Phys. **57**(2), 117 (1989)
31. P. Schewe, Phys. Today **52**(12), 9 (1999)
32. S.K. Blau, Phys. Today **55**(1), 15 (2002)
33. R. P. Feynman, R. B. Leighton, M. Sands, *The Feynman Lectures on Physics* (Addison-Wesley, Reading, Massachusetts 1963–1965)
34. N.D. Mermin, Phys. Today **46**(1), 9 (1993)
35. B.C. Reed, *The Bohr Atom: A Guide* (IOP Publishing, Bristol, 2020)
36. P.F. Schewe, Phys. Today **58**(7), 9 (2005)

Chapter 2
Schrödinger's Equation

Summary This chapter develops Schrödinger's equation, the fundamental law of quantum physics. Schrödinger's equation is a type of "wave" equation, so this chapter opens with a description of the "classical" wave equation, that which would be used to describe a traveling wave propagating along a string such as one encounters in elementary physics. Like Newton's laws of motion, there is no rigorous derivation or proof of Schrödinger's equation, but a plausibility argument based on the classical wave equation and energy concepts is presented. There are actually two versions of Schrödinger's equation, one time-dependent and one time-independent; while the focus in this book is on the latter, both are developed in this chapter. Solutions to Schrödinger's equation take the form of a mathematical function usually written as $\psi(x)$ and which is known as a "wavfunction". Associated with any wavefunction is also corresponding energy. The solution of Schrödinger's equation for any particular sitMap often leads to *multiple* possible wavefunctions and corresponding energies. The interpretation of this function is unusual in that it deals with probabilities. The last section in this chapter discusses what conditions must be applied to solutions of Schrödinger's equation and their interpretation.

From lines of evidence like those marshaled in Chap. 1, it became clear to physicists in the early part of the twentieth century that the laws of classical mechanics do not apply on atomic scales. The issue then arose of how to modify classical mechanics to produce a "quantum" mechanics consistent with the microscopic characteristics of atoms and molecules while remaining consistent with Newtonian mechanics on macroscopic scales. As the essence of classical physics is embodied in Newton's laws, the essence of quantum mechanics is embodied in Schrödinger's equation.

At this point we could simply write down Schrödinger's equation and say: "Here it is—let's see how to solve it in certain circumstances," just as you would solve for the motion of a classical point mass using $F = ma$ with an applied force and some specified initial conditions. This would be convenient, but it would not be very satisfying. Postulated physical laws, while they are derivable from nothing more fundamental, do not simply pop out of thin air: they are products of thinking

© The Author(s), under exclusive license to Springer Nature Switzerland AG 2022

B. C. Reed, *Quantum Mechanics*, https://doi.org/10.1007/978-3-031-14020-4_2

processes conditioned by past successes, empirical facts, intuition, trial and error, and inspired guesses. In an effort to smooth the transition, we introduce Schrödinger's equation via a plausibility argument. This is a fancy way of saying that, while no rigorous derivation exists, it is possible to arrive at Schrödinger's equation by deftly combining some classical and quantum concepts. Of course, any equation could be concocted in this way; whether or not the result is useful is a matter for experiment to arbitrate.

We develop Schrödinger's equation in Sects. 2.2 and 2.3 after first reviewing the classical wave equation in Sect. 2.1. In Sect. 2.4 we discuss what general constraints must be imposed on solutions of Schrödinger's equation, and how solutions are interpreted. We follow this procedure because the interpretation, like the equation itself, is fundamentally a postulate and it seems sensible to gather postulates together in one place, independent of any particular example. Examples of setting up and solving Schrödinger's equation are reserved for subsequent chapters.

2.1 The Classical Wave Equation

Any equation that purports to describe quantum physics must respect the incontrovertible fact that particles possess a fundamentally wavy nature. To physicists of the early twentieth century, this demand was familiar ground: wavelike disturbances propagating along strings, water waves, sound waves, electromagnetic waves and so forth were pervasive phenomena in classical physics. As a starting point, then, let us consider the classical wave equation and how one might modify it to accommodate the wave nature of individual particles.

As a classical model of a wave, consider a disturbance moving along a string. Figure 2.1a shows a sketch of such a disturbance at some particular time, say $t = 0$. Distance along the string from the origin is measured by x, while $y(x, t)$ measures the amplitude, that is, the up-or-down displacement of the string from its undisturbed position as a function of position and time. The pattern may in fact be quite complex, but we can describe it in principle by saying that at $t = 0$, $y = f(x)$.

The simplest wave pattern we can conceive is one that retains its shape as it propagates along the string. The speed with which the pattern flows through the string is determined by the tension T in the string and its mass per unit length μ: $v = \sqrt{T/\mu}$. In such a transverse wave, an individual point on the string oscillates up-and-down but does not shift left-to-right as the disturbance passes through. Sketch (b) shows the wave pattern after time t has elapsed; the whole pattern has advanced a distance vt along the string. The vertical offset between the two patterns is for clarity only.

Now imagine two observers, O_1 and O_2, standing at fixed positions on the x axis as shown; O_2 is at distance vt to the right of O_1. Distances measured by O_1 and O_2 along the x direction are referred to as x and x', respectively. Assuming that the

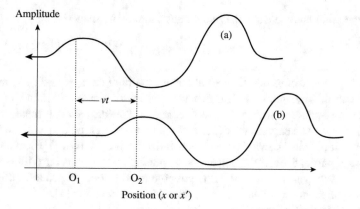

Fig. 2.1 A traveling wave at two different times

wave pattern retains its shape, each observer must describe the pattern with the same mathematical function of his or her distance coordinate:

$$y = f(x) = f(x').$$ (2.1)

In addition, when $x' = 0$, $x = vt$, so x and x' and are related according as

$$x' = x - vt.$$ (2.2)

Substituting (2.2) into (2.1) gives

$$y = f(x) = f(x - vt).$$ (2.3)

The interpretation of this expression is that for the wave pattern to maintain its shape, x, v, and t must always appear in the combination $(x - vt)$. Examples of acceptable wave patterns would be

$$y = \sin(x - vt),$$

$$y = e^{(x-vt)^{7/8}},$$

and

$$y = \text{Tanh}^{-1}[(x - vt)^{-4/3}].$$

On the other hand, a wave that would not maintain its shape is

$$y = \cos[x^2 - (vt)^2].$$

Although these considerations establish a constraint on x, v, and t if waves are to maintain their shape, it is clear that we could write down an infinite variety of possible wave patterns $y(x)$, or "wavefunctions" as they are called. Is there any more general way we can express the constraint without getting mixed up with an infinite number of functions? Inevitably, such generalized expressions of physical laws take the form of partial differential equations. Assuming that wave patterns must involve functions of the form $f(x - vt)$, it is possible to work out the general differential equation which any wave pattern must satisfy to be acceptable [1].

Now, we know that in general we can put

$$y(x, t) = f(x - vt) = f(z), \tag{2.4}$$

where we define $z = (x - vt)$. From this definition (the velocity is assumed to be constant)

$$\frac{\partial y}{\partial x} = \frac{\partial y}{\partial z}\frac{\partial z}{\partial x} = \frac{\partial y}{\partial z}, \tag{2.5}$$

and

$$\frac{\partial y}{\partial t} = \frac{\partial y}{\partial z}\frac{\partial z}{\partial t} = -v\frac{\partial y}{\partial z}. \tag{2.6}$$

From these we can form an identity:

$$\frac{\partial y}{\partial t} + v\frac{\partial y}{\partial x} = -v\frac{\partial y}{\partial z} + v\frac{\partial y}{\partial x} = 0. \tag{2.7}$$

Taking the partial derivative of this with respect to time gives:

$$\frac{\partial}{\partial t}\left(\frac{\partial y}{\partial t} + v\frac{\partial y}{\partial x}\right) = \frac{\partial}{\partial t}(0),$$

that is

$$\frac{\partial^2 y}{\partial t^2} + v\frac{\partial}{\partial t}\left(\frac{\partial y}{\partial x}\right) = 0,$$

or

$$\frac{\partial^2 y}{\partial t^2} + v\frac{\partial}{\partial x}\left(\frac{\partial y}{\partial t}\right) = 0, \tag{2.8}$$

where the last step followed by exchanging the order of the partial derivatives. From (2.6) we can replace $(\partial y/\partial t)$ by $-v(\partial y/\partial z)$, hence

$$\frac{\partial^2 y}{\partial t^2} + v\frac{\partial}{\partial x}\left(-v\frac{\partial y}{\partial x}\right) = 0$$

or

$$\frac{\partial^2 y}{\partial t^2} = v^2\frac{\partial^2 y}{\partial x^2}. \tag{2.9}$$

This is the *classical wave equation*. If you are familiar with some advanced mathematical physics, its solution in any particular case usually proceeds in terms of Fourier or Laplace transforms, but this will not concern us. The essential point for our purposes is that "classical wavefunctions" $y(x, t)$ must satisfy this equation.

Example 2.1 Show that the function $y = e^{(x-vt)^{7/8}}$ satisfies the classical wave equation.

We have a function of the form e^z where $z = (x - vt)^{7/8}$. It is helpful to recall that for any variable q,

$$\frac{d}{dq}\left[e^{z(q)}\right] = e^{z(q)}\frac{dz}{dq}.$$

Begin with the left side of the wave equation. Here we will have $q \equiv t$, and

$$\frac{\partial z}{\partial t} = \frac{7}{8}(x - vt)^{-1/8}(-v).$$

Hence

$$\frac{\partial y}{\partial t} = e^z\frac{\partial z}{\partial t} = -\frac{7}{8}(v)e^z(x - vt)^{-1/8}.$$

Remembering the chain rule of basic calculus, the second derivative goes as

$$\frac{\partial^2 y}{\partial t^2} = -\frac{7}{8}(v)\frac{\partial}{\partial t}\left[e^z(x - vt)^{-1/8}\right]$$

$$= -\frac{7}{8}(v)\left[e^z\frac{\partial z}{\partial t}(x - vt)^{-1/8} - \frac{1}{8}e^z(x - vt)^{-9/8}(-v)\right]$$

$$= -\frac{7}{8}(v)\left[-\frac{7}{8}(v)e^z(x - vt)^{-1/4} - \frac{1}{8}e^z(x - vt)^{-9/8}(-v)\right]$$

$$= (v^2)\left[\frac{49}{64}(x - vt)^{-1/4} - \frac{7}{64}(x - vt)^{-9/8}\right]e^z.$$

Now look at the right side of the wave equation. Here $q \equiv x$, and we have

$$\frac{\partial z}{\partial x} = \frac{7}{8}(x - vt)^{-1/8}.$$

Hence

$$\frac{\partial y}{\partial x} = e^z \frac{\partial z}{\partial x} = \frac{7}{8} e^z (x - vt)^{-1/8}.$$

The second derivative develops as

$$\frac{\partial^2 y}{\partial x^2} = \frac{7}{8} \frac{\partial}{\partial x} \left[e^z (x - vt)^{-1/8} \right]$$

$$= \frac{7}{8} \left[e^z \frac{\partial z}{\partial x} (x - vt)^{-1/8} - \frac{1}{8} e^z (x - vt)^{-9/8} \right]$$

$$= \frac{7}{8} \left[\frac{7}{8} e^z (x - vt)^{-1/4} - \frac{1}{8} e^z (x - vt)^{-9/8} \right]$$

$$= \left[\frac{49}{64} (x - vt)^{-1/4} - \frac{7}{64} (x - vt)^{-9/8} \right] e^z.$$

Comparing the two second derivatives shows that the wave equation is satisfied.

In developing the wave equation why didn't we stop at (2.6), $\frac{\partial y}{\partial t} = -v \frac{\partial y}{\partial x}$? The reason is that a wave equation should be space and time invariant, that is, insensitive to changes in the definition of the positive directions of x and t. Equation (2.9) has this property, but (2.6) does not.

Aside from spatial (and temporal) invariance, Eq. (2.9) has another important property: that of *linearity*. This means that wherever the dependent variable y appears, it does so only to the first power; there are no non-linear terms such as e^y or $\sin(y)$ or $\sqrt{y}$. The result of this is that if one has a number of independent solutions $y_1, y_2, \ldots y_n$ to the wave equation (all with the same speed v), then so too will be a function constructed by a linear sum of them, that is, a function of the form

$$y = a_1 y_1 + a_2 y_2 + \cdots + a_n y_n, \tag{2.10}$$

where the coefficients $a_1, a_2, \ldots a_n$ are constants.

This linearity property is of great value in the following sense. In discussions of harmonic motion in elementary physics texts, one often encounters solutions of the wave equation of the from

$$y = A \sin(kx - \omega t), \qquad (2.11)$$

where A is the amplitude of the wave, k is its wavenumber ($k = 2\pi/\lambda$, where λ is the wavelength), and ω is its angular frequency ($\omega = 2\pi\nu$, where ν is the frequency in cycles per second, or Hertz). This expression corresponds to a wave pattern moving in the direction of the $+x$ axis; for one moving in the $-x$ direction, replace the $-$ sign inside the brackets with a $+$ sign. Now, it might happen in some particular problem that the solution to the wave equation is in fact more complicated than this. However, since Fourier analysis will always let us represent any function as a linear sum of sines and cosines and since the wave equation is linear, we are guaranteed that such a sum will also be a solution. That is, we can always express any solution of the wave equation as a sum of sinusoidal functions of the from of (2.11). By using (2.11) as a prototype wave in our development of Schrödinger's equation, we can keep the mathematical manipulations fairly simple while sacrificing no generality.

2.2 The Time-Independent Schrödinger Equation

How can the classical wave equation be modified to accommodate the characteristics of quantum matter-waves? This is by no means obvious: strings exhibit wave phenomena as a whole; y and v in (2.9) refer to a pattern amplitude and speed. What we now desire is a wave equation applicable to individual particles. With no idea of what a quantum wave equation can be expected to look like, how should we proceed? It is at this point that our derivation becomes non-rigorous. What follows might at best be called a plausibility argument. Readers interested in a detailed account of the development of quantum ideas and fundamental relationships are urged to look at Max Jammer's masterful *The Conceptual Development of Quantum Mechanics* [2]. Other, more recent sources along these lines are David Lindley's *Uncertainty: Einstein, Heisenberg, Bohr, and the Struggle for the Soul of Science* and Olivier Darrigol's "A simplified genesis of quantum mechanics" [3, 4].

Schrödinger was led by the assumption that the putative quantum wave equation should possess two properties. Experience with classical waves argued strongly in favor of a linear, second-order differential equation, and the success of the Bohr model argued for retaining the concept of energy conservation. In classical notation, conservation of mechanical energy is expressed as

$$E = \frac{p^2}{2m} + V(x), \qquad (2.12)$$

where p is the momentum of the particle and where $V(x)$ denotes the potential energy of the particle as a function of position x: the so-called *potential function*. Momentum will of course also be a function of position: $p(x)$. We will derive Schrödinger's equation in one dimension, and later generalize the result to three dimensions.

To make the connection from classical wave to quantum-mechanical matter-waves, recall the de Broglie relation between the momentum of a particle and its wavelength:

$$\lambda = \frac{h}{p}. \tag{2.13}$$

Substituting this into the prototype sinusoidal wavefunction introduced in the last section, Eq. (2.11), along with $k = 2\pi/\lambda$ yields

$$y = A \sin\left(\frac{p}{\hbar}x - \omega t\right). \tag{2.14}$$

To impose the constraint of energy conservation without specifying any particular potential function $V(x)$, we extract an expression for the momentum from this wavefunction. Differentiating (2.14) twice with respect to x leads to

$$p^2 = -\frac{\hbar^2}{y}\left(\frac{\mathrm{d}^2 y}{\mathrm{d}x^2}\right), \tag{2.15}$$

which, upon substitution into the expression for conservation of energy, Eq. (2.12) leads to a differential equation:

$$-\frac{\hbar^2}{2m}\frac{\mathrm{d}^2 y}{\mathrm{d}x^2} + V(x)y = Ey. \tag{2.16}$$

Note how this derivation was carried out to yield a differential equation which bears no explicit trace of the form of the wavefunction that was assumed at the start! (2.16) is the nonrelativistic, one-dimensional, time-independent Schrödinger equation (also known as Schrödinger's wave equation) for a particle of mass m. This equation and some of its fundamental applications were put forth by Erwin Schrödinger in a series of papers in 1926 [5–8], and an English-language description of the theory was published in late 1926 [9]. Much of what we shall take up in the remainder of this book descends directly from these fundamental papers.

Mathematically, Eq. (2.16) can be classified as a second-order linear differential equation with independent variable x and dependent variable y. As derived here, this equation involves only one spatial dimension, x. This is fine for a pulse traveling along a string, but atoms and molecules live in three dimensions. By intuition, let us accept that (2.16) can be expressed in a three-dimensional form by adding second partial derivatives of the wave pattern with respect to y and z to the left side. This leads to a minor crisis of notation: confusion between the function representing the wave pattern and the coordinate y itself. Convention is to designate the time-independent wave pattern by the lower-case Greek letter psi (ψ). In this notation, Schrödinger's equation becomes

$$-\frac{\hbar^2}{2m}\left(\frac{\partial^2\psi}{\partial x^2}+\frac{\partial^2\psi}{\partial y^2}+\frac{\partial^2\psi}{\partial z^2}\right)+V\psi=E\psi, \tag{2.17}$$

where we have written partial derivatives because ψ may be a function of some or all of (x, y, z), and, in anticipation of the following section, also of time. Be sure to understand that both the potential function V and solution function ψ in this expression will in general be functions of all three coordinates (x, y, z).

An astute reader will have noticed three curious aspects regarding the foregoing derivation of Schrödinger's equation. The first is that the classical wave equation was in fact never used—only a prototype wavefunction. The second is that nowhere was the concept of quantization built into the development. Lastly, the total system energy E appears explicitly in the resulting differential equation. From the point of view of classical physics, this latter aspect is distinctly unusual. Newton's law of motion $F = ma$ makes no mention of energy: that is something to be specified by supplying the initial position and speed of the particle involved after solving the differential equation for a particular force law, a process known as supplying "boundary conditions." We will see that quantization is actually predicted by Schrödinger's equation on solving it for a given $V(x, y, z)$ and set of boundary conditions. Schrödinger's equation is of a form known to mathematicians as an *eigenvalue equation*. Speaking loosely, this means that for a given $V(x, y, z)$ and set of boundary conditions, there exists a certain restricted set of functions $\psi(x, y, z)$ that satisfy the equation, and that to each of these functions corresponds a particular value of E. This set of values of E are the quantized energy states of the system.

Equation (2.17) is expressed in Cartesian coordinates. Coordinate systems are, however, a human construct; any true physical law should be expressible in a form which makes no mention of any particular coordinate system. In the case of Schrödinger's equation we can satisfy this demand by writing it in the form

$$-\frac{\hbar^2}{2m}\nabla^2\psi(r)+V(r)\psi(r)=E\psi(r), \tag{2.18}$$

where ∇^2 is the Laplacian operator and where r designates a general (x, y, z) position.

This coordinate-system independence is actually a rather profound point of fundamental physics. Physical laws should transcend the choice of any particular coordinate system. To be sure, the Laplacian operator has different forms depending on the choice of coordinate system, but all of these forms express the same concept: a sum of second derivatives with respect to spatial coordinates. Ultimately, the results of physical calculations cannot depend upon the coordinate system used to develop them.

Example 2.2 Suppose that the function $\psi(x) = Ae^{-\iota kx}$ where $\iota = \sqrt{-1}$ and where A and k are constants is known to be a solution to the time-independent one-dimensional Schrödinger equation with energy E. What must be the potential function $V(x)$?

The time-independent Schrödinger equation in one dimension appears as

$$-\frac{\hbar^2}{2m}\frac{d^2\psi}{dx^2} + V(x)\psi = E\psi.$$

The second derivative of the given function is $d^2\psi/dx^2 = (\iota k)^2 Ae^{-\iota kx} = -k^2\psi$. Hence we have

$$-\frac{\hbar^2}{2m}(-k^2\psi) + V(x)\psi = E\psi.$$

We can cancel the common factor of ψ to leave

$$V(x) = E - \frac{\hbar^2 k^2}{2m},$$

that is, $V(x) = $ constant. This case corresponds to a so-called free particle, one whose energy is in fact unrestricted. We will see solutions of this form in Chap. 3. It is worth noting that the wavefunction $\psi = Be^{+\iota kx}$ would lead to the same result for $V(x)$. Usually, $V(x)$ is considered to be given and $\psi(x)$ is what is sought; for the case of $V(x) = $ constant we will see in Chap. 3 that the k's in both solutions will be the same. Hence the most general solution to this potential is $\psi = Ae^{-\iota kx} + Be^{+\iota kx}$. All second-order differential equations have two families of solutions.

2.3 The Time-Dependent Schrödinger Equation

Any sensible wave equation should be both space and time-dependent. In the derivation above, time dependence was overlooked by focusing on derivatives of y (or ψ) with respect to x. In doing so, any knowledge of the direction sense of the wave pattern was foregone. Since the vast majority of the problems that will be addressed in this book are time-independent, this represents no loss. It is, however, worthwhile presenting a brief pseudo-derivation of the time-dependent Schrödinger equation for sake of completeness, to touch on some historical issues, and to establish some results to be used in Chap. 3. Also, the time-dependent Schrödinger equation is necessary for a complete proof of Heisenberg's famous Uncertainty Principle (Appendix A).

That we do not consider solutions of the time-dependent Schrödinger equation should not be interpreted to mean that it is somehow of lesser importance than it's time-independent counterpart, which is referred to herein rather grandly as *the* Schrödinger equation. The time-independent Schrödinger equation can ultimately inform us only of the permitted energy levels corresponding to a potential, the "stationary states." On the other hand, particles that are changing position in time can only be modeled with solutions of the time-dependent Schrödinger equation, and an example of constructing such a "moving wave packet" is treated briefly in Sect. 4.8. Beyond this, one of the main applications of quantum mechanics is in calculating transition rates between stationary states, and for this one must also invoke the full machinery of time-dependence. The scope of the material presented here is restricted to time-independent solutions as the intent is an introductory treatment; developing and interpreting such solutions proves to be an area more than rich enough to fill several chapters.

In deriving the time-independent Schrödinger equation, the approach was to use an expression for momentum, Eq. (2.15), which was extracted from a prototype waveform, Eq. (2.14), and substitute it into conservation of energy as expressed in (2.12). To establish the time-dependent Schrödinger equation, we play a similar game by extracting an expression for the total energy E from the waveform and substituting it into the time-independent Schrödinger equation as follows. We start with Planck's energy-frequency relation for oscillators,

$$E = h\nu. \tag{2.19}$$

This, along with the definition of angular frequency $\omega = 2\pi\nu$, renders the prototype wave pattern as (in ψ notation)

$$\psi = A \sin\left[\frac{1}{\hbar}(px - Et)\right]. \tag{2.20}$$

We extract E by taking the time-derivative of this:

$$\frac{\partial \psi}{\partial t} = -\frac{E}{\hbar} A \cos\left[\frac{1}{\hbar}(px - Et)\right]. \tag{2.21}$$

However, this leads to a problem: if our wave equation is to represent a physical law, it should be independent of the form of any of its particular solutions: that is, it should involve ψ only, not any particular function for ψ. Since only a first derivative is necessary to extract a factor of E, we are left with a cosine function as opposed to a sine function. One way to rectify this is to recall that the sum of a squared sine and cosine of an argument is unity, that is,

$$A^2 \sin^2 \left[\frac{1}{\hbar}(px - Et) \right] + A^2 \cos^2 \left[\frac{1}{\hbar}(px - Et) \right] = A^2. \qquad (2.22)$$

Recognizing the first term in this expression as ψ^2 and rearranging to isolate the cosine term gives

$$A \cos \left[\frac{1}{\hbar}(px - Et) \right] = \pm\sqrt{A^2 - \psi^2}, \qquad (2.23)$$

where the sign ambiguity arises on taking the square root. Substituting this into (2.21) and solving for E gives

$$E = \frac{\mp\hbar}{\sqrt{A^2 - \psi^2}} \left(\frac{\partial\psi}{\partial t} \right), \qquad (2.24)$$

which, upon substitution into the right side of the time-independent Schrödinger equation renders it as

$$-\frac{\hbar^2}{2m}\frac{\partial^2\psi}{\partial x^2} + V(x,t)\psi = \frac{\mp\hbar\psi}{\sqrt{A^2 - \psi^2}} \left(\frac{\partial\psi}{\partial t} \right). \qquad (2.25)$$

Mathematically, this is a plausible approach, but serious objections attend this result. The first is that the amplitude of the wave should not appear in what is presumed to be a general physical law: amplitude, like energy, is to be dictated by the boundary conditions relevant to a given problem. Worse, because of the presence of the square root in the denominator of the right side, this differential equation is not linear: a sum of independent solutions would not itself be a solution. Also, there is a sign ambiguity.

One might argue that the amplitude objection could be circumvented by taking a second time derivative of the prototype waveform, which would return us to a sine function, that is, to the original ψ itself. However, this would lead to an expression involving E^2, with the result that the consequent differential equation would still be non-linear because E appears only to the first power in the energy-conservation expression, necessitating taking a square root of $\partial\psi^2/\partial t^2$. We are forced to conclude that no conventional prototype wave expression will lead to a sensible time-dependent wave equation.

Faced with this situation, Schrödinger hit on the idea of modifying the prototype wave to an imaginary exponential form,

$$\psi(x,t) = A \exp\left[\frac{\iota}{\hbar}(px - Et) \right]. \qquad (2.26)$$

The beauty of this assumption is that it leaves the time-independent Schrödinger equation unchanged (that is, we still have (2.16)—try it!), but leads to a more plausible time-dependent form. In this case we find

$$E = -\frac{\hbar}{\iota\psi}\left(\frac{\partial\psi}{\partial t}\right) = \frac{\iota\hbar}{\psi}\left(\frac{\partial\psi}{\partial t}\right), \tag{2.27}$$

where we have used the fact that $1/\iota = -\iota$. Upon substituting this result into the right side of the time-independent Schrödinger equation we find

$$-\frac{\hbar^2}{2m}\frac{\partial^2\psi}{\partial x^2} + V(x)\psi = \iota\hbar\left(\frac{\partial\psi}{\partial t}\right). \tag{2.28}$$

This is the time-dependent Schrödinger equation.

Despite our success in arriving at a linear time-dependent differential equation, a problem remains: we have played fast-and-loose with the form of the wavefunction in such a way that it now represents neither a wave moving to the right nor one moving to the left. Recalling Euler's identity for expressing a sine function in terms of exponentials,

$$\sin\theta = \frac{1}{2\iota}(e^{\iota\theta} - e^{-\iota\theta}), \tag{2.29}$$

we see that a right-moving "matter wave" should be written as

$$\psi(x,t) = A\sin\left[\frac{1}{\hbar}(px - Et)\right] = \frac{A}{2\iota}\left[e^{\frac{\iota}{\hbar}(px-Et)} - e^{-\frac{\iota}{\hbar}(px-Et)}\right]. \tag{2.30}$$

Curiously, a purely right (or left)-moving wave leads to an unacceptable time-dependent wave equation. On the other hand, either of the individual exponential components of which such a wave is composed does lead to a sensible differential equation, albeit with a change of sign.

Clearly, we cannot have *both* a sensible-looking wave equation and a classical left or right-moving wave; something has to give. And yet, if we are to speak of particle-waves, there presumably must be some concept of direction of travel. To resolve this impasse, Schrödinger proposed arbitrarily adopting one of the exponential terms in (2.30) as representing a right-directed (that is, $+x$ direction) matter wave, and the other to represent a left-moving matter wave. By historical convention these are taken to be the first and second terms, respectively:

$$\psi_{right}(x,t) = A\exp\left[\frac{\iota}{\hbar}(px - Et)\right], \tag{2.31}$$

and

$$\psi_{left}(x,t) = A\exp\left[-\frac{\iota}{\hbar}(px - Et)\right], \tag{2.32}$$

where the factor of $1/2\iota$ has been absorbed into the amplitude. Experiment has shown that one can consistently adopt these prescriptions.

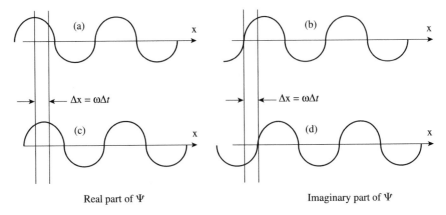

Real part of Ψ Imaginary part of Ψ

Fig. 2.2 Sketches of the real and imaginary parts of $\Psi(x, t)$ at $t = 0$ and $t = \Delta t$

Actually, the decision to adopt (2.31) and (2.32) as representing right and left-moving waves is not wholly arbitrary, and can be given some justification via the following argument.

Consider (2.28), the time-dependent Schrödinger equation. By comparing it to its time-*independent* counterpart, Eq. (2.16), it should be clear that the solution can in general be written as

$$\Psi(x, t) = \psi(x)e^{-\iota\omega t},$$

where $\psi(x)$ represents the solution to the time-independent Schrödinger equation and where $\omega = E/\hbar$. Now consider a wavefunction whose spatial part is of the form of the supposed right-moving wave, $\psi(x) = Ae^{\iota kx}$. The complete solution is then

$$\Psi(x, t) = Ae^{\iota(kx - \omega t)} = A\left[\cos\left(kx - \omega t\right) + \iota\sin\left(kx - \omega t\right)\right]. \tag{2.33}$$

Panels (a) and (b) of Fig. 2.2 respectively illustrate the real (cosine) and imaginary (sine) parts of Ψ at $t = 0$, and panels (c) and (d) illustrate them at some slightly later time Δt, presumed to be much less than the period of the wave. At the later time the maximum of the cosine function in the real part of Ψ is not reached until $x = \omega t/k$; likewise for the zero of the sine function in the imaginary part. The pattern for Ψ behaves as if it has advanced slightly to the right between $t = 0$ and $t = \Delta t$. A similar construction for $\psi(x) = Ae^{-\iota kx}$ reveals a pattern whose "motion" is to the left.

Bear in mind that the above derivations are not rigorous: while prototype waveforms and the de Broglie and Planck relations were used to set up Schrödinger's equation, we will find that solutions will in general look nothing like simple sinusoidal or exponential functions.

2.4 Interpretation of ψ: Probabilities and Boundary Conditions

For a wave on a string, the meaning of the wave pattern $y(x, t)$ is clear: the vertical displacement of the string from its equilibrium position. How are we to interpret a quantum-mechanical wavefunction $\psi(x, y, z)$? Does an electron orbiting a nucleus actually "wave" in some way? Does it require some medium within which to do so? An interpretation of ψ in terms of probabilities was hinted at in Sect. 1.4. The precise recipe was postulated by German physicist Max Born in 1926 [10, 11]. To state Born's result quantitatively requires an understanding of the concept of the complex conjugate of a quantity.

In general, the wavefunction $\psi(r)$ that satisfies Schrödinger's equation for some potential function $V(r)$ may be explicitly complex, that is, of the form $\psi = a + \iota b$, where a and b are real functions of (x, y, z). The complex conjugate of a function is formed by replacing ι by $-\iota$ wherever it appears, and designating the result with an asterisk: $\psi^* = a - \iota b$. The product of a function and its complex conjugate is always a purely real function:

$$\psi^*\psi = (a + \iota b)(a - \iota b) = a^2 + \iota ab - \iota ab - \iota^2 b^2 = a^2 + b^2,$$

because $\iota^2 = -1$. Thus, if ψ is complex (hence unphysical), we can construct a real quantity from it by computing $\psi^*\psi$. For brevity, $\psi^*\psi$ is often abbreviated as $|\psi|^2$ or simply ψ^2.

Born's interpretation of ψ can now be stated: The probability of finding the particle in a small volume of space dV located at position r is given by $\psi^*(r)\psi(r)dV$; note that dV is used here as an element of volume; do not confuse it with the potential function $V(r)$. This interpretation means that at any given time, we cannot say exactly where the particle is; we are restricted to specifying only that at some time, the probability of finding the particle between x and $x + dx$, y and $y + dy$, z and $z + dz$ is given by $\Psi^*(x, y, z, t)\Psi(x, y, z, t)dx\,dy\,dz$. Born is credited with coining the term "quantum mechanics"; see [12, 13].

This probabilistic interpretation of ψ is sometimes popularly misconstrued to imply that a particle can be in two or more places simultaneously. This is not so; particles are not assumed to lose their identity as (essentially) point objects. To make a macroscopic-scale analogy, suppose that a friend says "There is a ten percent probability that I will be at home this afternoon, a sixty percent probability that I will be at school, and a thirty percent probability that I will be at work." This does not mean that they will be at all three places simultaneously, but rather that they are more likely to be at school than at home or work—but they might be at any of the three locales. Until you do some experiment to find out where your friend actually is at some time, your knowledge is limited to these probabilities. For a very readable treatment on other common quantum misinterpretations, see [14].

A probabilistic interpretation of ψ implies a clear break with the sort of quantities one is familiar with in classical mechanics. Physicist and historian of science

Abraham Pais has remarked that the concept of probability as an inherent feature of fundamental physical law may be the most dramatic scientific change of the twentieth century. Instead of being able to specify explicit recipes for the position and velocity of a particle, we can only predict the probability of finding the particle in some region of space or of having momentum within some range. This probability will be greater in some places and less, perhaps even zero, in others. ψ itself has no physical reality: it is a mathematical construct used to keep account of probabilities. Particles are not actually "undulating" in space; the apparent "waviness" of matter is a consequence of its *probability distribution*.

Quantitatively, Born's interpretation can be expressed as

$$P(r)\mathrm{d}V = \psi^*(r)\psi(r)\mathrm{d}V. \tag{2.34}$$

Probabilities are dimensionless pure numbers. Inasmuch as an element of volume has units of (length)^3, $\psi^*\psi$ must have units of (length)$^{-3}$ to render $\psi^*\psi\mathrm{d}V$ dimensionless. Thus, $\psi^*\psi$ is known as a *probability density*; literally, this is the probability per cubic meter of space of finding the particle at position (x, y, z). The position at which the particle has the greatest probability of being found (that which maximizes $\psi^*\psi$) is known as the most probable position, r_{mp}.

There is an important corollary to Born's interpretation. If one adds up the probabilities of finding a particle over the entire space it could possibly occupy, the total must be unity. That is, the probabilities must add up to a "whole" particle. Mathematically, we can express this as

$$\int \psi^*\psi\mathrm{d}V = 1, \tag{2.35}$$

where, in practice, the limits of integration correspond to the limits of space the particle is restricted to. Since this is an integral over volume it is in general a triple integral. Any wavefunction for which this calculation can be performed is said to be *square-integrable*. In practice, this constraint represents a condition imposed on ψ which serves to pin down constants of integration that arise upon solving Schrödinger's equation for a given potential function $V(r)$. Once (2.35) has been imposed, a wavefunction is said to be *normalized*.

Schrödinger's equation is a second-order differential equation: Its solution for any particular $V(r)$ will in general lead to two constants of integration whose values become specified upon supplying boundary conditions. Boundary conditions are physical constraints applied to general mathematical solutions of differential equations; they are the means by which we specify the particular problem being solved. In general, one needs to supply as many boundary conditions as there are constants of integration. This is analogous to supplying the initial position and velocity of a particle in a Newton's law problem. In quantum-mechanical problems, however, these boundary conditions can be expressed very generally, quite independent of any particular example. While the description of them that follows below will make more sense in the context of particular solutions of Schrödinger's equation presented

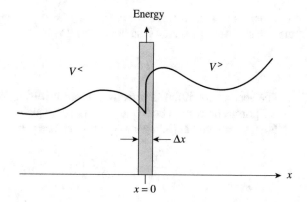

Fig. 2.3 Potential function with a discontinuity at $x = 0$

in subsequent chapters, this discussion appears here in order to connect them to its development.

Many quantum-mechanical problems involve different regions of space, each of which is described by its own potential function $V(r)$. In such cases, Schrödinger's equation is set up and solved independently for each region. These solutions may be very difficult to obtain, but this does not detract from the generality of the following argument. The boundary conditions can be thought of as recipes for "patching together" the resulting wavefunctions at the boundaries where the different regions of space meet. We can discern the nature of these conditions by considering a general boundary between two one-dimensional regions as shown in Fig. 2.3. For simplicity, we can imagine the boundary to be at $x = 0$: Functions referring to $x < 0$ are designated with superscript $<$, and those referring to $x > 0$ with superscript $>$. Thus, the potential function and wavefunction for $x < 0$ are designated $[V^<(x), \psi^<(x)]$, and those for $x > 0$ are designated $[V^>(x), \psi^>(x)]$. In general, there may or may not be a discontinuity in $V(x)$ at the boundary; one is shown in the figure for sake of generality. In reality, discontinuous potentials cannot exist (why?), but they can often be a very handy approximation for simplifying the description of a real potential.

Since a particle must have a unique energy, the two solutions to Schrödinger's equation must refer to the same energy (we consider only time-independent wavefunctions and potentials here):

$$-\frac{\hbar^2}{2m}\frac{d^2\psi^<}{dx^2} + V^<(x)\,\psi^< = E\psi^<, \tag{2.36}$$

and

$$-\frac{\hbar^2}{2m}\frac{d^2\psi^>}{dx^2} + V^>(x)\,\psi^> = E\psi^>. \tag{2.37}$$

The first general boundary condition is that the probabilistic interpretation of ψ can make sense only if each position in space has a unique probability associated with it. Thus, $\psi(x, y, z)$ *must be a continuous function.* In practice, this means forcing the two solutions to agree at the boundary, that is,

$$\psi^<(boundary) = \psi^>(boundary). \tag{2.38}$$

The second condition is somewhat more involved. What follows will not be a fully rigorous derivation but rather one based on physical grounds.

First, rearrange Schrödinger's equation to read

$$\frac{d^2\psi}{dx^2} = -\left(\frac{2m}{\hbar^2}\right)[E - V(x)]\psi(x).$$

Now imagine integrating this expression across the boundary:

$$\int\limits_<^> \left(\frac{d^2\psi}{dx^2}\right)dx = -\left(\frac{2mE}{\hbar^2}\right)\int\limits_<^> \psi(x)dx + \left(\frac{2m}{\hbar^2}\right)\int\limits_<^> V(x)\psi(x)dx, \tag{2.39}$$

where the limits of integration are to represent points within the $[<, >]$ domains. In general, the integral of the second derivative of ψ will be its first derivative. Suppose that the limits of integration straddle the boundary with a narrow extent Δx as shown in the diagram, that is, we are integrating from $x = -\Delta x/2$ to $x = +\Delta x/2$; we will eventually let $\Delta x \to 0$. Since ψ must be continuous across the boundary [that is, $\psi^<(x) \to \psi^>(x) \to \psi_{x=0}$ as $\Delta x \to 0$], we have

$$\left(\frac{d\psi^>}{dx}\right)_{\substack{\Delta x/2 \\ \Delta x \to 0}} - \left(\frac{d\psi^<}{dx}\right)_{\substack{-\Delta x/2 \\ \Delta x \to 0}} = -\left(\frac{2mE}{\hbar^2}\right)(\psi_0 \Delta x) \tag{2.40}$$

$$+ \left(\frac{2m}{\hbar^2}\right)[V^>(0) - V^<(0)]\left(\psi_0 \frac{\Delta x}{2}\right),$$

where ψ_0 denotes $\psi_{x=0}$ and where in the last term it is imagined that we will have $\Delta x \to 0$.

Now, as $\Delta x \to 0$, the first term on the right side of 2.40 will always vanish if E and ψ_0 remain finite. For the second term, two circumstances are possible: (i) If the potential is continuous or at worst finitely discontinuous across the boundary (that is, if $V^>(0)$ and $V^<(0)$ are equal or at most differ by a finite amount), then the second term will vanish as $\Delta x \to 0$. On the other hand, if the potential is infinitely discontinuous (that is, if $V^>(0)$ and $V^<(0)$ differ by an infinite amount), then the second term will be of the indeterminate form infinity times zero, in which case we can make no conclusion regarding the continuity (or not) of $d\psi/dx$. (Full disclosure: this statement is not true in a very pathological case involving a mathematical model for the potential known as a delta-function, but we will not encounter this function in this book.) It is crucial to appreciate that it is *differences* in potentials that are important, not their absolute magnitudes: forces are given by $F = -dV/dx$ and are unaffected if the potential is changed by an additive constant.

To repeat: if $V(x)$ is continuous or finitely discontinuous across a boundary, then the first derivative of ψ must be made continuous across the boundary. But if $V(x)$ is infinitely discontinuous across the boundary then no statement can be made regarding the continuity of $d\psi/dx$ across the boundary.

In summary, the three conditions that must be imposed on solutions of Schrödinger's equation are:

(1) ψ must be square integrable: $\int \psi^* \psi \, dV = 1$.
(2) ψ must be continuous everywhere.
(3) $d\psi/dx$ must be made continuous across a finite potential discontinuity, but cannot be specified across an infinite potential discontinuity.

All of these conditions will be used in examples studied in the next chapter.

In closing this chapter, a few quasi-philosophical remarks are appropriate. Schrödinger's equation is not in itself terribly imposing: classical physics is full of linear, second-order partial differential equations. It is the interpretation of the solutions and the constraints imposed upon them that are unlike anything that is familiar from Newtonian mechanics. The previously solid, deterministic Universe dissolves into fuzzy, intangible clouds of probability where there is no "yes" or "no" but only "might be." But in the end, experiment arbitrates the correctness of physical theories, and we will see that application of Schrödinger's equation to systems such as the hydrogen atom yields predictions entirely consistent with experiment.

Finally, if you find yourself uncomfortable with the notion of dealing with explicit complex functions, take heart: Historian of science Ricardo Karam has examined how Schrödinger himself struggled with the notion of complex functions, initially attempting to discard their imaginary parts. Apparently it took him the better part of a year to accept this aspect of his own theory [15]. Modern experiments confirm that quantum theory must inherently involve complex numbers [16]

Example 2.3 Presume that the wavefunction $\psi(x) = Ae^{-kx}$ applies over the domain $0 \leq x \leq \infty$; $\psi = 0$ otherwise. What must be the value of A in terms of k if this function is to be properly normalized? What is the probability of finding the particle between $x = 0$ and $x = 1/k$?

We must have $\int \psi^2 dx = 1$. In the present case this demands $A^2 \int_0^\infty e^{-2kx} dx = 1$. Because $\int e^{-ax} dx = -\frac{1}{a}e^{-ax}$, we have $a = 2k$ and hence

$$A^2 \int\limits_0^\infty e^{-2kx} dx = 1 \Rightarrow A^2 \left[-\frac{1}{2k} e^{-2kx} \right]_0^\infty = 1,$$

or

$$A^2 \left[0 - \left(-\frac{1}{2k} \right) \right] = 1,$$

that is, $A = \sqrt{2k}$.

The probability of finding the particle between $x = 0$ and $x = 1/k$ is given by the integral of ψ^2 over these limits:

$$P(0, k) = \int_0^{1/k} \psi^2 dx = A^2 \int_0^{1/k} e^{-2kx} dx = (2k) \left[-\frac{1}{2k} e^{-2kx} \right]_0^{1/k} = 1 - e^{-2} \sim 0.865.$$

The chance that the particle will be found between these limits is about 86.5%.

Summary

In as much as no derivation-from-first-principles of Schrödinger's equation exists, there is an almost infinite number of ways of concocting it; the development given in this chapter is a common one. In one dimension, Schrödinger's equation is

$$-\frac{\hbar^2}{2m} \frac{d^2 \psi}{dx^2} + V(x)\psi = E\psi.$$

This linear, second-order differential equation is to quantum mechanics what $F = ma$ is to Newtonian mechanics. In general, one solves this equation for the wavefunctions $\psi(x)$ and energy eigenvalues E for a system of mass m in an environment where the potential energy is given by $V(x)$; one also needs to apply the boundary conditions listed below. In itself, ψ has no physical meaning, but $\psi^* \psi dV$ is the probability of finding the system in a volume of space dV. More than anything else, it is this probabilistic interpretation of quantum mechanics that represents a distinct change from the sort of directly tangible quantities one is used to in Newtonian physics.

The boundary conditions that must be imposed on solutions of Schrödinger's equation are: (i) ψ must be square integrable; (ii) ψ must be continuous everywhere; and (iii) $d\psi/dx$ must be made continuous across a finite potential discontinuity, but cannot be specified across an infinite potential discontinuity.

Problems

2.1(E) Verify that if one has a number of independent solutions $y_1, y_2, \ldots y_n$ of the classical wave equation (all of pattern speed v), then a linear sum of them of the form (2.10) is also a solution.

2.2(E) Following Example 2.1, show that the wave pattern $y = \text{Tanh}^{-1}[(x - vt)^{-4/2}]$ satisfies the classical wave equation. $\text{Tanh}^{-1}(u)$ designates the inverse hyperbolic tangent of argument u. The derivative of this function with respect to any variable q is given by

$$\frac{d}{dq}\left\{\text{Tanh}^{-1}[u(q)]\right\} = \left(\frac{1}{1 - u^2}\right)\frac{du}{dq}.$$

2.3(I) Schrödinger's equation is not the only possible differential wave equation consistent with the law of conservation of energy and the prototype waveform (2.14). Derive one other.

2.4(I) Suppose that the solution to Schrödinger's equation for some potential gives rise to three wavefunctions, $\psi_1(x)$, $\psi_2(x)$, and $\psi_3(x)$ of the forms and domains

$$\begin{cases} \psi_1(x) = Ae^{kx} & (-\infty \le x \le 0) \\ \psi_2(x) = Bx^2 + Cx + D & (0 \le x \le L) \\ \psi_3(x) = 0 & (x \ge L). \end{cases}$$

What values must B, C, and D take in terms of A in order that these solutions satisfy the continuity conditions discussed in Sect. 2.4? In Chap. 3 we will see that a solution of the form $\psi_3(x) = 0$ requires the presence of an infinitely discontinuous potential at $x = L$; consequently, you do not need to apply the continuity condition on $d\psi/dx$ there. Derive also an expression that A, B, C, D, k, and L must satisfy in order that the overall solution be normalized over the domain $-\infty \le x \le \infty$.

2.5(E) Given the wavefunction $\psi(x) = Axe^{-kx}$ ($0 \le x \le \infty$; $k > 0$), what value must A take in terms of k in order that ψ be normalized? See Appendix C for the relevant integral.

2.6(E) Suppose that the wavefunction in the preceding problem is known to be a solution of Schrödinger's equation for some energy E. What is the corresponding potential function $V(x)$?

2.7(E) The wavefunction $\psi(x) = Ae^{-kx^2}$ ($-\infty \le x \le \infty$; $k > 0$) is known to be a solution of Schrödinger's equation for some energy E. What is the corresponding potential function $V(x)$? We will explore a potential like this in Chap. 5.

References

1. N. Gauthier, This derivation of the classical wave equation follows that given by Gauthier. Am. J. Phys. **55**(5), 477 (1987)
2. M. Jammer, *The Conceptual Development of Quantum Mechanics* (McGraw-Hill, New York, 1966)
3. D. Lindley, *Uncertainty: Einstein, Heisenberg, Bohr, and the Struggle for the Soul of Science* (Doubleday, New York, 2007)
4. O. Darrigol, Stud. Hist. Philos. Modern Phys. **40**, 151 (2009)
5. E. Schrödinger, Ann. Phys. **384**(4), 361 (1926)
6. E. Schrödinger, Ann. Phys. **384**(6), 489 (1926)
7. E. Schrödinger, Ann. Phys. **385**(13), 437 (1926)
8. E. Schrödinger, Ann. Phys. **386**(18), 109 (1926)
9. E. Schrödinger, Phys. Rev. **28**(6), 1049 (1926)
10. M. Born, Z. Phys. **37**(12), 863 (1926)
11. M. Born, Z. Phys. **38**(11–12), 803 (1926)
12. C.D. Galles, Phys. Today **54**(4), 94 (2001)
13. G. Holton, Phys. Today **54**(5), 90 (2001)
14. D.F. Styer, Am. J. Phys. **64**(1), 31 (1996)
15. R. Karam, Am. J. Phys. **88**(6), 433 (2020)
16. E. Conover, Sci. News **201**(2), 14 (2022)

Chapter 3
Solutions of Schrödinger's Equation in One Dimension

Summary When solving problems involving Newton's laws of motion, one usually begins by specifying a circumstance, say a particle moving in the Erath's gravitational field. Solving Schrödinger's equation is similar in that one must specify a particle mass and the environment within which it resides. The latter is specified via the so-called *potential function* $V(x)$, which specifies the potential energy of the particle at some position x. In this lengthy chapter, Schrödinger's equation is solved for several iconic potential functions. The first of these, the "infinite square well", is highly idealized but is analyzed as it reveals many characteristic qualities of such solutions, in particular how the imposition of *boundary conditions* leads to *quantized energy levels*. This is followed by solutions for the finite potential well, potential barriers, and an examination of how quantum mechanics was pivotal in understanding the nuclear process of alpha decay, which was otherwise incomprehensible based on concepts of classical physics.

The purpose of this chapter is to investigate solutions of Schrödinger's equation for a variety of one-dimensional potential functions. While such potentials have applications in areas such as modeling CCD chips (Example 3.3) and understanding alpha-decay (Sect. 3.9), our reason for starting with them as opposed to two-or-three dimensional problems is that they exhibit many uniquely quantum effects with a minimum of mathematical complexity. Solving them builds up experience for more involved problems in subsequent chapters.

This is a lengthy chapter, so an overview is given here to help orient readers. Part I of this chapter, comprising Sects. 3.1–3.6, is devoted to examining *potential wells*. In loose terms, this means situations where a particle is trapped in some region of space by force barriers which, classically, it does not possess sufficient energy to "leap over" and thus escape. Such situations lead to discrete bound-energy states analogous to those appearing in the Bohr model. We begin our study of potential wells in Sect. 3.1 by reviewing certain characteristics of classical potential-energy functions that make for valuable points of contact with quantum-mechanical problems. In Sect. 3.2 we solve Schrödinger's equation for the case of a potential known as an

© The Author(s), under exclusive license to Springer Nature Switzerland AG 2022
B. C. Reed, *Quantum Mechanics*, https://doi.org/10.1007/978-3-031-14020-4_3

infinite well. The solution of this highly idealized case illustrates how the presence of discrete wavefunctions and energies emerges upon application of the boundary conditions derived in Chap. 2. In Sects. 3.3 through 3.5 we will consider a similar potential, the finite well. This problem will help us understand the phenomenon of *quantum tunneling*. While this phenomenon has a classical analog for massless systems (penetration of electromagnetic waves through nonconducting media), we will see that, wave-mechanically, tunneling can also happen for material particles. Section 3.6 discusses how the information gleaned from the infinite and finite-well solutions can be used to develop guidelines for sketching approximate wavefunctions for any potential function.

Part II, comprising Sects. 3.7–3.10, is devoted to an analysis of *potential barriers*. Again speaking loosely, we can regard a potential barrier as a potential well that has been turned inside-out: Instead of being trapped within a well, an "incoming" particle encounters a barrier, through which it has some probability of penetrating. Much of the mathematics involved in studying wells and barriers is similar, but the motivations and results are different; this is why these concepts are brought under the roof of one chapter but are housed, as it were, in separate but adjacent apartments. Potential barriers are of interest, for example, to nuclear physicists who probe the structure and energy levels of nuclei by scattering accelerated subatomic particles from them. Results built up in Sects. 3.7 and 3.8 are applied in Sect. 3.9 to the phenomenon of *alpha decay*, a quantum tunneling effect whose explanation was one of the first great successes of quantum mechanics. Section 3.10 briefly discusses some further aspects of scattering experiments. This section could have been included with potential wells in Part I, but as the emphasis is on scattering we put it with Part II. Problems for all sections appear at the end of the chapter.

Part I: Potential Wells

3.1   Concept of a Potential Well

In Newtonian mechanics, the "object of the exercise" is often to establish expressions giving the position r and velocity v of a body of mass m as a function of time. To do this, one supplies the initial position and velocity and a description of the forces to which the mass is subject. The forces can be expressed either directly or in the form of a *potential energy function* $V(r)$; the force on the body is given by the negative gradient of V, $F = -\nabla V$, and the motion follows from Newton's second law, $F = ma$. In problems where quantum physics is applicable, the environment also enters through $V(r)$, but one solves for the wavefunctions and corresponding energy eigenvalues that satisfy Schrödinger's equation.

Two constructs which we shall use extensively are *energy-level diagrams* and *potential wells*. To introduce these concepts it is helpful to consider a simple Newtonian example. To this end, consider a car of mass m on a roller-coaster track as shown in Fig. 3.1.

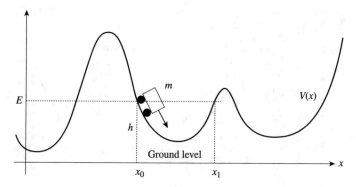

Fig. 3.1 Car on a roller-coaster track

If the car starts from height h above the ground, its potential energy is $V = mgh$. If the height of the track as a function of position is given by $h(x)$, then we can write $V(x) = mgh(x)$. $V(x)$ and $h(x)$ are equivalent but for a factor of mg; the curve in Fig. 3.1 can be regarded either as a sideways view of the track, or, more abstractly, as a plot of the potential energy of the car as a function of position. We could label the vertical axis interchangeably as "height in meters" or as "energy in Joules". If the car is released from rest at position x_0 as shown, its total energy would be $E = mgh x_0$. In the energy/position version of Fig. 3.1, this value defines a horizontal line at energy E. In the absence of friction or air resistance, E will remain constant at this level, which defines a "valley" or "well" between x_0 and x_1. When the car is released, it will accelerate, reach a maximum speed at the bottom of the valley, and then climb up the other side until it comes to rest momentarily at the same vertical height (or "energy level") from which it was released. Destined to slosh back-and-forth between x_0 and x_1, the car is constrained within a so-called *potential well*, and is said to be in a *bound energy state*. Technically, a bound state is one whose total energy is less than $V(x)$ as $x \to \infty$. Positions x_0 and x_1 are known as "classical turning points," positions at which the particle stops momentarily and reverses its direction of motion. If the car were released from x_0 with some nonzero speed, the total energy line would not cut the curve at x_0 because of the contribution of the initial kinetic energy: $E = mv_0^2/2 + mgh(x_0)$. In this event it would find itself in a new bound state, as illustrated in Fig. 3.2.

Another possibility is that if the track to the right of the release point is always lower than the vertical level of the release point, the car will eventually arrive at $x = +\infty$. Figure 3.3 illustrates such an *unbound* energy state.

The preceding comments possess broad applicability. Figures 3.1 through 3.3 illustrate what are known as *potential diagrams*. In these figures the x-axis always corresponds to a position coordinate and the y-axis to energy; in general, the vertical axes in these diagrams will not be equivalent to a vertical height. The potential energy function for the system is typically plotted as a solid curve or line in such diagrams.

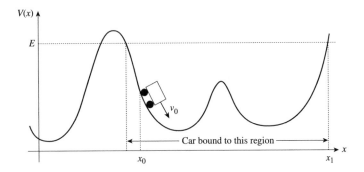

Fig. 3.2 Car in a bound energy state with initial velocity

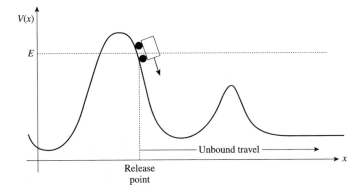

Fig. 3.3 Car in an unbound energy state

Classically, the total energy of a system is in principle unrestricted; E does not appear explicitly in Newton's second law. Quantum-mechanically, energy enters Schrödinger's equation explicitly, and solutions for a given potential will involve E as a parameter. With these general ideas in mind, let us now set up and solve Schrödinger's equation for some simple one-dimensional potentials.

3.2 The Infinite Potential Well

The simplest quantum environment imaginable for a particle is one in which it is trapped between walls so energetically high that it would require an infinite amount of energy to leap over them. The energy diagram for such a situation is shown in Fig. 3.4: a particle of mass m moves in an infinitely deep potential well between $x = 0$ and $x = L$. Such a construct is known variously as an infinite potential well, infinite rectangular well, or infinite square well.

Fig. 3.4 Infinite square well. The walls extend upwards to infinity

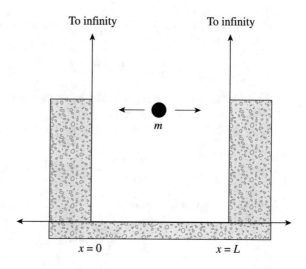

Fig. 3.5 Schematic diagram of an electron trapped in a one-dimensional box made of electrodes and grids in an evacuated tube. After Fig. 3.3 of *An Introduction to Quantum Physics* by A. P. French and E. F. Taylor

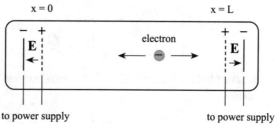

In Fig. 3.4 and subsequent similar diagrams, shading designates potential "walls." Analytically, the infinite potential well is defined by

$$V(x) = \begin{cases} 0, & 0 \le x \le L \\ \infty, & x < 0, \ x > L. \end{cases} \tag{3.1}$$

The walls need not (and in general will not) be physical barriers; Fig. 3.5 shows how one might construct an electronic potential well by trapping an electron between strong electric fields in an evacuated tube. In reality, the electric fields cannot be made infinitely strong, but the idea should be clear: as soon as the electron tries to penetrate to the left of $x = 0$ or to the right of $x = L$, it will experience a strong opposing force that turns it back. We refer to the regions defined by $x < 0$ and $x > L$ as being "outside" the well, and that defined by $0 < x < L$ as being "inside." We seek bound-state solutions with $E > 0$.

In the outside regions, Schrödinger's equation takes the form

$$-\frac{\hbar^2}{2m}\frac{d^2\psi}{dx^2} + (\infty)\psi = E\psi, \quad (x \leq 0, x \geq L). \tag{3.2}$$

This equation can only be satisfied by having $\psi = 0$ everywhere outside the well; this is so because E must presumably be finite.

Inside the well the story is different. Here $V(x) = 0$, and

$$-\frac{\hbar^2}{2m}\frac{d^2\psi}{dx^2} = E\psi \quad (0 \leq x \leq L). \tag{3.3}$$

Rearrange this equation to read

$$\frac{d^2\psi}{dx^2} = -k^2\psi, \tag{3.4}$$

where

$$k^2 = \frac{2mE}{\hbar^2}. \tag{3.5}$$

In general, the solution of (3.4) is a sum of exponential functions with imaginary arguments:

$$\psi(x) = Ae^{\iota kx} + Be^{-\iota kx}, \tag{3.6}$$

where $\iota = \sqrt{-1}$ and where A and B are constants of integration.

We saw in Chap. 2 that three boundary conditions must be applied to solutions of Schrödinger's equation: (i) they must be normalizable by being square-integrable; (ii) ψ must be continuous everywhere; and (iii) $d\psi/dx$ must be made continuous across potential discontinuities unless the discontinuity is infinite. It is generally easiest to consider the latter two conditions first. In the present case, condition (iii) cannot be used because we do in fact have infinite potential discontinuities. Condition (ii) demands

$$\psi(0) = 0 \tag{3.7}$$

and

$$\psi(L) = 0. \tag{3.8}$$

On substituting $x = 0$ into (3.6) we find that it reduces to $A + B = 0$, or $B = -A$. Thus the general solution inside the well can be cast as

$$\psi(x) = A(e^{\iota kx} - e^{-\iota kx}) = (2\iota A)\sin(kx). \tag{3.9}$$

At $x = L$ we must have

$$(2\iota A) \sin(kL) = 0. \tag{3.10}$$

There are two possibilities: either $A = 0$ or $\sin(kL) = 0$. If $A = 0$, the wave-function vanishes entirely, an unphysical solution because we cannot normalize it. The second possibility is much more interesting: $\sin(kL)$ will vanish if (and only if) $kL = n\pi$, where n is an integer. This restricts k to values such that $k = n\pi/L$; from (3.5) this sets a restriction on the permissible values for the total energy of the particle:

$$E_n = \left(\frac{\pi^2 \hbar^2}{2mL^2} \right) n^2. \tag{3.11}$$

Strictly, negative values of n also satisfy $\sin(n\pi) = 0$, but these are rejected on physical grounds: the boundary condition is $kL = n\pi$, and k must be positive for $E > 0$. As with the Bohr atom, n is known as the *principal quantum number* of the system; $n = 1, 2, 3, \ldots$ The *ground-state* energy corresponds to $n = 1$. (Note that $n = 0$ leads to a null solution; in this case $E = k = 0$, and (3.4) would imply $\psi = ax + b$. This solution can only be made to satisfy the boundary conditions if the constants a and b are set to zero.)

The important concept here is that *application of the quantum-mechanical boundary conditions has led to quantized energy levels.* This is characteristic of all bound-state solutions of Schrödinger's equation. It is customary to label the value of the energy corresponding to its quantum number n as has been done in (3.11). To each energy E_n corresponds a wavefunction ψ_n:

$$\psi_n(x) = (2\iota A) \sin \left(\frac{n\pi}{L} x \right), \tag{3.12}$$

where we have replaced k in (3.9) with $n\pi/L$.

What of the constant of integration A? There is one constraint we have not yet imposed: that ψ must be normalized, that is, that the probability distribution of the particle, when integrated over the well, must be unity. Thus, for every possible value of n we must demand

$$\int_0^L \psi_n^*(x)\psi_n(x)\mathrm{d}x = 1. \tag{3.13}$$

In the present case, $\psi_n^*(x)\psi_n(x) = 4A^2 \sin^2(kx)$, so normalization proceeds as

$$\int_0^L \psi_n^*(x)\psi_n(x)dx = 4A^2 \int_0^L \sin^2\left(\frac{n\pi}{L}x\right)dx$$

$$= 4A^2 \left[\frac{x}{2} - \frac{\sin(2n\pi x/L)}{4(n\pi/L)}\right]_0^L$$

$$= A^2 \left\{\left[\frac{L}{2} - \frac{\sin(2n\pi)}{4(n\pi/L)}\right] - \left[0 - \frac{\sin(0)}{4(n\pi/L)}\right]\right\}$$

$$= 4A^2 \left(\frac{L}{2}\right) = 2A^2 L.$$

Forcing this result to be equal to unity gives

$$A = \sqrt{\frac{1}{2L}}. \tag{3.14}$$

Combining (3.12) and (3.14) gives the normalized infinite rectangular well wave-functions:

$$\psi_n(x) = \iota\sqrt{\frac{2}{L}} \sin\left(\frac{n\pi}{L}x\right), \quad (0 \le x \le L), \quad n = 1, 2, 3, \ldots. \tag{3.15}$$

Note that A depends only on the length of the well and not on the quantum state n; this will prove to be an exception rather than the rule. Energy levels generally depend on the quantum state involved, the mass of the particle, the width of the well, and Planck's constant.

Because only $\psi_n^*(x)\psi_n(x)$ has physical meaning, we will usually drop the factor of ι when discussing infinite well wavefunctions and treat them as purely real.

It is worthwhile reiterating two key lessons from this solution of Schrödinger's equation:

(1) *Quantized energy levels* (also known as *energy eigenvalues*) arise naturally from the imposition of *boundary conditions* on solutions of Schrödinger's equation and lead to *quantum numbers* which can be used to label the energy eigenvalues. Compare the natural way that energy quantization emerges from Schrödinger's equation as opposed to its ad-hoc introduction by Bohr in his model of the hydrogen atom.

(2) To each energy eigenvalue corresponds a *wavefunction* (or *eigenfunction*) $\psi_n(x)$ which dictates the probability distribution of the system when it possesses total energy E_n.

Fig. 3.6 Infinite square well wavefunctions for $L = 1$ for the $n = 1$ (solid curve), $n = 2$ (short-dashed curve) and $n = 3$ (long-dashed curve) states

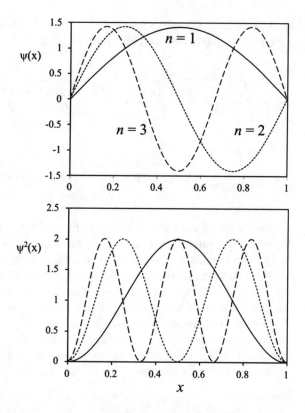

Figure 3.6 shows plots of ψ_n and ψ_n^2 for the three lowest-energy infinite-well wavefunctions for $L = 1$. The $n = 1$ state is known as the *ground state* or *normal state* of the system. In the plots of ψ_n^2, the peaks correspond to positions where there is a high probability of finding the particle, while the valleys correspond to positions of low probability. The number of extrema in the wavefunction is equal to n. There are a number of positions – so-called *nodes* – where $\psi = 0$, that is, where we would never expect to find the particle. Including the boundaries of the well, the number of nodes in state n is $n + 1$.

The wavefunctions plotted in Fig. 3.6 are reminiscent of standing waves on a string. In fact, the two problems are mathematically identical in that an integral number of half-wavelengths fit between the boundaries in each case. For a string, this results in a constraint on the possible vibrational frequencies; for a particle in the infinite well this translates to a constraint on possible energies via the de Broglie relation.

If we accept the probability-distribution interpretation of quantum mechanics, why do we not detect a wavy nature for matter in everyday life? The answer lies in the differing orders of magnitude of the quantum numbers involved in atomic-scale and macroscopic phenomena. To illustrate this, consider two cases. Case A is the quantum case: an electron of energy 20 eV (3.20×10^{-18} J) trapped in an atomic-scale potential well of width $L = 1$ Å (10^{-10} m). Case B is a mass of 1 kg and energy 1 J trapped in a potential well of width 1 m. Then

$$n_A = \sqrt{\frac{8m_A L_A^2 E_A}{h^2}} = \sqrt{\frac{8(9.11 \times 10^{-31} \text{ kg})(10^{-10} \text{ m})^2(3.20 \times 10^{-18} \text{ J})}{(6.626 \times 10^{-34} \text{ J-s})^2}} = 0.73$$

and

$$n_B = \sqrt{\frac{8m_B L_B^2 E_B}{h^2}} = \sqrt{\frac{8(1 \text{ kg})(1 \text{ m})^2(1 \text{ J})}{(6.626 \times 10^{-34} \text{ J-s})^2}} = 4.3 \times 10^{33}.$$

The n's should work out to be integers; the fact that n_A did not means that mutually incompatible values of m, L, and E were chosen. The important point is that in the atomic-scale problem, ψ would possess on the order of only one maximum over the length-scale of the problem, whereas in the macroscopic case there would be some 10^{33} maxima in the space of one meter. The distance between probability maxima in the latter case would be about 10^{-33} m. This is some 20 orders of magnitude smaller than the size of a nucleus; the peaks would be utterly unresolvable from each other. For practical purposes, the 1 kg mass can be located anywhere in the 1-m box with almost unlimited precision: on a macroscopic scale we do not notice the "bumpiness" of the probability distribution.

Example 3.1 Determine the probability of finding a particle of mass m between $x = 0$ and $x = L/10$ if it is in the $n = 3$ state of an infinite rectangular well.

We know that the probability of finding the particle in an infinitesimal region of space dx at position x is given by (2.34), $(\psi^*\psi)$dx. The answer to this question is given by integrating this probability over the desired range:

$$P(0, L/10) = \int_0^{L/10} \psi_3^*(x)\psi_3(x) \, dx = \frac{2}{L} \int_0^{L/10} \sin^2\left(\frac{3\pi}{L}x\right) dx$$

$$= \frac{2}{L}\left[\frac{x}{2} - \frac{\sin(6\pi x/L)}{(12\pi/L)}\right]_0^{L/10} = \frac{2}{L}\left[\frac{L}{20} - \frac{\sin(3\pi/5)}{(12\pi/L)}\right] = 0.0496,$$

or about 5%.

It is important to remember that probabilities have meaning only in a statistical sense. In practice, you would have to measure the positions of particles in a number of identically-prepared infinite wells and somehow determine what fraction of them were found between $x = 0$ and $x = L/10$.

In questions of the type of Example 3.1, a shortcut can often be used. In addressing that question we began with the fact that the probability of finding the particle between x and $x + dx$ when dx is infinitesimally small is given by

$$P(x, x + dx) = \psi_n^*(x)\psi_n(x)dx. \tag{3.16}$$

To compute the probability of finding the particle within some *finite* range Δx, it was necessary to integrate. However, if the desired range is small in comparison to the length scale over which ψ changes appreciably, one can assume that ψ remains approximately constant over Δx and write

$$P(x, x + \Delta x) \sim \psi_n^*(x)\psi_n(x)\Delta x.$$

For this case we have $\Delta x = L/10$. Evaluating ψ at the midpoint of the range $(x = L/20)$, gives

$$P(0, L/10) \sim \frac{2}{L} \sin^2 \left(\frac{3\pi}{20}\right) \left(\frac{L}{10}\right) \sim 0.041,$$

which compares favorably with the exact result obtained by integration.

3.3 The Finite Potential Well

We now consider a particle of mass m trapped in a potential well of finite depth. This is a somewhat more realistic problem than that of the infinite well, and can be imagined to correspond, for example, to an electron trapped near the surface of a metal in a photoelectric-effect experiment and needing only a few electron-volts of energy to escape. The situation is illustrated in Fig. 3.7. This arrangement is also referred to as a finite rectangular well or finite square well.

We take the well to have depth V_o Joules and width $2L$, symmetric about $x = 0$. (This arrangement is different from that used in the infinite well, but ultimately makes the mathematics easier). The well is described by

$$V(x) = \begin{cases} 0 & (-L \leq x \leq L) \\ V_o & (|x| > L). \end{cases} \tag{3.17}$$

We again divide space into inside ($|x| \leq L$) and outside ($|x| \geq L$) regions. Inside the well $V(x) = 0$, and Schrödinger's equation reduces to the same form as that inside the infinite rectangular well:

$$-\frac{\hbar^2}{2m}\frac{d^2\psi_{in}}{dx^2} = E\,\psi_{in}. \tag{3.18}$$

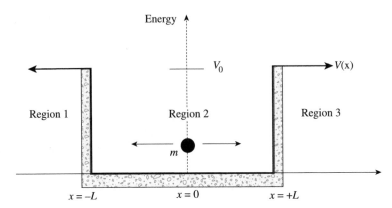

Fig. 3.7 Finite rectangular well

Outside the well, $V(x) = V_o$, and Schrödinger's equation becomes

$$-\frac{\hbar^2}{2m}\frac{d^2\psi_{out}}{dx^2} = (E - V_o)\psi_{out}.$$ (3.19)

Equation (3.19) expresses the first difference from the infinite-well case: ψ outside the well need not (and in general will not) be zero. To be precise, there are two outside regions, one with $x \leq -L$ and the other with $x \geq +L$. However, Schrödinger's equation is identical in both, and we can address this complication later. As a notational convenience, label the regions as 1 through 3 from left to right as shown in Fig. 3.7; these region numbers will be used as subscripts on the corresponding wavefunctions—they do not designate energy levels! Once again we are seeking bound energy states, that is, ones with $E < V_o$.

We can write Eqs. (3.18) and (3.19) more compactly as

$$\frac{d^2\psi_{in}}{dx^2} = -k_2^2\psi_{in}$$ (3.20)

and

$$\frac{d^2\psi_{out}}{dx^2} = k_1^2\psi_{out},$$ (3.21)

where

$$k_1^2 = \frac{2m}{\hbar^2}(V_o - E)$$ (3.22)

and

$$k_2^2 = \frac{2mE}{\hbar^2}.$$ (3.23)

Since the bottom of the well is at $V = 0$ and we are seeking bound states with $V_o > E$, both k_1 and k_1 are positive real; the subscripts on the k's refer to the regions to which they apply. Since the value of k in region 3 is equal to that in region 1, we can put $k_3 = k_1$.

The solution to (3.20) is of the same general form as that for the inside of the infinite rectangular well:

$$\psi_{in}(x) = A \exp(\iota k_2 x) + B \exp(-\iota k_2 x). \tag{3.24}$$

The general solution to (3.21) is

$$\psi_{out}(x) = C e^{k_1 x} + D e^{-k_1 x}, \tag{3.25}$$

where C and D are constants of integration.

As alluded to above, Eq. (3.25) actually applies to both outside regions; to be formally correct we should write down two versions of it, each with its own constants of integration. Hence we put

$$\begin{cases} \psi_1(x) = C e^{k_1 x} + D e^{-k_1 x} & (x \leq -L) \\ \psi_2(x) = A e^{\iota k_2 x} + B e^{-\iota k_2 x} & (-L \leq x \leq L) \\ \psi_3(x) = G e^{k_1 x} + F e^{-k_1 x} & (x \geq L), \end{cases} \tag{3.26}$$

where we subscript the wavefunctions according to the regions to which they apply. We do not use E as a constant of integration as that symbol is used to represent energy.

The problem now is to apply the boundary conditions derived in Chap. 2. Here the potential discontinuities at $x = -L$ and $x = +L$ are not infinite, so we must demand continuity of both the wavefunctions and their derivatives at the boundaries. That is, we must demand

$$\psi_1(x) = \psi_2(x) \quad at \ x = -L, \tag{3.27}$$

$$\psi_2(x) = \psi_3(x) \quad at \ x = +L, \tag{3.28}$$

$$\frac{d\psi_1}{dx}\bigg|_{-L} = \frac{d\psi_2}{dx}\bigg|_{-L}, \tag{3.29}$$

and

$$\frac{d\psi_2}{dx}\bigg|_{+L} = \frac{d\psi_3}{dx}\bigg|_{+L}. \tag{3.30}$$

Before applying these, we can simplify matters somewhat by considering the third boundary condition: that ψ be normalizable by being square integrable. In practice, this means that $\psi^*\psi$ must integrate to unity. This demands

$$\int\limits_{-\infty}^{-L} \psi_1^* \psi_1 \, \mathrm{d}x + \int\limits_{-L}^{L} \psi_2^* \psi_2 \mathrm{d}x + \int\limits_{L}^{\infty} \psi_3^* \psi_3 \mathrm{d}x = 1.$$

Note that the square of each wavefunction is integrated only over the region within which it applies.

This square integrability demand presents a problem. Consider the wavefunction in region 1, ψ_1, the first of (3.26). As $x \to -\infty$, $\psi_1(x)$ diverges due to the $De^{-k_1 x}$ term, with the result that the integral of $\psi_1^* \psi_1$ would never converge! A similar problem arises with the first term of ψ_3 as $x \to \infty$. The only way to alleviate this catastrophe is to insist that if the domain of a wavefunction includes $\pm\infty$, then we must insist that $\psi \to 0$ asymptotically:

$$\psi_1(x) \to 0 \quad \text{as } x \to -\infty \tag{3.31}$$

and

$$\psi_3(x) \to 0 \quad \text{as } x \to +\infty. \tag{3.32}$$

These conditions require $D = G = 0$ in (3.26). This reduces the solution to

$$\begin{cases} \psi_1(x) = Ce^{k_1 x} & (x < -L) \\ \psi_2(x) = \ Ae^{\imath k_2 x} + Be^{-\imath k_2 x} & (-L \leq x \leq L) \\ \psi_3(x) = Fe^{-k_1 x} & (x > L). \end{cases} \tag{3.33}$$

All seems well: we have four constants to determine (A, B, C, F), and four conditions to apply: (3.27) through (3.30). But this is not the whole story: after applying (3.27) through (3.30) we still have to actually apply the square-integrability condition, so we have an overdetermined system of five conditions on four constants! The resolution of this apparent difficulty will become clear below [see following (3.53)]. Applying conditions (3.27) through (3.30) yields:

$$Ae^{-\imath k_2 L} + Be^{\imath k_2 L} = Ce^{-k_1 L} \tag{3.34}$$

$$\imath k_2 Ae^{-\imath k_2 L} - \imath k_2 Be^{\imath k_2 L} = k_1 Ce^{-k_1 L} \tag{3.35}$$

$$Ae^{\imath k_2 L} + Be^{-\imath k_2 L} = Fe^{-k_1 L} \tag{3.36}$$

$$ \iota k_2 A e^{\iota k_2 L} - \iota k_2 B e^{-\iota k_2 L} = -k_1 F e^{-k_1 L}. \tag{3.37} $$

Equations (3.34) and (3.35) derived from (3.27) and (3.29); similarly, Eqs. (3.36) and (3.37) derived from Eqs. (3.28) and (3.30). The dimensionless products $k_1 L$ and $k_1 L$ appear frequently in these equations, which suggests defining two new constants:

$$ \eta = k_1 L = \sqrt{\frac{2m(V_o - E)}{\hbar^2}} L \tag{3.38} $$

and

$$ \xi = k_2 L = \sqrt{\frac{2m E}{\hbar^2}} L. \tag{3.39} $$

With these definitions, we can write (3.34) through (3.37) as

$$ A e^{-\iota \xi} + B e^{\iota \xi} = C e^{-\eta}, \tag{3.40} $$

$$ \iota A e^{-\iota \xi} - \iota B e^{\iota \xi} = (\eta/\xi) C e^{-\eta}, \tag{3.41} $$

$$ A e^{\iota \xi} + B e^{-\iota \xi} = F e^{-\eta}, \tag{3.42} $$

$$ \iota A e^{\iota \xi} - \iota B e^{-\iota \xi} = -(\eta/\xi) F e^{-\eta}. \tag{3.43} $$

We treat these as four equations in four unknowns. Solving (3.40) for A gives

$$ A = e^{\iota \xi} (C e^{-\eta} - B e^{\iota \xi}). \tag{3.44} $$

Substituting this result into (3.41) and isolating B gives

$$ B = (\iota/2) C e^{-\iota \xi} e^{-\eta} (\eta/\xi - \iota). \tag{3.45} $$

Back-substituting (3.45) into (3.44) gives A in terms of C alone:

$$ A = \frac{1}{2} C e^{-\eta} e^{\iota \xi} [1 - \iota(\eta/\xi)]. \tag{3.46} $$

Eliminating A and B in (3.42) and (3.43) via (3.45) and (3.46) gives, after some manipulation,

$$ F = C [\cos(2\xi) + (\eta/\xi) \sin(2\xi)], \tag{3.47} $$

and

$$ F = -C [\cos(2\xi) - (\xi/\eta) \sin(2\xi)], \tag{3.48} $$

where we have used the Euler identities. Equating these two expressions gives a condition on η and ξ:

$$2\cos(2\xi) + (\eta/\xi - \xi/\eta)\sin(2\xi) = 0. \tag{3.49}$$

A, B, C, and F have disappeared. The meaning of (3.49) can be elucidated by noting that η and ξ are not independent. From (3.38) and (3.39), we have

$$\xi^2 + \eta^2 = \frac{2mV_oL^2}{\hbar^2} = \text{constant} = K^2. \tag{3.50}$$

K is known as the *strength parameter* of the well. Expressing η as $\sqrt{K^2 - \xi^2}$, we can cast (3.49) into a constraint on ξ alone:

$$f(\xi, K) = (K^2 - 2\xi^2)\sin(2\xi) + 2\xi\sqrt{K^2 - \xi^2}\cos(2\xi) = 0. \tag{3.51}$$

For a given value of K, only certain values of ξ will satisfy (3.51); from (3.38) or (3.39), these will correspond to particular values of the energy of the particle. *The roots of* (3.51) *correspond to the energy eigenvalues of the finite potential well.* If we label the roots of (3.51) as ξ_n, then the energy levels are given by $E_n = \xi_n^2\hbar^2/2mL^2$. Further, if either of (3.47) or (3.48) is squared and (3.51) is used to eliminate $\sin(2\xi)$ or $\cos(2\xi)$, the result is $F^2 = C^2$, that is, $F = \pm C$. If $F = +C$, then (3.40) and (3.42) reduce to $A = B$; in this case $\psi_2(x) = (2A)\cos(k_2x)$, and the overall wavefunction has *even parity*: $\psi(x) = \psi(-x)$. On the other hand, if $F = -C$, Eqs. (3.41) and (3.43) reduce to $A = -B$ and one finds $\psi_2(x) = (2\iota A)\sin(k_2x)$, that is, the overall wavefunction is of *odd parity*: $\psi(x) = -\psi(-x)$.

From the above description, there evidently exist two separate families of solutions to the finite rectangular well problem. These are referred to as the even and odd solution sets:

$$\left.\begin{array}{l} B = A \\ F = C \end{array}\right\} \text{even solutions,} \tag{3.52}$$

$$\left.\begin{array}{l} B = -A \\ F = -C \end{array}\right\} \text{odd solutions.} \tag{3.53}$$

The value of making the well symmetric about $x = 0$ is that it causes the solution to separate naturally into families of different parity. However, Eq. (3.51) is valid for *all* of the energy eigenvalues. Odd and even-parity solutions to Schrödinger's equation will always be the case whenever a symmetric potential is involved; this is discussed further in Sect. 3.6.

The presence of two solution-families resolves the dilemma alluded to in the paragraph following (3.33): not all of the boundary conditions can be satisfied simultaneously.

Fig. 3.8 Finite rectangular well wavefunctions. Not to scale

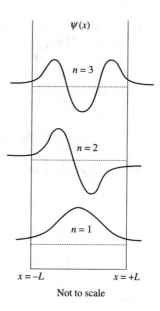

Not to scale

An important aspect of the finite potential well is that there is some probability of finding the particle beyond the walls of the well, where, classically, it cannot be found. This behavior, known as *tunneling* or *barrier penetration*, is, like quantized energies, a purely quantum-mechanical phenomenon that arises due to the wave nature of particles. Physicist Eugen Merzbacher has written that "…among all the successes of quantum mechanics as it evolved in the third decade of the 20th century, none was more impressive than the understanding of the tunnel effect—the penetration of matter waves and the transmission of particles through a high potential barrier." [1] A *characteristic tunneling length* for a given energy is given by the reciprocal of k_1 of (3.22).

Figure 3.8 shows sketches of the three lowest finite-well wavefunctions; again, we usually plot $|\psi|$, suppressing any explicitly complex nature of the wavefunction. Note the alternating parities and tunneling phenomena; the number of maxima in the wavefunctions again corresponds to n.

The finite well illustrates a number of concepts which apply generally for any potential well. Inside a well, that is, wherever $E > V(x)$, the wavefunction will always have the oscillatory form of the second of (3.26). In regions where $E < V(x)$, the solution will have the form of an exponential decay, corresponding to the tunneling effect. The wavefunction and its first derivative must be continuous everywhere, and ψ must often be forced to vanish at the limits of its domain if those limits go to $\pm\infty$. The number of maxima in the wavefunction is equal to the quantum number of the system, and can be used to label the various possible states.

Before closing this section, it is worthwhile to elaborate briefly on the energy-eigenvalue condition, Eq. (3.51). As it stands, this expression conations no quantum number n to index the bound states in the way (3.11) does for the infinite well. But such a quantum number can be introduced through the following manipulation [2, 3].

First, extract a factor of K^2 from both terms in (3.51) to cast the eigenvalue condition as

$$[1 - 2(\xi/K)^2]\sin(2\xi) + 2(\xi/K)\sqrt{1 - (\xi/K)^2}\cos(2\xi) = 0. \tag{3.54}$$

Now define the quantity Φ by

$$\Phi = \mathrm{Sin}^{-1}(\xi/K). \tag{3.55}$$

From this definition it follows that

$$\cos^2\Phi = 1 - (\xi/K)^2, \tag{3.56}$$

so that

$$\cos(2\Phi) = \cos^2\Phi - \sin^2\Phi = 1 - 2(\xi/K)^2 \tag{3.57}$$

and

$$\sin(2\Phi) = 2\sin\Phi\cos\Phi = 2(\xi/K)\sqrt{1 - (\xi/K)^2}. \tag{3.58}$$

Now consider the quantity

$$\sin(2\xi + 2\Phi) = \sin(2\xi)\cos(2\Phi) + \cos(2\xi)\sin(2\Phi). \tag{3.59}$$

Substituting (3.57) and (3.58) into (3.59) gives

$$\sin(2\xi + 2\Phi) = [1 - 2(\xi/K)^2]\sin(2\xi) + 2(\xi/K)\sqrt{1 - (\xi/K)^2}\cos(2\xi). \tag{3.60}$$

Compare (3.54) and (3.60): they are identical. This means that we can write the eigenvalue equation for the finite well as

$$\sin(2\xi + 2\Phi) = 0. \tag{3.61}$$

The sine of any angle will be zero only when that angle is equal to an integer number of π radians, that is, we must have

$$2(\xi + \Phi) = n\pi,$$

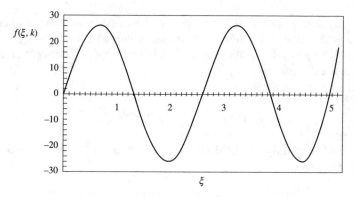

Fig. 3.9 Example 3.2

or

$$\xi + \sin^{-1}\left(\frac{\xi}{K}\right) = \frac{n\pi}{2}. \tag{3.62}$$

This expression is the recast eigenvalue condition for the finite well: in this considerably simplified version, an integer quantum number appears explicitly.

Example 3.2 Given an electron in a finite well of depth 100 eV and $L = 1\,\text{Å}$, find the bound energy states.

From (3.50),

$$K^2 = \frac{2mV_0L^2}{\hbar^2} = \frac{2\,(9.1095 \times 10^{-31}\,\text{kg})(1.6022 \times 10^{-17}\text{J})(10^{-10}\text{m})^2}{(1.0546 \times 10^{-34}\,\text{J-s})^2} = 26.2462.$$

Figure 3.9 shows $f(\xi, K)$ [see (3.51)] vs. ξ for this value of K^2; note that, from (3.50) and (3.51), ξ cannot exceed K since both η and ξ must be positive real. By solving either (3.51) or (3.62) above, one finds that there are four bound states; these lie at $\xi = 1.3118, 2.6076, 3.8593$, and 4.9630, corresponding to energies of $6.56, 25.91, 56.75$, and 93.85 eV, respectively; these are calculated from (3.39).

3.3.1 A Matrix Approach to the Finite Potential Well

This section, which can be considered optional, examines an alternate way of solving for the energy eigenvalues of the finite potential well via matrix algebra.

To approach this method, it is helpful to cast the middle member of (3.33), that for the wavefunction in the interior of the well, into an equivalent trigonometric form. The purpose in doing this is to avoid having to deal with complex quantities in numerical calculations.

The interior wavefunction is

$$\psi_2(x) = Ae^{\iota k_2 x} + Be^{-\iota k_2 x}.$$

Now, Euler's identity tells us that $e^{\pm \iota \theta} = \cos(\theta) \pm \iota \sin(\theta)$. Applying this to ψ_2 with $\theta \equiv k_2 x$ gives

$$\psi_2(x) = \iota(A - B)\sin(k_2 x) + (A + B)\cos(k_2 x).$$

Now redefine $\iota(A - B)$ as A and $(A + B)$ as B; these are just new constants of integration. Either or both of these might turn out to be complex in practice, but this will not matter for what follows. We can now write ψ_2 as

$$\psi_2(x) = A\sin(k_2 x) + B\cos(k_2 x).$$

The overall solution is then of the form

$$
\begin{cases}
\psi_1(x) = Ce^{k_1 x} & (x < -L) \\
\psi_2(x) = A\sin(k_2 x) + B\cos(k_2 x) & (-L \le x \le L) \\
\psi_3(x) = Fe^{-k_1 x} & (x > L).
\end{cases}
\tag{3.63}
$$

k_1 and k_2 still have the same meanings as in (3.22) and (3.23); similarly for ξ and η with (3.38) and (3.39). In this formulation, $A = 0$ corresponds to the even-parity solutions and $B = 0$ to the odd-parity ones.

Applying the standard boundary conditions to the various ψ functions and their derivatives give

$$-A\sin\xi + B\cos\xi = Ce^{-\eta}, \tag{3.64}$$

$$\xi A\cos\xi + \xi B\sin\xi = \eta Ce^{-\eta}, \tag{3.65}$$

$$A\sin\xi + B\cos\xi = Fe^{-\eta}, \tag{3.66}$$

$$\xi A\cos\xi - \xi B\sin\xi = -\eta Fe^{-\eta}, \tag{3.67}$$

In (3.64) through (3.67), bring the terms on the right sides of the equalities to the left sides. If you are familiar with matrix notation, the four resulting equations can be expressed as

$$\begin{bmatrix} -\sin\xi & \cos\xi & -e^{-\eta} & 0 \\ \xi\cos\xi & \xi\sin\xi & -\eta e^{-\eta} & 0 \\ \sin\xi & \cos\xi & 0 & -e^{-\eta} \\ \xi\cos\xi & -\xi\sin\xi & 0 & \eta e^{-\eta} \end{bmatrix} \begin{bmatrix} A \\ B \\ C \\ F \end{bmatrix} = \begin{bmatrix} 0 \\ 0 \\ 0 \\ 0 \end{bmatrix}. \tag{3.68}$$

The only way to satisfy this expression is to demand that the determinant of the four-by-four square matrix on the left side be zero:

$$\begin{vmatrix} -\sin\xi & \cos\xi & -e^{-\eta} & 0 \\ \xi\cos\xi & \xi\sin\xi & -\eta e^{-\eta} & 0 \\ \sin\xi & \cos\xi & 0 & -e^{-\eta} \\ \xi\cos\xi & -\xi\sin\xi & 0 & \eta e^{-\eta} \end{vmatrix} = 0. \tag{3.69}$$

Many spreadsheet packages can quickly evaluate determinants; finding the energy eigenvalues then reduces to isolating, by trial and error, those values of ξ and η [which are related by (3.50)] that satisfy (3.69). It is helpful to know in advance how many solutions to expect.

How can we determine the constants of integration A, B, C, and F by this approach? To do this, leave one of the right-hand sides in (3.64) through (3.67) in the right side, say that in (3.66). In this case, Eq. (3.68) becomes

$$\begin{bmatrix} -\sin\xi & \cos\xi & -e^{-\eta} & 0 \\ \xi\cos\xi & \xi\sin\xi & -\eta e^{-\eta} & 0 \\ \sin\xi & \cos\xi & 0 & 0 \\ \xi\cos\xi & -\xi\sin\xi & 0 & \eta e^{-\eta} \end{bmatrix} \begin{bmatrix} A \\ B \\ C \\ F \end{bmatrix} = \begin{bmatrix} 0 \\ 0 \\ Fe^{-\eta} \\ 0 \end{bmatrix}. \tag{3.70}$$

Call the four-by-four matrix appearing here by the name W (for "Well"); let W_{ij}^{-1} denote the element in the i'th column and j'th row in the inverse of W. From the mathematics of matrix inversion, we can then solve (3.70) for A, B, C, and F:

$$\begin{cases} A = F \left(W_{13}^{-1}\right)\left(e^{-\eta}\right) \\ B = F \left(W_{23}^{-1}\right)\left(e^{-\eta}\right) \\ C = F \left(W_{33}^{-1}\right)\left(e^{-\eta}\right) \\ F = F \left(W_{43}^{-1}\right)\left(e^{-\eta}\right). \end{cases} \tag{3.71}$$

Equations (3.71) leave F undetermined; it can only be pinned down after the fact from knowledge of the ratios A/F, B/F, C/F and the demand that the solution be normalized. The last of (3.71) provides a check on the method: if $\left(W_{43}^{-1}\right)\left(e^{-\eta}\right) \neq 1$, then something has gone wrong with the calculation.

A final observation on the manipulation that led to (3.70): we could have left any one of the right-sides of (3.64) through (3.67) on the right side in (3.70), but chose to do so with that for (3.66). You should find the same ξ-values if any of the other three conditions had been so manipulated. What you do not want to do, however, is leave an A or B on the right side of (3.70). The reason for this is that, depending on the parity of the solution, either A or B will be zero, and the check condition analogous to $\left(W_{43}^{-1}\right)\left(e^{-\eta}\right) = 1$ will be unworkable. Neither C nor F will ever be zero in this problem, so the formulation of (3.70) and (3.71) will work for all of the energy eigenvalues.

As an exercise, try reproducing the ξ-values for of Example 3.2 by this method.

3.4   Finite Potential Well-Even Solutions

In this section we explore the even-parity solutions to the finite rectangular well in more detail. With $B = A$ and $F = C$, Eq. (3.33) becomes

$$
\begin{cases}
\psi_1(x) = Ce^{k_1 x} & (x < -L) \\
\psi_2(x) = 2A \cos(k_2 x) & (-L \leq x \leq L) \\
\psi_3(x) = Ce^{-k_1 x} & (x > L).
\end{cases}
\tag{3.72}
$$

Similarly, Eqs. (3.40) through (3.43) reduce to two conditions:

$$
2A \cos \xi = Ce^{-\eta}
\tag{3.73}
$$

and

$$
2A \sin \xi = (\eta/\xi)\, Ce^{-\eta}.
\tag{3.74}
$$

Normalization demands

$$
\int\limits_{-\infty}^{-L} \psi_1^* \psi_1 dx + \int\limits_{-L}^{L} \psi_2^* \psi_2 dx + \int\limits_{L}^{\infty} \psi_3^* \psi_3\, dx = 1.
\tag{3.75}
$$

Again note that each wavefunction is integrated only over the domain wherein it applies. Since we are dealing with even-parity states, the first and third integrals are equal, and reduce to

$$
\int\limits_{-\infty}^{-L} \psi_1^* \psi_1 dx = \int\limits_{L}^{\infty} \psi_3^* \psi_3\, dx = C^2 \int\limits_{L}^{\infty} e^{2k_1 x}\, dx = \frac{C^2 e^{-2\eta}}{2k_1}.
\tag{3.76}
$$

The middle integral evaluates as

$$\int_{-L}^{L} \psi_2^* \psi_2 dx = 4A^2 \int_{-L}^{L} \cos^2(k_2 x) \, dx = 4A^2 \left[\frac{x}{2} + \frac{\sin(2k_2 x)}{4k_2} \right]_{-L}^{L}$$

$$= 4A^2 \left[\frac{\sin(2\xi)}{2k_2} + L \right].$$

(3.77)

Hence we have

$$\frac{C^2 e^{-2\eta}}{k_1} + 4A^2 \left[\frac{\sin(2\xi)}{2k_2} + L \right] = 1.$$

(3.78)

Eliminating C via either of (3.73) or (3.74) gives

$$4A^2 \left[\frac{\cos^2 \xi}{k_1} + \frac{\sin(2\xi)}{2k_2} + L \right] = 1,$$

(3.79)

which, on eliminating k_1 and k_2 in favor of ξ and η, becomes

$$4A^2 = \frac{\eta \xi}{L \left[\xi \cos^2 \xi + \eta \sin \xi \cos \xi + \eta \xi \right]}.$$

(3.80)

Hence, from (3.73),

$$C^2 = \frac{\eta \xi e^{2\eta} \cos^2 \xi}{L[\xi \cos^2 \xi + \eta \sin \xi \cos \xi + \eta \xi]} = \frac{\eta \xi e^{2\eta}}{L \left[\xi + \eta \tan \xi + \eta \xi \sec^2 \xi \right]}.$$

(3.81)

Equations (3.72), (3.80), and (3.81) specify the even-parity wavefunctions of the finite rectangular well: once a given energy state ξ_n is specified, A and C can be computed. Unlike the case of the infinite well, the normalization constants depend on the energy eigenvalue involved.

The probability of finding the particle outside the well is given by

$$P_{out}^{even} = \int_{-\infty}^{-L} \psi_1^* \psi_1 dx + \int_{L}^{\infty} \psi_3^* \psi_3 dx = \frac{C^2 e^{-2\eta}}{k_1} = \frac{\xi}{[\xi + \eta \tan \xi + \eta \xi \sec^2 \xi]}.$$

(3.82)

Note that the probability is dimensionless.

The probability distribution of electrons trapped in an atomic-scale two-dimensional potential well have actually been measured via Scanning Tunneling Microscopy (STM), a technique invented in 1981 by Gerd Binning and Heinrich Rohrer of IBM's

Fig. 3.10 Ellipsoidal quantum corral of cobalt atoms on a copper surface. The major axis of the ellipse is about 15 nm wide. *Source* Public domain; https://commons.wikimedia.org/wiki/File: Co_ellipse.png from J. A. Stroscio, R. J. Celotta, Steven R. Blankenship, and Frank M. Hess, https:// www.nist.gov/programs-projects/atom-manipulation-scanning-tunneling-microscope

Zurich Laboratory and for which they shared the 1986 Nobel Prize for Physics. In this technique, a very tiny metallic probe tip (ideally, only one atom wide at the tip) is positioned within a few atomic diameters of a conducting surface. A small voltage applied between the surface and the tip sets up a potential barrier through which electrons can tunnel. The resulting "tunneling current" is very sensitive to the barrier width (see Problem 3.18, but read Sect. 3.7 first); consequently, by adjusting the tip/surface separation to maintain a constant current, the surface can be mapped at the atomic level; a resolution of about 0.2 nm is possible. A related technique, *atomic force microscopy* (AFM), makes it possible to move individual atoms on a substrate to create nano-scale structures. Figure 3.10 shows a scanning-tunneling microscope "image" of a "quantum corral" constructed by atomic force microscopy; the wavelike ripples within the corral reflect the probability distribution of surface-state electrons. This striking image is direct evidence for the atomic-scale probability distributions predicted by quantum mechanics. We will have more to say about tunneling in Sects. 3.7–3.9.

A good introduction to the physics of atomic force microscopy can be found in [4].

3.5 Number of Bound States in a Finite Potential Well

In Example 3.2 we saw that a finite rectangular well of strength $K = 5.12$ is capable of binding four energy states, the most energetic of which was close to the top of the well. From the form of the eigenvalue condition (3.51) as exemplified in Fig. 3.9, one might infer, correctly as it turns out, that a well with a larger K would likely bind more states. Is there any way to tell in advance how many states a well of a given K can bind without having to solve the eigenvalue equation in detail? This is indeed possible.

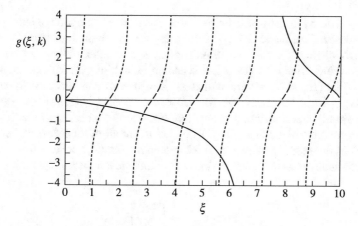

Fig. 3.11 Tangent and eigenvalue functions for $K = 10$ finite well

To see this, it is convenient to rearrange the eigenvalue equation into the form

$$\tan(2\xi) = -\frac{2\xi\sqrt{K^2 - \xi^2}}{K^2 - 2\xi^2} = g(\xi, K).\tag{3.83}$$

This expression indicates another way one might solve for the eigenvalues: by finding the intersections of the function $g(\xi, K)$ with branches of the function $\tan(2\xi)$. This is illustrated in Fig 3.11 for the case of $K = 10$.

A plot of $\tan(2\xi)$ consists of a number of nearly vertical branches with zeros at $\xi = 0, \pi/2, \pi, 3\pi/2, \ldots$. These branches approach positive infinity as ξ approaches $\pi/4, 3\pi/4, 5\pi/4, \ldots$, and increase toward zero from negative infinity when ξ is just past $\pi/4, 3\pi/4, 5\pi/4, \ldots$. The function $g(\xi, K)$ has two branches. Initially, as ξ increases from zero (recall that the range of ξ is $0 \le \xi \le K$), $g(\xi, K)$ decreases from zero to negative infinity as ξ approaches $K/\sqrt{2}$, and then decreases from positive infinity just past $\xi = K/\sqrt{2}$ down to zero at $\xi = K$. While the figure shows six intersections (eigenvalues!) between the branches of $\tan(2\xi)$ and the two branches of $g(\xi, K)$, there are actually seven: one lies below the bottom of the plot at $\xi \sim 7.07$.

It should be clear that no matter what value of K one is dealing with, $g(\xi, K)$ will always be of the form shown in Fig. 3.11: two descending branches separated by a break at $K/\sqrt{2}$, one of negative values and one of positive, with $g(\xi, K) \to 0^+$ as $\xi \to K$.

As you look at Fig. 3.11, imagine decreasing K: the branches of $g(\xi, K)$ will shrink toward the left edge of the figure while the branches of $\tan(2\xi)$ will remain unchanged. No matter how small the value of K, the descending branch of the $g(\xi, K)$ function will always cut at least the first branch of the $\tan(2\xi)$ function, so we can conclude that a finite rectangular well will always support at least one bound state.

Now, the branches of $\tan(2\xi)$ have zeroes at $\xi = 0, \pi/2, \pi, 3\pi/2, \ldots$. Imagine increasing K so that the $g(\xi, K)$ curve begins to stretch out to the right. If

K is less than $\pi/2$, there will be only one eigenvalue, that corresponding to the positive-descending branch of $g(\xi, K)$ cutting the first branch of $\tan(2\xi)$ (that is, the branch which begins at $\xi = 0$). As soon as K exceeds $\pi/2$ there will be a second eigenvalue, where the positive-descending branch of $g(\xi, K)$ cuts the second branch of $\tan(2\xi)$. The pattern is that as K increases and successively exceeds $2\pi/2, 3\pi/2, 4\pi/2, 5\pi/2, \ldots$, we will have 3, 4, 5, 6, ... eigenvalues. Study the figure carefully and convince yourself that because the branches of $\tan(2\xi)$ go to $\pm\infty$, we are guaranteed that $g(\xi, K)$ is bound to cut all of them that lie below the terminus of the $g(\xi, K)$ curve and which extend into $\tan(2\xi) > 0$.

We can summarize these conclusions by writing the number of eigenvalues N for a given value of K as

$$N(K) = 1 + \left[\frac{2K}{\pi}\right], \tag{3.84}$$

where $[2K/\pi]$ designates the largest integer less than or equal to $2K/\pi$. For $K = 10$, this gives $N = 7$ as expected.

Finite potential wells find practical use in a number of everyday devices. Digital cameras obtain their images with a two-dimensional grid of potential wells known as a charge-coupled-device or "CCD" chip. Made from light-sensitive materials, these devices respond to incident photons by liberating electrons, which are trapped in electronic potential wells. Each well is known as a "picture element or" "pixel." After the exposure is complete, the image is read out by lowering and raising the potential "walls" to march the electrons off the chip for counting and thus clearing it for the next exposure.

Example 3.3 A manufacturer of CCD cameras for amateur astronomers advertises that one of its products uses a chip with "9-micron pixels" that each have a "full well" of 60,000 electrons. This means that each tiny potential well is $9 \times 9\,\mu m$ ($1\,\mu m = 10^{-6}$ m) in size and that they can hold 60,000 electrons before becoming full. (Each energy level can hold two electrons, one each of opposite spin—see Chap. 8.) Supposing for simplicity that the pixels are one-dimensional finite rectangular wells, what must be their depth V_o?

With $N = 30{,}000$, the factor of unity on the right side of (3.84) can be neglected:

$$K \sim \frac{N\pi}{2}.$$

Combining this with (3.50) and solving for V_o gives

$$V_o = \frac{(N\pi\hbar)^2}{8mL^2},$$

where L is the half-width of the well, $4.5\,\mu$m. Putting in the numbers,

$$V_o = \frac{\left[30,000\pi\,(1.055 \times 10^{-34}\,\text{J-s})\right]^2}{8(9.109 \times 10^{-31}\,\text{kg})(4.5 \times 10^{-6}\,\text{m})^2} = 6.70 \times 10^{-19}\,\text{J},$$

or about 4.18 eV. Such a device can easily be battery-powered, a necessity for astronomers at remote sites with no access to power.

3.6   Sketching Wavefunctions

The finite potential well is a simple, symmetric potential, yet its solution is not trivial. It is easy to imagine that more complex potentials could involve intractable algebra. Indeed, closed analytic solutions to Schrödinger's equation are obtainable for only a very limited number of potentials; in many cases brute-force series solutions, approximation methods, or numerical integration must be appealed to. It is, however, possible to make *qualitative* sketches of plausible wavefunctions for complicated potentials based on a few simple rules. This is instructive in its own right, and can be helpful as a guide as to what form an analytic solution might take.

Figure 3.12 shows a hypothetical example. We wish to get an idea of the form of the wavefunction(s) corresponding to a particle of mass m and total energy E moving in the potential shown. The points where the total-energy line cuts $V(x)$ are the classical turning points of the motion; these points serve to delimit the inside and outside regions wherein the solutions will be of sinusoidal and exponential-decay form, respectively.

At any position, we can divide the total energy into potential and kinetic contributions as shown. At positions where the well is deep, the kinetic energy will be relatively large. According to de Broglie's hypothesis, this will correspond to a smaller wavelength since $\lambda = h/p$. This is the first rule for sketching wavefunctions:

Fig. 3.12 A hypothetical potential well

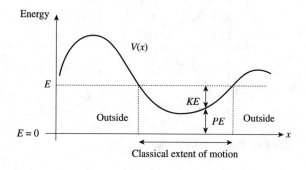

deeper well → shorter λ. Further, wavefunctions with more nodes (that is, points where $\psi(x) = 0$) exhibit shorter characteristic wavelengths, meaning higher energies. The ground state wavefunction has no nodes, the first excited state one, the second excited state two, and so on.

What of the amplitude of ψ as the well shallows or deepens? This question can also be addressed from an energy argument. At a position where the kinetic energy is large, the velocity of the particle will be high, hence it will spend less time in deeper parts of the well than shallower regions, that is, there should be a lower chance of finding it where the well is deep. Since the probability of finding the particle at some position is proportional to ψ, the amplitude of ψ must therefore decrease where the well deepens. This is the second rule for sketching wavefunctions: deeper well → smaller amplitude for ψ.

To get a more complete sense of how solutions of Schrödinger's equation behave in the various regions, it is helpful to look at some of its purely mathematical aspects. The discussion here is adapted from that in Eisberg and Resnick [5].

We can write Schrödinger's equation as

$$\frac{d^2\psi}{dx^2} = \frac{2m}{\hbar^2}[V(x) - E]\psi(x).$$

Consider an outside region, where $[V(x) - E] > 0$. If $\psi > 0$ at some point, then $d^2\psi/dx^2$ must be > 0, whereas if $\psi < 0$, then $d^2\psi/dx^2$ will be < 0. For an inside region, $d^2\psi/dx^2$ must be $< 0 (> 0)$ if $\psi > 0 (< 0)$ because $[V(x) - E]$ is < 0. From elementary calculus, we know that functions with $d^2\psi/dx^2 > 0$ are concave upwards, while those with $d^2\psi/dx^2 < 0$ are concave downwards. Such functions are sketched for the inside and outside regions in Fig. 3.13.

Imagine that you are inside the well at the point where the inside and left-outside regions meet, and that $\psi > 0$ with $d\psi/dx < 0$ [or > 0] at this point. If you proceed leftward and enter the outside region, ψ must have the form typical of such regions:

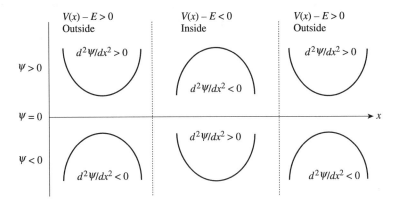

Fig. 3.13 Nature of ψ functions

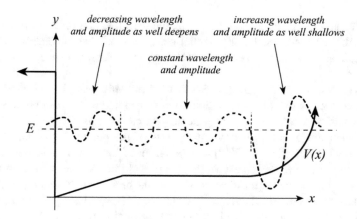

Fig. 3.14 Hypothetical $n = 9$ wavefunction

a sum of positive and negative exponentials as in (3.25). ψ would initially decrease [or increase] in value, but will eventually begin to turn away from the $\psi = 0$ line since $d^2\psi/dx^2 > 0$. If the domain of the potential goes to $x \to \infty$, the divergent part of ψ will have to be arbitrarily suppressed to prevent an infinite divergence. Similar logic applies to the right-outside region. If on the other hand you proceed rightward from the outside toward the inside of the well, ψ will always tend to turn towards the $\psi = 0$ line whether the initial values of ψ and its first derivative be positive or negative. This behavior will always ensure the oscillatory-form solutions characteristic of the infinite and finite wells.

Inasmuch as they derive from considerations of fundamental quantum relationships and the nature of Schrödinger's equation, the above comments regarding the wavelength, number of nodes, and curvatures of wavefunctions are quite general for any potential. Other factors to keep in mind are that that the number of extrema in the wavefunction is equal to the quantum number of the state involved, and that if the well is symmetric the wavefunctions will alternate parity. Figure 3.14 illustrates an application of these rules.

When sketching such figures, it is conventional to superimpose the sketch of the wavefunction over the sketch of the well at the vertical level of the energy of the wavefunction.

Although the preceding argument for the amplitude of ψ based on potential and kinetic energy sounds plausible, it can on occasion lead to erroneous predictions regarding what ψ should look like. An example is provided by the ground state of the parabolic-shaped harmonic oscillator potential $V(x) = kx^2/2$ discussed in Chap. 5. In this case, $\psi(x)$ in fact has its greatest amplitude where the potential well is deepest; see Figs. 5.1 and 5.2. This is because the above reasoning is based on classical concepts. The deeper well $\to$ smaller amplitude reasoning is correct for high-n energy levels where the results of classical and quantum computations must agree, but fails completely for the lowest energy levels. "High," however, need not mean "terribly high": as shown in Fig. 5.2, the wavefunction for the fourth excited state of the harmonic oscillator potential concurs with what the energy argument would indicate. One must always be careful in applying classical concepts to quantum situations.

Finally, it was remarked in Sect. 3.3 that a symmetric potential will always lead to odd and even-parity solutions to Schrödinger's equation. This can be understood on the basis of a some simple arguments. We can always define the coordinate system such that $x = 0$ is the symmetry axis of the potential; that is, such that the potential is described by $V(x) = V(-x)$. Since the two sides of the potential are indistinguishable, we can expect that the particle should have no preference for being found on one side or the other, that is, that the probability of finding the particle should also be symmetric about $x = 0$: $\psi^2(x) = \psi^2(-x)$. Taking the square root of this gives $\psi(x) = \pm\psi(-x)$, that is, $\psi(x)$ must be of either even or odd parity. If it is of even parity then it must have either a local maximum or minimum at the symmetry point, with $d\psi/dx = 0$ there. If $\psi(x)$ is of odd parity, then we must have $\psi = 0$ at $x = 0$ and $d\psi/dx \neq 0$ there. By sketching a few even and odd-parity functions you should be able to convince yourself of some further conclusions. One is that the lowest energy state for a symmetric potential must be of even parity. For an $n = 1$ state we expect one maximum; no odd-parity function can have only one maximum and still satisfy the boundary conditions $\psi \to 0$ as $x \to \pm\infty$, whereas a single-maximum even-parity function must always behave in this way. Ground-state wavefunctions of symmetric potentials will thus always have an appearance similar to that for the $n = 1$ state of a finite rectangular well as sketched in Fig. 3.8.

As to the higher-energy states of symmetric potentials, consider first those corresponding to odd quantum numbers, $n = 3, 5, 7 \ldots$. To achieve both the desired symmetry and have an odd number of maxima, ψ must have an extremum at $x = 0$. This can only be satisfied by an even-parity function. (An odd-parity function must have $d\psi/dx \neq 0$ at $x = 0$; it is impossible to satisfy that requirement and the demands of symmetry and an odd number of maxima simultaneously.) For states with even quantum numbers $n = 2, 4, 6 \ldots$, one can always easily sketch a suitable odd-parity solution since a symmetric odd-parity function satisfying $\psi \to \infty$ as $x \to \pm\infty$ must have a total number of maxima that is even. Even-parity functions are possible in this circumstance, but only if they have a minimum at $x = 0$. This cannot be demanded, however, as this would be asking to impose a further boundary condition on the solution. For these reasons, solutions to symmetric potentials will alternate parity in the sequence even, odd, even, odd, and so forth. The harmonic-oscillator potential of Chap. 5 is an example of a symmetric potential.

Part II: Potential Barriers and Scattering

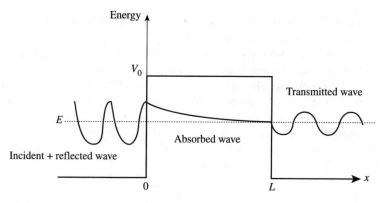

Fig. 3.15 Rectangular barrier penetration

3.7 Potential Barriers and Scattering

We now turn our attention to potential barriers.

The general situation is depicted in Fig. 3.15, where a matter-wave of energy E propagating from the left encounters a rectangular potential barrier of height V_0 Joules and thickness L. A simple classical analog is a ball rolling up a hill of height h: if the energy of the ball is greater than mgh (that is, if $E > V_0$), then the ball will scale the hill and roll down the other side; if not, it will roll back toward the direction from which it came. The presence of tunneling phenomena in potential-well problems, however, hints that the quantum-mechanical situation could be quite different: there may be some probability of the matter wave penetrating through the barrier even if $E < V_0$.

As a result of such a reflection process, the region to the left of the barrier would presumably contain a combination of incident and partially reflected waves, that to the right a transmitted or scattered wave, and within the barrier itself will be an absorbed component comprising both exponential growth and decay terms of the form of (3.25). Intuitively, we might expect the amplitude of the transmitted wave to depend on E, V_0, L, and the amplitude of the incident wave.

Potential barrier problems possess considerable practical application. Essentially everything we know of atomic and nuclear structure derives from experiments wherein some incident particle strikes an atom or nucleus. From the intensity and spatial distribution of the scattered particles, much can be inferred as to the size and structure of the target object. In reality these experiments are of course three-dimensional, but many of the important features can be gleaned from a one-dimensional analysis.

In principle, a number of situations are possible here. One could have particles with $E > V_0$ or $E < V_0$. The incident energy could be in the form of a spatially isolated wave packet or a continuous, time-independent stream of monoenergetic particles. The barrier could be thick or thin compared to the de Broglie wavelength of the incident particles. Since we will be analyzing the situation via time-independent

wavefunctions like those used in our studies of potential wells, we are restricted to
dealing with a steady-state situation, that is, the case of a steady stream of particles
of energy $E < V_0$ incident on the barrier. (Aside: intuitively appealing as they are,
it is misleading to invoke terms such as "incident", "reflected", and "transmitted" as
these imply that some particles are time-dependently burrowing their way through
the barrier to emerge on the other side. What we will have is a time-independent
probability distribution whose amplitude varies across the barrier.) We will first
work out the general situation for a barrier of arbitrary thickness and then specialize
our results to a few particularly interesting cases.

The potential for the case depicted in Fig. 3.15 is

$$V(x) = \begin{cases} V_0 & (0 \le x \le L) \\ 0, & \text{otherwise.} \end{cases} \tag{3.85}$$

We divide space into three regions as shown in the figure. In the first region where
$x \le 0$, we have $V = 0$. The solution to Schrödinger's equation will be of the same
form as that inside an infinite or finite rectangular well:

$$\psi_1(x) = A_0 e^{ik_1 x} + A e^{-ik_1 x}, \tag{3.86}$$

where

$$k_1^2 = \frac{2mE}{\hbar^2}, \tag{3.87}$$

and where A_0 and A are constants of integration.

Equation (3.86) can be interpreted as the sum of two waves: an incident matter-
wave of amplitude A_0 propagating to the right, and a reflected one of amplitude A
propagating to the left, corresponding to any partial reflection arising from the barrier
at $x = 0$. Amplitude A_0 would be set experimentally.

For $0 \le x \le L$ where $E < V_0$, the wavefunction will be of the form

$$\psi_2(x) = B e^{-k_2 x} + C e^{k_2 x}, \tag{3.88}$$

where

$$k_2^2 = \frac{2m(V_0 - E)}{\hbar^2}. \tag{3.89}$$

Neither B nor C can be eliminated from (3.88) as the barrier does not extend to
infinity; both terms must be retained for complete generality.

For $x \ge L$ we again have $V = 0$, but there is nothing for the transmitted wave
to reflect from. In this region, then, we take the solution to be of the form of a
matter-wave propagating to the right:

$$\psi_3(x) = D e^{ik_1 x}. \tag{3.90}$$

Note that the wave number k is the same in both outside regions since V is the same in both. That fraction of the original incident wave which emerges from the barrier does so with the same energy that it entered with, but with a different, presumably smaller, amplitude. (Radio signals generated by your favorite station do not differ in frequency inside and outside of your house!)

Experimentally, one would measure the transmitted amplitude D as a function of the amplitude of the incident wave, A_0, presumably over a range of incident energies k_1. To get D in terms of A_0 it is necessary to apply the usual boundary conditions to (3.86) through (3.90), with the exception that we cannot force $\psi \to 0$ as $x \to \infty$ since the (incoming + reflected) and transmitted waves propagate to infinity in both directions; the dimensions of a laboratory are likely to be large in comparison with the size of a nucleus. Demanding the continuity of the wavefunctions and their derivatives at $x = 0$ and $x = L$ gives four conditions:

$$A_0 + A = B + C \tag{3.91}$$

$$\imath k_1 A_0 - \imath k_1 A = -k_2 B + k_2 C \tag{3.92}$$

$$B e^{-k_2 L} + C e^{k_2 L} = D e^{\imath k_1 L} \tag{3.93}$$

$$-k_2 B e^{-k_2 L} + k_2 C e^{k_2 L} = \imath k_1 D e^{\imath k_1 L}. \tag{3.94}$$

Taking A_0 to be fixed, we can treat (3.91) through (3.94) as four equations in four unknowns: A, B, C, and D. Solving for D in terms of A_0 is tedious but straightforward, and follows the same sort of logic as was applied in the study of the finite rectangular well in Sect. 3.3. The result is

$$D = \frac{4 \imath k_1 k_2 e^{-\imath k_1 L}}{[(k_2 + \imath k_1)^2 e^{-k_2 L} - (k_2 - \imath k_1)^2 e^{k_2 L}]} A_0. \tag{3.95}$$

D/A_0 is the amplitude of the scattered-wave probability density as a fraction of that of the incident wave. The squared magnitude of this number, $(D/A_0)(D/A_0)^*$, is then a measure of the fraction of matter waves that penetrate through the barrier to emerge on the other side, and is known as the *transmission coefficient* T:

$$T = D^* D. \tag{3.96}$$

It is customary to set $A_0 = 1$ for simplicity. To expedite the algebra involved in computing T, it is helpful to extract a factor of $e^{k_2 L}$ from the denominator of (3.95) and place it in the numerator:

$$D = \frac{4\iota k_1 k_2 e^{-\iota k_1 L} e^{-k_2 L}}{[(k_2 + \iota k_1)^2 e^{-2k_2 L} - (k_2 - \iota k_1)^2]}.$$

Computing T is another exercise in algebra. The result is

$$T = \frac{16 k_1^2 k_2^2 e^{-2k_2 L}}{[(k_1^2 + k_2^2)^2 (1 + e^{-4k_2 L}) - e^{-2k_2 L}(2k_1^4 - 12k_1^2 k_2^2 + 2k_2^4)]}. \tag{3.97}$$

This result is quite general for our $E < V_0$ scenario, but rather awkward. In what follows we consider two cases where simplifying approximations can be made. If you are familiar with hyperbolic trigonometric functions, Eq. (3.97) can be expressed as

$$T = \left[1 + \left(\frac{k_1^2 + k_2^2}{2k_1 k_2} \right)^2 \sinh^2(2k_2 L) \right]^{-1},$$

but (3.97) will suffice for our purposes as it is.

Thick Barrier Approximation

In cases where the width of the barrier is large compared to the wavelength $\lambda = 2\pi k_2$, we can put $k_2 L \gg 1$, or

$$\frac{2mL^2(V_o - E)}{\hbar^2} \gg 1. \tag{3.98}$$

In this event, the $e^{-4k_2 L}$ and $e^{-2k_2 L}$ terms in (3.97) approach zero; the second term in the denominator drops away to give

$$T_{THICK} \approx \frac{16k_1^2 k_2^2 e^{-2k_2 L}}{(k_1^2 + k_2^2)^2},$$

or, on replacing k_1 and k_2 via (3.87) and (3.89),

$$T_{THICK} \approx 16 \left(\frac{E}{V_0} \right) \left(1 - \frac{E}{V_0} \right) \exp \left(-\frac{2L}{\hbar} \sqrt{2m(V_0 - E)} \right). \tag{3.99}$$

In this result, the term being exponentiated will be large and negative, which will make T_{THICK} exceedingly small. This is demonstrated in the following example.

Example 3.4 Electrons of energy $E = 1$ eV are incident on a barrier of height $V_0 = 50$ eV and width $L = 10$ Å. What is the transmission coefficient?

We first verify that the thick barrier approximation is valid:

$$k_2 L = \frac{L\sqrt{2m(V_0 - E)}}{\hbar} = \frac{(10^{-9} \text{ m})\sqrt{2(9.109 \times 10^{-31} \text{ kg})(7.850 \times 10^{-18} \text{ J})}}{1.055 \times 10^{-34} \text{ J-s}}$$

$$= 35.85,$$

which we regard as $\gg 1$ for the purpose of this example. Hence

$$T \sim 16 \left(\frac{1}{50}\right)\left(\frac{49}{50}\right) \exp(-71.7) \sim 2.29 \times 10^{-32}.$$

The chance of an electron penetrating the barrier is remote. If 10^6 electrons are incident per second, one would have to wait, on average, some 1.4×10^{18} years for a penetration to occur. This is some eight orders of magnitude greater than the estimated age of the Universe! However, such calculations are not without practical value: If the electrons have higher energy, or if the barrier is thinner, tunneling is much more likely. An electronic device known as a Josephson junction depends on quantum tunneling to control tiny currents very precisely by raising and lowering potential barriers.

A variant of the thick barrier scenario is the so-called semi-infinite potential step. In this situation, the thickness of the barrier in Fig. 3.15 as allowed to become infinite: $L \to \infty$. The transmission coefficient in this case goes identically to zero: A particle may penetrate into the barrier, but will eventually be turned back.

Thin Barrier Approximation

In this case the thickness of the barrier is assumed to be small with respect to the wavelength of the absorbed wave, that is, $k_2 L \ll 1$. This means that the incoming wave will hardly notice the barrier at all; we can expect $T \to 1$. Since $e^{-2k_2 L} \sim e^{-4k_2 L} \sim 1$, the denominator of (3.97) reduces to $16(k_1^2 K_2^2)$, leaving

$$T_{THIN} \sim e^{-2k_2 L}. \tag{3.100}$$

Inasmuch as $k_2 L \ll 1$, T_{THIN} will be only slightly less than 1. In itself, this result is not particularly noteworthy; we will put it to much more interesting use in the following section.

3.8 Penetration of Arbitrarily-Shaped Barriers

Equation (3.100) gives the transmission coefficient for a matter-wave incident on a thin rectangular barrier of height V_0 and thickness L. In reality, an experiment is not likely to involve such a simple configuration; the potential presented by a nucleus to an incoming test particle is likely to have some more complex shape. However, it is possible to derive an expression for the penetration probability in the case of a particle encountering an arbitrarily-shaped barrier by considering the barrier to be comprised of a number of thin barriers placed back-to-back.

Figure 3.16 illustrates such a case. A particle of mass m and energy E is incident on some potential barrier described by the function $V(x)$.

We divide the barrier into a large number of back-to-back rectangular barriers, each of width Δx. Since each thin barrier has a different height $V(x)$, it will have a different k in (3.100):

$$k(x) = \sqrt{\frac{2m[V(x) - E]}{\hbar^2}}. \qquad (3.101)$$

If the entire barrier is divided into n sub-barriers, the probability of traversing the entire barrier is just the product of penetrating each sub-barrier in turn:

$$P \sim (e^{-2k_1 \Delta x})(e^{-2k_2 \Delta x}) \cdots (e^{-2k_n \Delta x}), \qquad (3.102)$$

where the subscripts now refer to particular sub-barriers. This can be expressed more compactly as

$$P \sim e^{-2(k_1 + k_2 + \cdots + k_n \Delta x]} \sim e^{-2 \sum_{i=1}^{n} k_i(x) \Delta x}. \qquad (3.103)$$

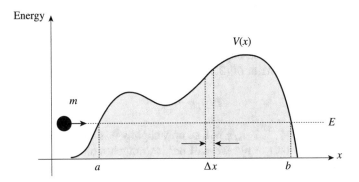

Fig. 3.16 General barrier penetration

The k's are written as functions of x as a reminder to use (3.101) for each sub-barrier. In the limit of $n \to \infty$, Eq. (3.103) can be replaced by an integral over the entire barrier:

$$P_{PENETRATE} \sim e^{-2\int k(x)dx},$$

or, on using (3.101),

$$P_{PENETRATE} \sim e^{-\frac{2\sqrt{2m}}{\hbar}\int_a^b \sqrt{V(x)-E}dx}. \tag{3.104}$$

The limits of integration are taken to be the points at which E cuts the $V(x)$ curve, "a" and "b" in Fig. 3.16. In reality, an incoming particle would sense the barrier before striking it, so these limits are an approximation, a complication we ignore.

The presence of the square root in (3.104) limits the number of cases that can be handled analytically; one often has to resort to numerical integration, as demonstrated in the following example.

Example 3.5 A electron of energy $E = A/2$ is incident on a potential barrier given by

$$V(x) = Ae^{-(x/x_0)^2}.$$

If $A = 1$ eV and $x_0 = 1$ Å, what is the probability of it penetrating the barrier? See Fig. 3.17.

Equation (3.104) gives

$$\ln P \sim -\frac{2\sqrt{2m}}{\hbar}\int \sqrt{Ae^{-(x/x_0)^2} - E}\,dx.$$

Substituting $E = A/2$ and extracting a common factor of A gives

$$\ln P \sim -\frac{2\sqrt{2mA}}{\hbar}\int \sqrt{e^{-(x/x_0)^2} - 1/2}\,dx.$$

For a reason that will become clear momentarily, it is helpful to make the change of variable $z = x/x_0$, giving

$$\ln P \sim -\frac{2x_0\sqrt{2mA}}{\hbar}\int \sqrt{e^{-z^2} - 1/2}\,dz.$$

The limits of integration correspond to the points at which E cuts the $V(x)$ curve, namely $x = \pm x_0\sqrt{\ln 2}$, or $z = \pm\sqrt{\ln 2}$. Recognizing that the integral is symmetric about $z = 0$, we have

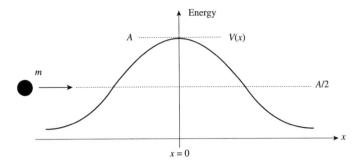

Fig. 3.17 Example 3.5

$$\ln P \sim -\frac{4x_0\sqrt{2mA}}{\hbar} \int\limits_{0}^{\sqrt{\ln 2}} \sqrt{e^{-z^2} - 1/2}\, dz.$$

The advantage of changing variables from x to z is that the integral has been rendered dimensionless, a significant convenience inasmuch as it can only be solved numerically. The reader should verify that it evaluates to 0.4416, giving

$$\ln P \sim -2.498\frac{x_0\sqrt{mA}}{\hbar}.$$

For $A = 1$ eV and $x_0 = 1$ Å for an electron, $P \sim 0.4047$, a distinctly non-negligible result.

3.9 Alpha-Decay as a Barrier Penetration Effect

Historically, one of the first great successes of quantum mechanics was the explanation of alpha-decay as a barrier penetration process of the type formulated in the previous section. This explanation was developed independently by George Gamow and the team of Ronald Gurney and Edward Condon in 1928 [6, 7].

Alpha-decay is a process wherein a nucleus spontaneously decays by emission of an alpha-particle, a synonymous term for a helium nucleus. The resulting *daughter nucleus*, having two fewer protons than the original nucleus, is thus an isotope of an element two places lower in the periodic table than the *parent nucleus*. Half-lives for this process vary from microseconds to billions of years, a range of over 20 orders of magnitude. The emitted alpha-particles range in energy from about 4 to 9 MeV, with more energetic decays being accompanied by shorter half-lives. A good example of this process is the decay of nuclei of the common isotope of uranium, U-238, into thorium-234:

$$^{238}_{92}U \rightarrow ^{234}_{90}Th + ^4_2He.$$

The half-life for this decay is about 4.47 billion years; it was this particular decay that led Henri Becquerel to the discovery of radioactivity in early 1896. Although we will refer to this particular case for sake of specificity, our analysis will be quite general and will ultimately be applied in detail to decay of thorium isotopes.

In papers published in 1911 and 1912, Hans Geiger and John Nutall found an empirical relation between the logarithm of the half-life and the decay energy [8, 9],

$$\ln\left(t_{1/2}\right) = a + \frac{b}{\sqrt{E_\alpha}}, \tag{3.105}$$

where the constants a and b depend on the element involved. Treating α-decay as a barrier penetration process provided an understanding of this otherwise purely empirical result.

The dilemma in trying to understanding alpha-decay was that to hold nuclei together against the tremendous repulsive forces between protons required positing the existence of short-range ($\sim 10^{-15}$ m) but very strongly attractive *nuclear forces* that act between nucleons; the resulting potential can be modeled as a very deep potential well with a spatial extent on the scale of the size of the nucleus [10]. This is sketched in Fig. 3.18, where the nuclear potential is represented by a rectangular well of width R; an alpha-particle of charge $+2e$ is trapped within the well. In reality, nuclei live in three dimensions, so the x-axis in Fig. 3.18 represents a radial coordinate measured from the center of the original nucleus; x is not defined for $x < 0$. We represent the potential as an infinite wall at $x = 0$, but this will not affect the argument that follows.

The circles at the base of the figure represent the post-decay nucleus and emitted alpha-particle: we take R to be defined by their center-to-center distance when they are just in contact. After the alpha is emitted, it will be strongly repelled by the inverse-square Coulomb force of the daughter nucleus; this is represented by the decreasing curved part of the potential function; recall that the Coulomb potential behaves as $1/r$.

The difficulty in understanding alpha-decay was that nuclear potentials were known to be some tens of MeV deep (see below). How could an alpha-particle of an energy of only a few MeV possibly jump out of such as well? Gamow and Gurney & Condon modeled alpha-decay as a penetration through the nuclear-plus-Coulomb barrier.

To get a sense of the distance scale involved, it is helpful to know that nuclei can be modeled as spheres whose radii depend on their number of nucleons A: $r \sim r_0 A^{1/3}$, where r_0 is empirically known from scattering experiments to be ~ 1.2 fm (1 fm = 1 femtometer = 10^{-15} m; a femtometer is also known as a Fermi, in honor of Enrico Fermi.) For our thorium-alpha system, for example,

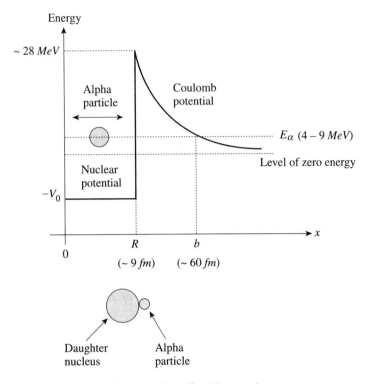

Fig. 3.18 Alpha-decay as a barrier penetration effect. Not to scale

$$R \sim r_0(A_{Th}^{1/3} + A_{\alpha}^{1/3}) \sim (1.2\,\text{fm})(234^{1/3} + 4^{1/3}) \sim 9.3\,\text{fm}.$$

The potential energy of this system when the nuclei are just in contact is given by the Coulomb potential. If Z is the atomic number of the post-decay daughter nucleus, then

$$V(R) = \frac{Q_\alpha\,Q_Z}{4\pi\varepsilon_o R} = \frac{2Ze^2}{4\pi\varepsilon_o R}, \tag{3.106}$$

because the alpha-particle has a charge of $Q_\alpha = 2e$. The combination of constants can be put into more convenient units as

$$\beta = \frac{e^2}{4\pi\varepsilon_o} = \frac{(1.6022 \times 10^{-19}\,\text{C})}{4\pi(8.8542 \times 10^{-12}\,\text{C}^2/\text{Nm}^2)} \tag{3.107}$$

$$= 2.3071 \times 10^{-28}\,(\text{J-m}) = 1.440\ \text{MeV-fm}.$$

For our post-decay thorium-alpha system, the potential at contact is then

$$V(R) = \frac{2\beta Z}{R} \sim 27.9 \text{ MeV},$$

where we have used $Z = 90$ for the thorium nucleus. The height of the (nuclear + Coulomb) barrier at R is consequently about 28 MeV, as indicated in Fig. 3.18.

To reinforce your appreciation of why alpha-decay could not be understood on the basis of classical energetics, imagine running the process in reverse: Fire an α-particle towards a fixed thorium nucleus; if the alpha has kinetic energy of only 4.2 MeV, it would be classically impossible for it to scale the 28-MeV high Coulomb barrier to the point where it would drawn into and fuse with the nucleus via nuclear forces. Quantum-mechanically, however, there is a slight probability that the alpha could penetrate through the Coulomb barrier. An actual alpha-decay is the reverse of this process, but this makes no difference to understanding the situation: any fundamental physical process should be time-reversible; whether we imagine the α-particle as incoming or escaping makes no difference to the calculation.

Classically, if the incoming alpha has kinetic energy E_α, then it should be able to approach no more closely to a target nucleus of charge Z than distance b given by

$$b = \frac{2\beta Z}{E_\alpha}. \tag{3.108}$$

For the thorium-alpha system with $E_\alpha = 4.2$ MeV this evaluates (check it!) to $b \sim 62$ fm, or $b \sim 6.6R$; this distance is sketched in Fig. 3.18.

We are now ready to proceed with the penetration probability calculation. The potential barrier is described by (3.106). For algebraic convenience, we work with the natural logarithm of the penetration probability:

$$\ln P_\alpha \sim -\frac{2\sqrt{2m_\alpha}}{\hbar} \int_R^b \sqrt{\frac{2Ze^2}{4\pi\varepsilon_o x} - E_\alpha}\, dx. \tag{3.109}$$

Because R is fairly small compared to b, we invoke the approximation $R \sim 0$. Extracting a factor of E_α from within the radical and defining the change of variable $y = x/b$ casts (3.109) into the form

$$\ln P_\alpha \sim -\frac{2\sqrt{2m_\alpha E_\alpha}b}{\hbar} \int_0^1 \left[\frac{\sqrt{y-1}}{y^2}\right] dy.$$

Substituting for b from (3.108) and defining another change of variable to $y = \sin^2 w$ gives

Table 3.1 Thorium alpha-decays

Mass number	E_α	Half-life
A	(MeV)	(sec)
220	8.95	10^{-5}
222	8.13	2.24×10^{-3}
224	7.31	1.04
226	6.45	1854
228	5.52	6.0×10^7
230	4.77	2.5×10^{12}
232	4.08	4.4×10^{17}

$$\ln P_\alpha \sim -\frac{8\beta Z \sqrt{2m_\alpha}}{\hbar \sqrt{E_\alpha}} \int\limits_0^{\pi/2} \cos^2 w \; dw.$$

This integral evaluates to $\pi/4$, giving

$$\ln P_\alpha \sim -\frac{2\beta Z \sqrt{2m_\alpha}\,\pi}{\hbar} \frac{1}{\sqrt{E_\alpha}} \sim -\frac{K}{\sqrt{E_\alpha}}. \tag{3.110}$$

This constant K is known as the *Gamow factor* for alpha-decay.

To connect this result to the Geiger-Nutall law, we make the following argument. Imagine that the alpha-particle is trapped within the parent nucleus, making N escape attempts per second. The number of successful escapes per second would then be $N P_\alpha$, and the average time τ between escapes would be the reciprocal of this, $1/(N P_\alpha)$, or $\tau \sim (1/N)e^{(K/\sqrt{E_\alpha})}$. If τ is in some way identifiable with the half-life, we would predict

$$\ln t_{1/2} \sim \ln\left(\frac{1}{N}\right) + \frac{K}{\sqrt{E_\alpha}}. \tag{3.111}$$

This expression is of exactly the same form as the Geiger-Nutall law! Strictly speaking, we should imagine a large number of nuclei, each containing an alpha that is rattling around inside it, attempting to escape; τ would then represent an average time-to-escape.

While reproducing the form of the Geiger-Nutall law is impressive enough, we can also show that this calculation yields predicted values of K in remarkably good accord with experiment.

Table 3.1 lists half-lives and decay energies for seven isotopes of thorium; this element makes for a good test case in view of the tremendous range of half-lives involved, from 10^{-5} seconds to nearly 14 billion years. Thorium decays to radium, which has $Z = 88$. These data are plotted in Fig. 3.19.

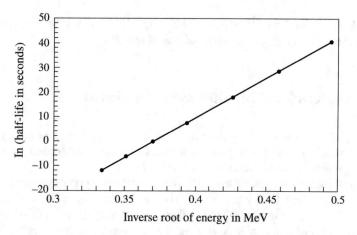

Fig. 3.19 Derived and predicted thorium $t_{1/2}$ versus E_α relationship

The straight line in Fig. 3.19 is a least-squares fit to the data (E_α is taken to be in MeV and $t_{1/2}$ in seconds):

$$\ln t_{1/2} \sim -199.84 + \frac{323.87}{\sqrt{E_\alpha}}. \tag{3.112}$$

The units of K are MeV$^{1/2}$. For comparison, we can get a *predicted* value of K from (3.110). To do this, it is helpful to write the mass of an α-particle as an energy equivalent, $m_\alpha = \mu_\alpha/c^2$, where $\mu_\alpha = 3727.38$ MeV. Hence

$$K = \frac{2\beta Z \sqrt{2m_\alpha}\pi}{\hbar} = \frac{2\beta Z \sqrt{2\mu_\alpha}\pi}{\hbar c}.$$

Now, $\hbar c = 3.1616 \times 10^{-26}$ J-m $\equiv 197.33$ MeV-fm. With the value of β from (3.107) and $Z = 88$, we predict

$$K = \frac{2(1.440)(88)\sqrt{2(3727.38)}\pi}{197.33} = 348.38 \text{ MeV}^{1/2},$$

only 7.5% high compared to the value based on the data. On considering the assumptions that went into this model (modeling a three-dimensional phenomenon as one-dimensional, ignorance of the precise shape of nuclear potentials, and taking $R \sim 0$), this is remarkably good agreement. The key point is that treating alpha-decay as a barrier-penetration effect provides a natural explanation of the Geiger-Nutall law. Few physical theories give results even qualitatively consistent with experiment over so many orders of magnitude. The Gamow-Gurney-Condon analysis provided a convincing example of the veracity of wave mechanics and also showed that it is valid *within* nuclei. The factor of -199.8 in (3.112) can only be established by "calibrating" (3.111) with the half-life of one of the decays.

3.10 Scattering by One-Dimensional Potential Wells

The last phenomenon we consider in this chapter is the scattering of matter waves from potential wells, that is, a situation wherein a matter-wave encounters a region of lower potential. The situation is illustrated in Fig. 3.20.

For convenience, we put the top of the well at zero energy; if it is of depth V_0, the bottom of the well will be at $V(x) = -V_0$. The well has width L, with one side located at $x = 0$. A matter-wave of energy $E > 0$ is incident from the left, and experiences sudden changes of potential at $x = 0$ and $x = L$, leading to partial reflection and transmission at those positions. In regions 1 and 2 there will be waves propagating to both the right and left, whereas only a transmitted rightward-propagating wave will be present in region 3. The analysis parallels that of Sect. 3.7. We can write the wavefunctions as

$$\psi_1(x) = A_0 e^{\iota k_1 x} + A e^{-\iota k_1 x}, \tag{3.113}$$

$$\psi_2(x) = B e^{\iota k_2 x} + C e^{-\iota k_2 x}, \tag{3.114}$$

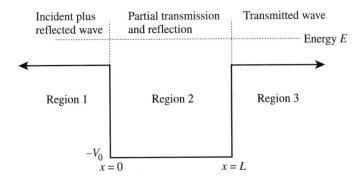

Fig. 3.20 Scattering by a one-dimensional potential well

and

$$\psi_3(x) = De^{\iota k_1 x}, \tag{3.115}$$

with

$$k_1^2 = \frac{2mE}{\hbar^2} \tag{3.116}$$

and

$$k_2^2 = \frac{2m(E + V_0)}{\hbar^2}. \tag{3.117}$$

The difference in sign between (3.117) and (3.89) is due to the fact that the top of the well is now at zero energy. Classically, the incoming particle has sufficient energy to climb back out of the well once it has fallen in. Quantum-mechanically, we will see that there is a non-zero probability of the particle becoming "trapped" inside the well in that the transmission probability need not be unity.

A_0 is again presumed to be experimentally set; the goal is to express D in terms of A_0 and then compute the transmission coefficient $T = D^*D$. Applying the usual boundary conditions on the continuity of ψ and its first derivative at $x = 0$ and $x = L$ gives

$$A_0 + A = B + C \tag{3.118}$$

$$\iota k_1 A_0 - \iota k_1 A = \iota k_2 B - \iota k_2 C \tag{3.119}$$

$$B e^{\iota k_2 L} + C e^{-\iota k_2 L} = D e^{\iota k_1 L} \tag{3.120}$$

$$\iota k_2 B e^{\iota k_2 L} - \iota k_2 C e^{-\iota k_2 L} = \iota k_1 D e^{\iota k_1 L}. \tag{3.121}$$

Taking $A_0 = 1$ and solving for D yields, after some algebra,

$$D = \frac{4k_1 k_2 e^{-\iota k_1 L}}{[(k_1 + k_2)^2 e^{-\iota k_2 L} - (k_2 - k_1)^2 e^{\iota k_2 L}]}. \tag{3.122}$$

The transmission coefficient $T = D^*D$ is given by

$$T = \frac{16 k_1^2 k_2^2}{(k_1 + k_2)^4 - (k_2^2 - k_1^2)^2 (e^{2\iota k_2 L} + e^{-2\iota k_2 L}) + (k_2 - k_1)^4}.$$

This expession can be simplified as follows. The first and last terms in the denominator can be expanded and combined to give $2(k_1^4 + 6k_1^2 k_2^2 + k_2^4)$. The sum of the exponential functions in the middle term can be expressed as $2\cos(2k_2L)$ via Euler's identity. Hence

$$T = \frac{16k_1^2k_2^2}{2(k_1^4 + 6k_1^2k_2^2 + k_2^4) - 2(k_2^2 - k_1^2)^2 \cos(2k_2L)}.$$

On substituting $\cos(2k_2L) = 2\cos^2(k_2L) - 1$, further simplification results, leaving a fairly compact final result:

$$T = \frac{4k_1^2k_2^2}{(k_1^2 + k_2^2)^2 - (k_2^2 - k_1^2)^2 \cos^2(2k_2L)}. \qquad (3.123)$$

There are three interesting cases, plus a general one:

(a) Low-Energy Scattering
Here the idea is that the incoming particle is of very low energy compared to the depth of the well: $E \ll V_0$, that is $k_1 \ll k_2$. With $E \sim 0$, we can regard k_1 as effectively zero; k_2 will then have only a weak dependence on E. Setting $k_1 \sim 0$ in the denominator of (3.123) gives

$$T(E \ll V_0) \sim \frac{4k_1^2}{k_2^2 \sin^2(2k_2L)}. \qquad (3.124)$$

The energy-dependence of this result resides primarily in k_1. Because $k_1 \propto \sqrt{E}$, we can conclude that at low energies, the transmission coefficient is proportional to the energy of the incident wave. As $E \to 0$, the incoming particles are captured by the well.

(b) High-Energy Scattering
In the case of a high-energy incident wave, we have $E \gg V_0$, that is, $k_1 \sim k_2$ in (3.123), which reduces to

$$T(E \gg V_0) \sim 1. \qquad (3.125)$$

This is not surprising: high energy particles barely notice the well.

(c) Resonance Scattering
It might seem from the above approximations that perfect transmission ($T = 1$) is never possible. However this is in fact not the case; there are certain energies for which $T = 1$ exactly.

Consider a case where $k_2L = n\pi$, with n an integer. In this event, $\cos^2(2k_2L) = 1$, and the denominator of (3.123) reduces to exactly $4k_1^2k_2^2$, giving

$$T(k_2L = n\pi) = 1, \qquad (3.126)$$

This is known as a *resonance condition*. At certain energies, the well becomes transparent to the incident matter-wave. The necessary condition is that the width of the well correspond exactly to an integral number of half-wavelengths of the matter-wave (recall that $k = 2\pi/\lambda$). This effect, known as Ramsauer-Townsend scattering, manifests itself in the form of a minimum in the scattering cross-section presented by noble-gas atoms to incident electrons [11]. Since noble-gas atoms consist of closed electron shells (Chap. 8), they present a fairly sharply defined force to a bombarding particle.

Combining the condition $k_2 L = n\pi$ with the definition of k_2 gives an explicit expression for the resonance-scattering energies:

$$E_{resonance} = \frac{1}{2m}\left(\frac{n\pi\hbar}{L}\right)^2 - V_0, \quad (n = 1, 2, 3, \ldots). \tag{3.127}$$

As an example, consider $n = 1$, $m = m_e$, $L = 2\text{Å}$, and $E_{res} = 1$ eV; these give $V_0 = 8.41$ eV, a quite reasonable number on an atomic scale. The increase of T with E predicted by (3.123) is, however, so sharp that T remains very close to unity after the first maximum is reached.

(d) General Scattering

To get a sense of what (3.123) looks like in general, it is helpful to define two new variables:

$$\alpha = E/V_0 \tag{3.128}$$

and

$$\beta = \sqrt{2mV_0}\left(\frac{L}{\hbar}\right). \tag{3.129}$$

With these, Eq. (3.123) becomes

$$T = \frac{4\alpha(\alpha + 1)}{(2\alpha + 1)^2 - \cos^2(\beta\sqrt{1 + \alpha})}, \tag{3.130}$$

For sake of explicitness, we consider a specific case: an electron scattering against a well of depth $V_0 = 10$ eV and width $L = 20$ Å. For these values,

$$\beta = \sqrt{2mV_0}\left(\frac{L}{\hbar}\right)$$

$$= \sqrt{2(9.1094 \times 10^{-31}\,\text{kg})(1.6022 \times 10^{-18}\,\text{J})}\left(\frac{20 \times 10^{-10}\,\text{m}}{1.0546 \times 10^{-34}\,\text{Js}}\right)$$

$$= 32.40.$$

Figure 3.21 shows the run of T versus E/V_0 for $0 \le E/V_0 \le 2$; seven resonance-scattering energies are apparent, with another just off the right edge of the plot. Because T is a non-linear function of α, the resonance peaks are not equally spaced in energy. Note that as E/V_0 increases, $T \to 1$ as discussed in (b) above.

Fig. 3.21 Resonance
scattering of an electron
against a well with
$\beta = E/V_0 = 32.4$

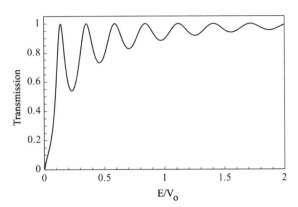

Summary
The energy levels of the infinite rectangular well of width L are given by

$$E_n = \left(\frac{\pi^2 \hbar^2}{2mL^2} \right) n^2.$$

In general, in any region where $V = $ constant and $E > V$, solutions of Schrödinger's equation take the form

$$\psi(x) = A e^{\iota k x} + B e^{-\iota k x},$$

where k is given by

$$k^2 = \frac{2m}{\hbar^2}(E - V).$$

When $V = $ constant and $E < V$, then

$$\psi(x) = C e^{k x} + D e^{-k x},$$

with k^2 above replaced by $-k^2$. If $V(x)$ is a symmetric function, there will be two families of solutions whose energy levels interleave, one family each of even and odd parity.

The energy levels for a particle of mass m moving in a finite rectangular well of width $2L$ and depth V_0 are given by solving

$$f(\xi, K) = (K^2 - 2\xi^2) \sin(2\xi) + 2\xi \sqrt{K^2 - \xi^2} \cos(2\xi) = 0,$$

where K is the strength parameter of the well,

$$K^2 = \frac{2m V_0 L^2}{\hbar^2},$$

and where

$$\xi = \sqrt{\frac{2mE}{\hbar^2}} L.$$

The number of bound states that can be supported by such a well is given by

$$N(K) = 1 + \left[\frac{2K}{\pi}\right].$$

Solutions to Schrödinger's equation when $E < V$ give rise to the phenomenon of barrier penetration. The probability of finding the particle a distance x within the barrier decreases exponentially with a characteristic decay length given by

$$\lambda_{tunneling} \sim \frac{\hbar}{2\sqrt{2m[V(x) - E]}}.$$

The probability of a particle of mass m and energy E penetrating through a rectangular barrier of height V and thickness L depends in a complicated way on all these parameters. For a barrier of arbitrary shape defined by $V(x)$, the penetration probability is given by

$$P_{PENETRATE} \sim \exp\left[-\frac{2\sqrt{2m}}{\hbar} \int_a^b \sqrt{V(x) - E}\, dx\right],$$

where the limits of integration are given by where E cuts the $V(x)$ curve. The half-life of various isotopes to alpha-decay may be understood as a barrier penetration effect.

For a particle of mass m incident on a rectangular potential well of with L and depth $V = -V_0$, 100% transmission will occur at resonance energies given by

$$E_{resonance} = \frac{1}{2m}\left(\frac{n\pi\hbar}{L}\right)^2 - V_0, \quad (n = 1, 2, 3, \ldots).$$

Problems

3.1(E) A possible alternate version of Schrödinger's equation is (see Problem 2.3)

$$-\left(\frac{\hbar^2}{2m}\right)\left(\frac{d\psi}{dx}\right)^2 + V\psi^2 = E\psi^2.$$

Show that this admits no plausible solution for the case of the infinite potential well.

3.2(I) Write a computer program to calculate values for E_n, ψ_n, and ψ_n^2 for an electron in an infinite rectangular well with $L = 10$ Å. Plot separately, to scale, ψ_n and ψ_n^2 for $n = 1$ to 6. What are the energies of these states?

3.3(E) Using classical arguments, derive an expression for the speed v of a particle in a one-dimensional infinite potential well. Apply your result to the case of an electron in a well with $L = 1$ Å. For what value of n does v exceed the speed of light? To what energy does this correspond?

3.4(I) An experimental physicist submits a proposal to a granting agency requesting support to construct an infinite potential well analogous to Fig. 3.5. Specifically, the proposal is to build a well with $L = 1$ mm, inject some electrons into it, and measure the wavelengths of photons emitted during low-n transitions (say, $n = 2 \rightarrow 1$) via optical spectroscopy. As an expert on quantum mechanics, you are asked to evaluate the proposal. What is your recommendation?

3.5(E) A particle in some potential is described by the wavefunction $\psi(x) = Axe^{-kx}$, $(0 \leq x \leq \infty; k > 0)$. If $k = 0.5$ Å^{-1}, what is the probability of finding the particle between $x = 2.0$ and 2.1 Å?

3.6(I) Verify the derivation of (3.51) starting from the wavefunctions given in (3.33) and the appropriate boundary conditions.

3.7(I) Verify the assertions leading to (3.52) and (3.53).

3.8(I) By sketching a few even and odd-parity functions, convince yourself of the assertions in Sect. 3.6 regarding the parity of solutions for symmetric potentials.

3.9(E) A proton is moving within a nuclear potential of depth 25 MeV and full width $2L = 10^{-14}$ m. If the potential can be modeled as a finite rectangular well, how many energy states are available to the proton?

3.10(I) Write a computer program to calculate and plot $\tan(2\xi)$ and $g(\xi, K)$ as a function of ξ for various values of K as in Fig. 3.11. Verify by comparing against Example 3.2. Run your program for a variety of combinations of m, V_0, and L; can you discover any trends?

3.11(I) Consider a particle of mass m moving in the semi-infinite potential well illustrated in Fig. 3.22. Set up and solve Schrödinger's equation for this system; assume $E < V_0$. Apply the appropriate boundary conditions at $x = 0$ and $x = L$ to derive an expression for the permissible energy eigenvalues. With $V_0 = 10$ eV, $L = 5$ Å and $m = m_{electron}$, compute the values of the energy (in eV) for the two lowest bound states.

3.12(I) Derive an expression analogous to (3.82) for the odd-parity states of the finite rectangular well. For convenience, write the interior solution as $\psi_2(x) = A \sin(k_2 x)$.

3.13(I) Show that the eigenvalue condition for the even-parity states of a finite square well can be written as $\xi^2 \tan^2 \xi = K^2(1 - \epsilon)$, where K is as defined in (3.50) and $\epsilon = E/V_0$. For low-energy states, ξ will be very small. From your result, derive an

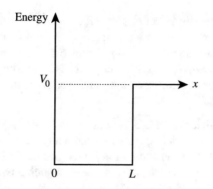

Fig. 3.22 Problem 3.11: semi-infinite potential well

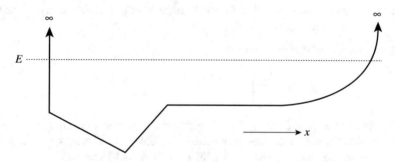

Fig. 3.23 Problem 3.15

expression for the energy of such states. Apply your result to a case where $K = 1.5$; compare your result with the exact value.

3.14(E) For the system discussed in Example 3.2, determine the probability of finding the particle outside the well for each of the four possible bound states.

3.15(I) For the potential well illustrated in Fig. 3.23, sketch a plausible form for a wavefunction corresponding to energy E.

3.16(I) Verify the expressions given for D and T in the derivation of the rectangular potential barrier, Eqs. (3.95) and (3.97).

3.17(E) Electrons of energy 10 eV are incident on a rectangular potential barrier of height 20 eV and width 50 Å. What is the transmission coefficient?

3.18(E) In quantum tunneling, the penetration probability is sensitive to slight changes in the height and/or width of the barrier. Consider an electron with $V(x) - E = 10$ eV incident on a barrier of width 20 Å. By what factor does the penetration probability change if the width is increased to 21 Å?

3.19(I) Write a computer program to numerically integrate the barrier penetration probability (3.104) for any m, E, and $V(x)$. Verify by comparing with Example 3.5. For this same problem, generate a plot of P vs. E for $E = 0.1, 0.2, \ldots 0.9$ Å.

3.20(I) A potential barrier is defined by $V(x) = \sqrt{A^2 - \alpha^2 x^2}$, $|x| \leq (A/\alpha)$. (a) Sketch $V(x)$. (b) A particle of mass m and energy $E < A$ is incident on this barrier from the left. Derive an integral expression for the penetration probability in terms of $\beta = E/A$ and $\sin(y) = \alpha x/A$. (c) Evaluate your expression numerically for $\beta = 0.1$ to 1.0 in steps of 0.1 for an electron striking such a barrier with $A = 10$ eV and $\alpha = 2$ eV/Å. Plot your result.

3.21(I) A potential barrier is defined by $V(x) = V_0(1 - x/A)$ for for $(0 \leq x \leq A)$; $V(x) = 0$ otherwise. A particle of mass m and energy $E < V_0$ is incident on this barrier from the left. Derive an expression for the penetration probability. Evaluate your result numerically for an electron incident on such a barrier with $V_0 = 5$ eV, $E = 2$ eV, and $A = 12$ Å. Potentials of this form are used to model the spontaneous escape of electrons from metal surfaces subjected to electric fields, a process known as cold emission. The expression for the transmission probability is known as the *Fowler-Nordheim* formula.

3.22(I) A potential barrier is defined by

$$V(x) = \begin{cases} (V_o/L^2)\, x^2 & 0 \leq x \leq L \\ 0 & x < 0;\ x > L. \end{cases}$$

A particle of mass m and energy $V_0/2$ is incident on this barrier from $x < 0$. Derive an expression for the penetration probability. Evaluate your expression numerically for an electron striking such a barrier with $L = 10$ Å and $V_0 = 5$ eV.

3.23(E) A sprinter of mass 70 kg running at 5 m/s does not have enough kinetic energy to leap over a wall of height 5 m, even if all of that kinetic energy could be directed into a vertically upward jump. If the wall is 0.2 m thick, estimate the probability of the sprinter being able to quantum tunnel through it rather than attempting to leap over it.

3.24(E) From the information given in Sect. 3.9, compute the approximate number of escape attempts per second for an alpha particle trapped within a thorium nucleus.

3.25(I) Table 3.2 gives a list of alpha-decay energies and half-lives for various isotopes of radium, element 88. Does radium follows the Gamow model for α-decay?

Table 3.2 Radium alpha-decays

Mass number A	E_α (MeV)	Half-life
214	7.136	2.46 s
220	7.46	0.019 s
221	6.608	29 s
222	6.656	38 s
223	5.716	11.435 days
224	5.686	3.66 days
226	4.784	1599 years

3.26(E) Some nuclei decay by beta-decay, the emission of an ordinary electron. By examining the development of the Gamow factor, would you expect the beta-decay probability for a given nucleus to be greater or less than that for alpha decay, all other factors being equal?

3.27(I) Verify the expression for the transmission coefficient for scattering by a potential well, Eq. (3.123).

3.28(E) In an experiment involving electron scattering from a finite rectangular well of depth 4 eV, it is found that electrons of energy 5 eV are completely transmitted. What must be the width of the well? At what next higher energy can one expect to again observe $T = 1$?

References

1. E. Merzbacher, Phys. Today **55**(8), 44–49 (2002)
2. D.L. Aronstein, C.L. Stroud, Am. J. Phys. **68**(10), 943 (2000)
3. D.W.L. Sprung, H. Wu, J. Martorell, Eur. J. Phys. **13**(1), 21 (1992)
4. D. Rugar, P. Hansma, Phys. Today **43**(10), 23–30 (1990)
5. R. Eisberg, R. Resnick R., *Quantum Physics of Atoms, Molecules, Solids, Nuclei and Particles* (Wiley, New York, 1974), Sections 5–7
6. G. Gamow, Z. Phys. **51**, 204 (1928)
7. R.W. Gurney, E.U. Condon, Nature **122**(3073), 439 (1928)
8. H. Geiger, J.M. Nutall, Phil. Mag. **22**, 613 (1911)
9. H. Geiger, J.M. Nutall, Phil. Mag. **23**, 439 (1912)
10. K. Heyde, *Basic Ideas and Concepts in Nuclear Physics*, 3rd edn. (Institute of Physics Publishing, Bristol, 2004)
11. C.W. Ramsauer, Ann. Phys. **64**, 513 (1921); J.S. Townsend, Phil. Mag. **43**, 593 (1922)

Chapter 4
Operators, Expectation Values, and Various Quantum Theories

Summary This chapter is a mathematical intermission devoted to some of the analytic formalisms of quantum theory. A central concept in this chapter is how physical quantities such as energy and momentum can be represented by mathematical constructs known as "operators". This leads to a formulation of how predictions of the behavior of quantum systems are expressed as integrals of these operators acting on the wavefuction(s) $\psi(x)$ for some system, a development which leads to the famous *Heisenberg Uncertainty Principle*. This chapter also gathers together some theorems regarding wavefunctions which will be useful in later chapters: the orthogonality theorem, the superposition theorem, the virial theorem, and the behavior of time-dependent wave "packets" constructed from "superposition states".

In this chapter we consider various aspects of Schrödinger's equation and matter waves pertinent to the interpretation of quantum mechanics. The purpose here is not to explore new solutions of Schrödinger's equation but rather to broaden understanding of how solutions are interpreted. As many of our points are postulates or require more background to prove rigorously, we will sometimes resort to intuitive justifications.

We begin in Sect. 4.1 by considering the concept of a mathematical *operator*. By showing how Schrödinger's equation can be expressed in *operator notation*, we gain insight as to what is meant by an *eigenvalue equation*. This leads to a discussion of how mathematical operators relate to physically observable quantities. In Sect. 4.2 we discuss *expectation values*. The purpose of doing any calculation is, presumably, to compute something that can be compared with experiment, and it is via expectation values that quantum mechanics informs us of its predictions. In Sects. 4.3 and 4.4 we give a semi-intuitive justification of the famous *Heisenberg uncertainty principle* and elucidate the recipe by which one can determine whether or not two quantities will be linked by an uncertainty relation. Section 4.5 is devoted to *Ehrenfest's theorem*, which shows how suitably-defined average values of quantities in quantum mechanics agree with the predictions of Newtonian mechanics. Section 4.6 develops the *orthogonality theorem*, which will prove to be of great value in the development of perturbation theory in Chap. 9. In Sect. 4.7 we take up the *superposition theorem*,

B. C. Reed, *Quantum Mechanics*, https://doi.org/10.1007/978-3-031-14020-4_4

which will also prove to be of value in developing *perturbation theory* in Chap. 9. As an example of an application of this theorem, in Sect. 4.8 we construct a representation of a moving particle trapped in an infinite rectangular potential well via a time-dependent summation of two infinite-well states. In Sect. 4.9 we derive the *virial theorem*, a relationship between the expectation values of kinetic and potential energy for any potential. Section 4.10, which can be considered optional, deals with how wavefunctions can be transformed to "momentum-space" representations in order to explore their momentum probability distributions.

4.1 Properties of Operators

Consider again the one-dimensional time-independent Schrödinger equation:

$$\left(-\frac{\hbar^2}{2m}\frac{d^2}{dx^2} + V(x)\right)\psi(x) = E\psi(x). \tag{4.1}$$

On the left side appears the second derivative of $\psi(x)$ times a constant, plus $V(x)\psi(x)$. The result of this calculation is $E\psi(x)$. The term in large brackets on the left side can be said to perform an operation on $\psi(x)$ and return $E\psi(x)$. Inasmuch as we interpret E as the energy corresponding to $\psi(x)$, this term acts as a *total energy operator*. As in classical physics, this is known as the *Hamiltonian operator*, designated by $\boldsymbol{H}$:

$$\boldsymbol{H} = -\frac{\hbar^2}{2m}\frac{d^2}{dx^2} + V(x). \tag{4.2}$$

In *operator notation*, Schrödinger's equation appears as

$$\boldsymbol{H}\psi(x) = E\psi(x). \tag{4.3}$$

An *operator* is an entity that does something to a function and returns a result, not unlike a subroutine in a computer program which processes an input value or function and returns an output value. The operation performed could be to multiply the input function by a constant, differentiate it, integrate it, exponentiate it, add something to it, or so on. There are many possible everyday examples of operators; for example, you can imagine an operator called "tax" that operates on the value of a purchase to return the amount of tax to be paid. In isolation, an operator is meaningless; only an expression wherein an operator acts on a function has meaning. Do not make the mistake of thinking that the ψ's in (4.3) can be canceled out.

We already know that only certain *eigenfunctions* $\psi(x)$ will satisfy Schrödinger's equation for a given potential $V(x)$, each with its own energy *eigenvalue* or *eigenenergy*. The general form of an eigenvalue equation is

$$(Operator)(Eigenfunction) = (Eigenvalue)(Eigenfunction). \qquad (4.4)$$

As in conventional algebra, the operator acts on whatever quantity or function appears immediately to the right of it, but only if the operator is acting on one of its eigenfunctions will the form of (4.4) be valid.

Returning to (4.1), we can consider the total energy operator to be the sum of individual kinetic-energy and potential-energy operators:

$$KE_{op} = -\frac{\hbar^2}{2m}\frac{d^2}{dx^2} \qquad (4.5)$$

and

$$PE_{op} = V(x). \qquad (4.6)$$

An example of operator algebra is provided by the infinite-well wavefunctions:

$$\psi_n(x) = \sqrt{\frac{2}{L}}\sin\left(\frac{n\pi x}{L}\right).$$

In this case $V(x) = 0$, so

$$\begin{aligned} H\psi_n(x) &= -\frac{\hbar^2}{2m}\frac{d^2\psi_n(x)}{dx^2} \\ &= -\frac{\hbar^2}{2m}\frac{d^2}{dx^2}\left(\sqrt{\frac{2}{L}}\sin\left(\frac{n\pi x}{L}\right)\right) = \frac{\hbar^2}{2m}\frac{n^2\pi^2}{L^2}\left(\sqrt{\frac{2}{L}}\sin\left(\frac{n\pi x}{L}\right)\right) \\ &= E_n\psi_n(x), \end{aligned}$$

were we have used (3.11).

Aside from energy, there are other *dynamical variables* whose values we may be interested in computing. These are quantities such as position, momentum, or angular momentum. What are the operators corresponding to these quantities? That there is an operator corresponding to every observable quantity is a fundamental postulate of quantum mechanics. In simple cases such as position (x), the operator is simply x itself (in three dimensions: $r = xx + yy + zz$, where x, y, and z are the Cartesian-coordinate unit vectors). The operator for linear momentum can be deduced by considering the expression given in Chap. 2 for a right or left-propagating matter wave:

$$\psi(x, t) = A\exp\left[\pm\frac{\iota}{\hbar}(px - Et)\right],$$

Extracting a factor of p by differentiating gives

$$\frac{\partial \psi}{\partial x} = \pm \left(\frac{\iota}{\hbar}\right) p \left\{ A \exp\left[\pm\frac{\iota}{\hbar}(px - E t)\right] \right\} = \pm \left(\frac{\iota}{\hbar}\right) p\psi,$$

or

$$p\psi \equiv \pm \left(\frac{\hbar}{\iota}\right) \frac{\partial \psi}{\partial x} = \left(\mp \iota\hbar \frac{\partial}{\partial x}\right) \psi,$$

where we have used $1/\iota = -\iota$. The sign ambiguity can be resolved by arguing that, in a wave to the right (or left), we choose the same sign (upper/lower) as is used in the description of the wave. Hence we have

$$(p_x)_{op} = -\iota\hbar \frac{\partial}{\partial x}. \tag{4.7}$$

Similarly, for the y and z directions, we infer

$$(p_y)_{op} = -\iota\hbar \frac{\partial}{\partial y} \tag{4.8}$$

and

$$(p_z)_{op} = -\iota\hbar \frac{\partial}{\partial z}. \tag{4.9}$$

Notice that we have switched to partial-derivative notation as each of p_x, p_y, and p_z take derivatives with respect to only one coordinate. These results can be summarized in a convenient vector form which makes no explicit mention of any particular coordinate system:

$$p_{op} = -\iota\hbar \nabla, \tag{4.10}$$

where ∇ is the gradient operator. Some useful operators are summarized in Table 4.1.

Table 4.1 Useful operators

Physical quantity	Operator
Position	x
Linear momentum	$-\iota\hbar\nabla$
Kinetic energy	$-\frac{\hbar^2}{2m}\nabla^2$
Potential energy	$V(x)$
Any function $f(x)$ of x	$f(x)$

Just as one can multiply together a string of numbers, it is possible to successively treat a function by the action of more than one operator. There is, however, an important difference between ordinary algebra and operator algebra: In the latter, the order of operations can make a difference to the final outcome, that is, operator algebra is not necessarily commutative. An example utilizing the position, momentum, and kinetic energy operators will illustrate this point. Consider some arbitrary function $f(x)$ operated on first by the position operator and then by the momentum operator:

$$(p_x)_{op}[(x)_{op} f(x)] = (p_x)_{op}[xf(x)] = -\iota\hbar \frac{\partial}{\partial x}[xf(x)]$$

$$= -\iota\hbar \left[f(x) + x\frac{df}{dx} \right]. \quad (4.11)$$

On the other hand, if the order of the operators is reversed, we find

$$(x)_{op}[(p_x)_{op} f(x)] = (x)_{op}\left[-\iota\hbar\frac{\partial}{\partial x} f(x) \right] = -\iota\hbar x \frac{df}{dx}. \quad (4.12)$$

Only if $f(x) = 0$ will the operators for p and x and x commute, but we would then be left with a null problem. In other instances, however, the overall result is insensitive to the order of the operators. For example:

$$(p_x)_{op}[(KE)_{op} f(x)] = (p_x)_{op}\left[-\frac{\hbar^2}{2m}\frac{d^2 f}{dx^2} \right] = \frac{\iota\hbar^3}{2m}\frac{d^3 f}{dx^3}. \quad (4.13)$$

and

$$(KE)_{op}[(p_x)_{op} f(x)] = (KE)_{op}\left[-\iota\hbar\frac{df}{dx} \right] = \frac{\iota\hbar^3}{2m}\frac{d^3 f}{dx^3}. \quad (4.14)$$

At present, the fact that the order of operations may or may not have an influence on the result of a calculation is academic. Later we will see that this feature is closely linked to Heisenberg's Uncertainty Principle.

Some comments on notation are appropriate here. Normally, bold type is reserved to denote vectors or vector operators such as that for linear momentum. An exception to this is to use H to designate the important Hamiltonian operator, a scalar quantity. Also, we will henceforth often drop the superfluous "*op*" notation when there is little possibility of confusion arising.

4.2 Expectation Values

We have already seen that some questions in quantum contexts will yield exact answers ("What is the energy of a particle in the $n = 3$ state of an infinite well?"), whereas others can only be answered in terms of a probability distribution ("What is

Fig. 4.1 A hypothetical
distribution of heights

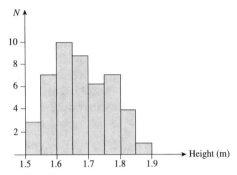

the position of the particle?"). In these latter cases, however, further information can
be gleaned by rephrasing the question. While it may be meaningless to ask where
the particle is at any particular time, it is meaningful to ask what its *average* position
is. Experimentally, one might answer such a question by preparing a large number
of identical infinite wells containing electrons in the $n = 3$ state, somehow simultaneously measure their positions, and then average the results. The theoretically
predicted value of this average is known as the *expectation value* of the position,
designated by $\langle x \rangle$. Our goal in this section is to devise a procedure for expressing
the expectation value of any function of x in terms of a given wavefunction $\psi(x)$.

To establish the relationship between expectation values and probability distributions, it is helpful to start with the familiar concept of an arithmetic average. To this
end, suppose that we take a census of the heights of N individuals; if we designate
the height of individual i to be x_i, then the average height is

$$\langle x \rangle = \frac{1}{N} \sum_{i=1}^{N} x_i. \tag{4.15}$$

Now imagine binning the results, that is, counting up, histogram-style, the number
of individuals in various height ranges. This is shown schematically in Fig. 4.1.

Designate bins by the index j. If x_j designates the height corresponding to the
middle of the range of bin j and n_j is the number of people whose heights fall in bin
j, then the average height can be expressed as

$$\langle x \rangle = \frac{1}{N} \sum_{bins\ j} (\text{height corresponding to bin j}) (\text{number of people in bin j}),$$

or

$$\langle x \rangle = \frac{1}{N} \sum_{bins\ j} x_j n_j = \sum_{bins\ j} x_j \left(\frac{n_j}{N} \right). \tag{4.16}$$

Now, n_j/N is just the ratio of the number of people whose heights fall in bin j to the total sample size. This is exactly the probability p_j of a randomly-chosen individual falling in bin j:

$$\langle x \rangle = \sum_{bins\ j} x_j\ p_j. \tag{4.17}$$

Summing over all possible bins is equivalent to summing over all possible heights. With this understanding, (4.17) can then be cast into a more suggestive form:

$$\langle x \rangle = \sum_{all\ x} x\ p(x). \tag{4.18}$$

In words, the average of some quantity x is just the sum over all x of x itself times the probability of a given member of the sample having as x its value for the quantity under consideration.

A simple example with a discretely-valued variable can be used to illustrate (4.18). Suppose that a census of 30 students reveals 3 who are each enrolled in 3 courses, 18 who are each enrolled in 4 courses, 8 who are each enrolled in 5 courses, and 1 who is enrolled in 6 courses. The variable "x" is the number of courses that a student is enrolled in, and the index of summation "j" runs from 3 to 6 (why?) The probability that a given student in the group is enrolled in (3, 4, 5, 6) courses is given by (3/30, 18/30, 8/30, 1/30) = (0.100, 0.600, 0.267, 0.033); these numbers are the n_j/N of (4.16). Note that the probabilities add up to unity. The average number of courses that any student is enrolled is then computed from (4.18) as 3(0.100) + 4(0.600) +5(0.267) + 6(0.033) = 4.233. This could also be computed form the total number of student-courses involved, 127, divided by the number of students. Note that no student is actually enrolled in the average number of courses! This last feature is analogous to what happens in what are known as quantum mechanical superposition states (see Sects. 4.8 and 9.2): a measurement of the number of courses that any one student is enrolled in will yield a discrete "eigenvalue" of 3, 4, 5, or 6, but the average is not and in general will not be one of the eigenvalues.

In the context of a quantum-mechanical average, we know that the probability of finding a particle between x and $x + dx$ is $\psi^*(x)\psi(x)dx$. On realizing that position is a continuous variable and not a discrete one, we can write the average or *expectation value* of position as

$$\langle x \rangle = \int x\psi^*(x)\psi(x)dx. \tag{4.19}$$

The limits of integration in any particular case correspond to the physical limits of space that the particle could occupy. Sometimes these are written purely formally as $\pm\infty$.

What if one wishes to compute the expectation value of a quantity that is a function of position or involves derivatives of position, such as momentum or energy? The general rule for calculating expectation values is:

$$\langle \text{observable} \rangle = \int \psi^*(x) \left[(\text{Operator}) \psi(x) \right] dx, \tag{4.20}$$

where (Operator) denotes the operator corresponding to the observable of interest. The order of computation is important here. First, [(Operator $\psi(x)$)] is computed. The result of this is then multiplied by ψ^*, and then the integral is carried out.

A formal proof of (4.20) is rather involved. For the moment, let us suffice it to say that this form for computing expectation values is the only one consistent with the intuitive notion that the results of quantum calculations should correspond to their Newtonian counterparts when one averages over the system. We discuss this point further in Sect. 4.5.

It is worth emphasizing that eigenvalues (for example, the possible "stationary energy states" of a system) and expectation values are very different quantities. As explained further in Sect. 4.8, a single measurement on a system will always yield one of the possible eigenvalues, but not necessarily the expectation value; only if the system is in a time-independent stationary state will a measurement result in the expectation value. A time-dependent phenomenon such as the motion of a particle can only be modeled via what are known as superposition states (see Sects. 4.7 and 4.8), in which case the expectation value of some quantity is a sort of average of its possible eigenvalues.

Example 4.1 Use (4.20) to compute $\langle x \rangle$, $\langle p \rangle$, $\langle E \rangle$, $\langle x^2 \rangle$ and $\langle p^2 \rangle$ for the infinite-well wavefunctions.

We have

$$\psi_n(x) = \sqrt{\frac{2}{L}} \sin\left(\frac{n\pi x}{L}\right).$$

According to (4.20) and setting $a = n\pi/L$,

$$\langle x \rangle = \int_0^L \psi_n^*(x) \{ x \, \psi_n(x) \} dx = \frac{2}{L} \int_0^L \sin(ax) \{ x \, \sin(ax) \} dx$$

$$= \frac{2}{L} \int_0^L x \sin^2(ax) dx = \frac{2}{L} \left[\frac{x^2}{4} - \frac{x \sin(2ax)}{4a} - \frac{\cos(2ax)}{8a^2} \right]_0^L$$

$$= \frac{L}{2}.$$

This result means that the average of many measurements of the position of a particle in an infinite well would be $x = L/2$. Notice that this result is independent of the state number n. This is consistent with the fact that, since the well is symmetric, the particle should have no reason to prefer one side to the other no matter what state it is in.

When computing $\langle p \rangle$, care must be taken to respect the differential nature of $(p_x)_{op}$:

$$\langle p \rangle = \int_0^L \psi_n^*(x) \left[-\imath\hbar \frac{d}{dx} \psi_n(x) \right] dx = -\frac{2\imath\hbar}{L} \int_0^L \sin(ax) \left[\frac{d}{dx} \sin(ax) \right] dx$$

$$= -\frac{2\imath a\hbar}{L} \int_0^L \sin(ax) \cos(ax) dx = -\frac{\imath\hbar}{L} [\sin^2(ax)]_0^L = 0.$$

This result is also not surprising: the symmetry of the well is such that the particle should have no preference for traveling one way or the other.

The expectation value of the energy proceeds similarly:

$$\langle E \rangle = \int_0^L \psi_n^*(x) [H\psi_n(x)] dx = \frac{2}{L} \int_0^L \sin(ax) \left[-\frac{\hbar^2}{2m} \frac{d^2}{dx^2} \sin(ax) \right] dx$$

$$= \frac{\hbar^2 \pi^2 n^2}{2mL^2} = E_n.$$

The expectation value of the energy for infinite well state n is just the energy eigenvalue of that state.

For $\langle x^2 \rangle$ we have

$$\langle x^2 \rangle = \int_0^L \psi_n^*(x) \left[x^2 \psi_n(x) \right] dx = \frac{2}{L} \int_0^L x^2 \sin^2(ax) dx$$

$$= \frac{2}{L} \left[\frac{x^3}{6} - \left(\frac{x^2}{4a} - \frac{1}{8a^3} \right) \sin(2ax) - \frac{x \cos(2ax)}{4a^2} \right]_0^L = \frac{2}{L} \left[\frac{L^3}{6} - \frac{L^3}{4\pi^2 n^2} \right]$$

$$= \frac{L^2}{3} \left[1 - \frac{3}{2\pi^2 n^2} \right].$$

Note that $\langle x \rangle^2$ is not equal to $\langle x^2 \rangle$; they are different quantities having different interpretations. $\langle x \rangle^2$ is the square of the mean value of the position, whereas $\langle x^2 \rangle$ is the mean value of (position squared).

The operator for any power of the linear momentum can be derived by operating p_{op} successively on itself. Working in one dimension,

$$p_x^2 = (p_x)(p_x) = \left(-\iota\hbar\frac{\partial}{\partial x}\right)\left(-\iota\hbar\frac{\partial}{\partial x}\right) = -\hbar^2\frac{\partial^2}{\partial x^2} = 2m(KE).$$

Hence, for the infinite-well wavefunctions,

$$\langle p^2\rangle = 2m\langle KE\rangle = \frac{\pi^2\hbar^2 n^2}{L^2}.$$

Note that $\langle x^2\rangle$ and $\langle p^2\rangle$ have units of (length)^2 and (momentum)^2, respectively. These results will be used later (Example 4.3) to establish measures of the spreads of p and x about their mean values.

Example 4.2 In the study of the hydrogen atom in Chap. 7 we will often encounter wavefunctions of the form $\psi(x) = Ax^n e^{-kx/2}$, $(0 < x < \infty)$, with k positive and n being a positive integer. Such a function always exhibits a maximum at some value of x followed by an infinitely long exponential tail. If this function is to be normalized, what must be the value of A? At what value of x does ψ reach a maximum? Finally, what is $\langle x\rangle$, and how does it compare to the value of x at which ψ reaches a maximum?

Normalizing demands

$$A^2\int_0^\infty x^{2n} e^{-kx}\,dx = 1,$$

This integral can be evaluated with the help of Appendix C, and emerges as $(2n)!/k^{2n+1}$. Hence

$$A^2 = \frac{k^{2n+1}}{(2n)!}.$$

To find where ψ reaches its maximum value, set $d\psi/dx = 0$:

$$\frac{d\psi}{dx} = A\left(nx^{n-1} - kx^n/2\right)e^{-kx/2} = 0$$

$$\Rightarrow x_{\text{max}} = \frac{2n}{k}.$$

Now, the expectation value of x is given by

$$\langle x \rangle = \int \psi(x\psi)\,dx = A^2 \int_0^\infty x^{2n+1}e^{-kx}\,dx.$$

Appendix C is again useful, giving

$$\langle x \rangle = A^2 \frac{(2n+1)!}{k^{2n+2}} = \frac{k^{2n+1}}{(2n)!}\frac{(2n+1)!}{k^{2n+2}} = \frac{2n+1}{k}.$$

For all n, the mean value of x exceeds the value of x at which ψ reaches its maximum value:

$$\frac{\langle x \rangle}{x_{\text{max}}} = \frac{2n+1}{2n}.$$

It is not always necessarily true that $\langle x \rangle > x_{max}$; this depends on the nature of the wavefunction. It is possible to have, for example, a wavefunction with multiple maxima, with the maximum of greatest amplitude at the $x \to \infty$ end of the range and consequently resulting in $\langle x \rangle < x_{max}$; many of the hydrogen atom wavefunctions of Chap. 7 illustrate this behavior (see Fig. 7.7).

Integrals of the general form of (4.20) arise frequently in quantum-mechanical calculations. To reduce the tedium of writing these out, it is convenient to adopt a shorthand known as *Dirac notation*, sometimes known as *Dirac braket notation*; this is not a mis-spelling. In this notation, the expectation value of $f(x)$ is written as $\langle \psi^*(x) | f(x) | \psi(x) \rangle$:

$$\left\langle \psi^*(x) \,|\, f(x) \,|\, \psi(x) \right\rangle = \int \psi^*(x)\{Op[f(x)]\}\psi(x)\,dx, \qquad (4.21)$$

where $|f(x)|$ denotes $Op[f(x)]$, the operator corresponding to $f(x)$. Dirac called $\langle \psi^*(x)|$ a "bra" and $|\psi(x)\rangle$ a "ket." A further simplification is to drop the wave-functions and simply write $\langle f(x) \rangle$. In this system of notation, there need not even be an operator involved. For example, the "overlap integral" of two functions $\psi_1(x)$

and $\psi_2(x)$ (which are presumed to be valid over the same domain) can be written in Dirac notation as

$$\langle \psi_1 | \psi_2 \rangle = \int \psi_1(x)\, \psi_2(x) dx. \tag{4.22}$$

This is also known as the *inner product* of $\psi_1(x)$ and $\psi_2(x)$. These various forms of Dirac notation will occasionally be utilized throughout the remainder of this book in order to simplify the writing of integrals.

Before closing this section, an interesting corollary can be made to the argument above that $\langle p \rangle = 0$ for the infinite-well wavefunctions is not surprising in view of the symmetry of the situation. One can in fact show that $\langle p \rangle = 0$ for any time-independent wavefunction. In general, the momentum operator is given in three dimensions by $\mathbf{p} = -\imath \hbar \nabla$. Hence

$$\langle \mathbf{p} \rangle = -\imath \hbar \int_V \psi(\nabla \psi)\, dV,$$

where the integral extends over the volume domain of ψ; dV is an element of volume. Since $\psi(\nabla \psi) = \frac{1}{2}\nabla(\psi^2)$, we can write this as

$$\langle \mathbf{p} \rangle = -\frac{\imath \hbar}{2} \int_V \nabla(\psi^2)\, dV$$

Now, the divergence theorem (also known as Gauss's theorem and Green's theorem) from vector calculus tells us that the volume integral of the gradient of a function can be replaced with the integral of the function over the surface that bounds that volume:

$$\langle \mathbf{p} \rangle = -\frac{\imath \hbar}{2} \int_S \psi^2 d\mathbf{S}, \tag{4.23}$$

where $d\mathbf{S}$ denotes an outwardly-directed element of surface area bounding the volume V. Since the limits of any potential extend in principle be extended to the edge of the Universe, we are forced to conclude that ψ must always vanish asymptotically in order for it to be to be normalizable. That is, $[\psi^2]_{boundaries} = 0$ in general, and so we must have $\langle \mathbf{p} \rangle = 0$ always.

That we have found $\langle \mathbf{p} \rangle = 0$ is not terribly surprising: intuitively, we should expect a time-independent wavefunction to have a time-independent expectation value for its position (that is, $\langle \mathbf{r} \rangle = $ constant) because the position operator $\langle \mathbf{r} \rangle$ has no time-dependence; it follows that $\langle \mathbf{p} \rangle = 0$. Thus, a quantum particle with a time-independent ψ can be considered as analogous to a classical particle that is stationary. This result, while simple, points to a curious aspect of the quantum momentum operator. In classical mechanics, momentum is given by a time derivative: $p = mv = m(dr/dt)$, but in quantum mechanics it is given by a spatial derivative operator: $p \equiv -\iota\hbar\nabla$. Classically, taking a spatial derivative to find momentum would be considered completely wrong, but, quantum mechanically, it works. Even if the spatial derivative is non-zero for a time-independent ψ, the boundary conditions guarantee $\langle p \rangle = 0$. This strange behavior of the p operator is yet another example of quantum counterintuitiveness.

4.3 The Uncertainty Principle

Because particles possess a wave nature, it is impossible to specify their positions precisely. A quantitative statement of this uncertainty is embodied in the *Heisenberg Uncertainty Principle*. We begin with a qualitative argument to illustrate this important principle before moving on to a formal analysis.

Figure 4.2 shows two hypothetical matter waves. Pattern (a) is a segment of a sine or cosine curve of wavelength λ which otherwise extends from $-\infty$ to $+\infty$, while pattern (b) represents a matter wave where the bulk of the particle is found within some spatial extent Δx. (Such a "wave packet" can be constructed using a technique known as superposition—See Sects. 4.7 and 4.8).

Pattern (a) can be described by a well-defined wavelength λ. Via the de Broglie wavelength, we then have precise knowledge of the momentum of this matter wave: $p = h/\lambda$. In case (b), the particle is essentially confined in a region of space of extent Δx. However, in localizing this wave packet we sacrifice precise knowledge of its momentum, since the pattern is described by no unique wavelength. If you are familiar with Fourier analysis you will know that any wave pattern can be constructed by addition of a number of waves of appropriate amplitudes and wavelengths. If the desired pattern is highly localized like pattern (b), then a great number of sine and cosine terms must be summed, that is, there will be a large number of contributing momenta. By adding more and more terms, Δx can be made arbitrarily small, but at the expense of the particle's momentum becoming less and less well-known.

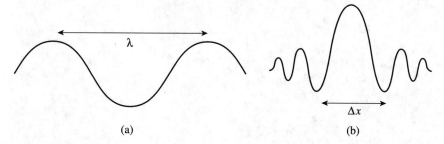

(a)   (b)

Fig. 4.2 Hypothetical matter-wave patterns

Nature imposes limits on how precisely we can simultaneously know position and momentum.

A formal statement of the uncertainty principle is as follows. If we define the uncertainty in x and p to be their variances

$$\Delta x = \sqrt{\langle x^2\rangle - \langle x\rangle^2}, \tag{4.24}$$

and

$$\Delta p = \sqrt{\langle p^2\rangle - \langle p\rangle^2}, \tag{4.25}$$

then the product $\Delta x \Delta p$ must obey the restriction

$$\Delta x \Delta p \geq \hbar/2. \tag{4.26}$$

Example 4.3 Use the results of Example 4.1 to calculate $\Delta x \Delta p$ for a particle moving in an infinite rectangular well.

From Example 4.1 we have

$$\langle x\rangle = L/2,$$

$$\langle p\rangle = 0,$$

$$\langle x^2\rangle = \frac{L^2}{3}\left[1 - \frac{3}{2\pi^2 n^2}\right],$$

and

$$\langle p^2\rangle = \frac{\pi^2 \hbar^2 n^2}{L^2}.$$

These give

$$\Delta x = L\sqrt{\frac{1}{12} - \frac{1}{2\pi^2 n^2}}$$

and

$$\Delta p = \frac{\pi \hbar n}{L}.$$

Therefore

$$\Delta x \Delta p = \sqrt{\frac{n^2\pi^2}{12} - \frac{1}{2}}\,\hbar.$$

The minimum value of $\Delta x \Delta p$ obtains for $n = 1$:

$$(\Delta x \, \Delta p)_{n=1} = 0.568 \, \hbar,$$

which satisfies the Uncertainty Principle.

For large values of n, we find

$$(\Delta x \, \Delta p)_{n \to \infty} \sim \frac{n\pi \hbar}{\sqrt{12}} \sim 0.907 n\hbar.$$

Parenthetically, one might wonder if there is any potential having a wavefunction that realizes the minimum possible $\Delta x \, \Delta p$. There is: the ground-state of the simple harmonic oscillator potential; the wavefunction in this case has a Gaussian form, $\psi \sim \exp(-kx^2)$. We will study this potential and its solutions in Chap. 5.

An important practical aspect of the uncertainty principle lies in its use as a means to estimate ground state energies for a particle moving in a given potential. Consider again an infinite well of width L. Knowing that the particle must be confined within the well gives $\Delta x \leq L$. From symmetry, we can argue that $\langle p \rangle = 0$, hence $\Delta p = \sqrt{\langle p^2 \rangle} = \sqrt{2m \langle E \rangle}$. Hence we have

$$L\sqrt{2m \, \langle E \rangle} \geq \hbar/2,$$

or

$$\langle E \rangle \geq \frac{\hbar^2}{8mL^2}.$$

Comparing this result with the exact value, (3.11), shows that we are low by a factor of $4\pi^2$. Uncertainty-based arguments for ground state energies will not in general yield exact results, but are useful for establishing approximate estimates and for determining how the energy eigenvalues of a system depend on the parameters involved. The following example demonstrates this.

Example 4.4 The harmonic oscillator potential is given by $V(x) = kx^2/2$, $-\infty \leq x \leq \infty$. Use the uncertainty principle to estimate the ground-state energy for a particle of mass m moving in this potential.

This potential is symmetric about $x = 0$, so we must have $\langle x \rangle = \langle p \rangle = 0$. We can write the total energy of the system as

$$E = \frac{p^2}{2m} + V(x) = \frac{p^2}{2m} + \frac{kx^2}{2}.$$

We can write the average value of the total energy as the sum of the average value of the kinetic and potential energies:

$$\langle E \rangle = \frac{\langle p^2 \rangle}{2m} + \frac{k}{2}\langle x^2 \rangle.$$

Hence

$$\Delta x = \sqrt{\langle x^2 \rangle - \langle x \rangle^2} = \sqrt{\langle x^2 \rangle}$$

and

$$\Delta p = \sqrt{\langle p^2 \rangle - \langle p \rangle^2} = \sqrt{\langle p^2 \rangle}.$$

These give

$$< E > = \frac{(\Delta p)^2}{2m} + \frac{k}{2}(\Delta x)^2.$$

The uncertainty relation demands

$$\Delta p \geq \frac{\hbar}{2(\Delta x)},$$

hence

$$\langle E \rangle = \frac{\hbar^2}{8m(\Delta x)^2} + \frac{k}{2}(\Delta x)^2.$$

The minimum possible value of E is found by minimizing this result with respect to Δx. Setting $d\langle E \rangle / d(\Delta x)$ to zero gives

$$\frac{d\langle E \rangle}{d(\Delta x)} = -\frac{\hbar^2}{4m(\Delta x)^3} + k(\Delta x) = 0,$$

or

$$(\Delta x)^4_{\min} = \frac{\hbar^2}{4km}.$$

On back-substituting this result in the expression for $\langle E \rangle$, we get

$$\langle E \rangle \geq \hbar\omega/2,$$

where ω is the classical angular frequency of the oscillator, $\sqrt{k/m}$. We will see in Chap. 5 that this energy is exactly what is predicted for the ground state by a formal solution of Schrödinger's equation for this potential.

A proof of the uncertainty principle is somewhat lengthy, but worth looking at in view of its historical importance. So as not to disturb the flow of the text, the proof appears in Appendix A1.

4.4 Commutators and Uncertainty Relations

How can one know *a priori* whether or not two observable quantities will obey an uncertainty relation? Equivalently, one might ask under what circumstances is it possible to measure simultaneous eigenvalues for two operators? We can answer this in terms of a generalized uncertainty relation. Consider any two operators A and B, and their corresponding variances ΔA and ΔB, defined as in (4.24). It is possible to show that these operators obey the general uncertainty relation [1]

$$(\Delta A)(\Delta B) \geq \frac{1}{2}\sqrt{-[A, B]^2},$$
(4.27)

where $[A, B]$ is itself an operator, known as the *commutator* of operators A and B, defined as

$$[A, B] = A_{op}B_{op} - B_{op}A_{op}.$$
(4.28)

If $[A, B]$ is non-zero, then there will be an uncertainty relation linking A and B. Pairs of physical variables that are linked by an uncertainty relation are known as *conjugate* or *complementary* variables. From the form of (4.28), it is obvious that if a wavefunction is an eigenfunction of both A and B, then the order of operations will not matter and we will have $[A, B] = 0$, that is, the eigenvalues of A and B will be measurable simultaneously to arbitrary precision.

We can verify the uncertainty principle in one dimension for $(\Delta x \Delta p)$ via these general relations. Recall

$$x_{op} = x,$$
(4.29)

and

$$(p_x)_{op} = -\imath\hbar\frac{d}{dx}.$$
(4.30)

Using a dummy wavefunction ψ to keep the order of differentiations straight, we have, dropping the "*op*" notation,

$$\psi = (xp - px)\psi = \{x(p\psi) - p(x\psi)\}$$

$$= \left\{ x\left(-\iota\hbar\frac{d\psi}{dx}\right) + \iota\hbar\frac{d}{dx}(x\psi)\right\} = \iota\hbar\left\{-x\frac{d\psi}{dx} + x\frac{d\psi}{dx} + \psi\right\}$$

$$= (\iota\hbar)\psi.$$

Hence

$$[x, p_x] \equiv \iota\hbar,$$

and

$$(\Delta x)(\Delta p_x) \geq \frac{1}{2}\sqrt{-(\iota\hbar)^2},$$

or

$$(\Delta x)(\Delta p_x) \geq \frac{\hbar}{2},$$

precisely the Uncertainty Principle.

A considerable amount of very muddy philosophy has been written around the Uncertainty Principle. Pais summarizes the situation elegantly in the following quotation [2]:

I have often felt that the expression uncertainty relation is unfortunate since it has all too often invoked imagery in popular writings utterly different from what Heisenberg very clearly had in mind, to wit, that the issue is not: what don't I know? but rather: what can't I know? In common language, "I am uncertain" does not exclude "I could be certain." It might therefore have been better had the term unknowability relations been used. Of course one neither can nor should do anything about that now.

Before closing this section, we establish a property of commutators that will be useful in Chap. 7. Suppose that we have three operators, A, B, and C, and that we wish to evaluate the commutator $[AB, C]\psi$. Tempting as it might be to factor the operator A outside the bracket and write $A[B, C]\psi$, this would not be a correct manipulation as it would change the order of the operations. A form of factoring can be achieved, however, by the following argument. The correct expansion of the commutator is

$$[AB, C]\psi = AB(C\psi) - C(AB\psi).$$

Now, it is quite legal to add to the right side of this expression the quantity $AC(B\psi) - AC(B\psi)$, since the net effect of such an addition is to add nothing:

$$[AB, C]\psi = AB(C\psi) - C(AB\psi) + AC(B\psi) - AC(B\psi).$$

Rearrange the right side of this expression to bring the first and fourth and second and third terms together:

$$[AB, C]\psi = A(BC - CB)\psi + (AC - CA)(B\psi).$$

Convince yourself that this is a legitimate rearrangement. The first bracketed term on the right side is just the commutator of B and C, while the second is the commutator of A and C:

$$[AB, C]\psi = A[B, C]\psi + [A, C](B\psi),$$

that is, we can factor the original commutator into the form

$$[AB, C] \equiv A[B, C] + [A, C]B. \tag{4.31}$$

Example 4.5 A wavefunction ψ has an energy eigenvalue E given by $H\psi = E\psi$, where H is the Hamiltonian operator $H = -\epsilon(d^2\psi/dx^2) + V(x)$, with $\epsilon = (\hbar^2/2m)$. Taking $A \equiv x$, $B \equiv p_x$, and $C \equiv H$, evaluate $[xp_x, H]$ using (4.31).

First compute $A[B, C]\psi$:

$$A[B, C]\psi = x(p_x H - Hp_x)\psi = x\left\{p_x(E\psi) - H(-\iota\hbar\psi')\right\},$$

where we use a prime to denote differentiation with respect to x. Applying the Hamiltonian operator gives

$$\begin{aligned}
A[B, C]\psi &= x\left\{(-\iota\hbar)(E\psi') - \left(-\varepsilon\frac{d^2}{dx^2} + V\right)(-\iota\hbar\psi')\right\} \\
&= x\left\{(-\iota\hbar)(E\psi') + (\iota\hbar)(-\varepsilon\psi''' + V\psi')\right\} \\
&= (\iota\hbar)\left\{xV\psi' - Ex\psi' - \varepsilon x\psi'''\right\}.
\end{aligned}$$

The other term in the right side of (4.31) is

$$[A, C] (B\psi) = [x, H](p_x \psi) = (-\iota\hbar)[x, H]\psi' = (-\iota\hbar)\left\{x(H\psi') - H(x\psi')\right\}$$

$$= (-\iota\hbar)\left\{x\left(-\varepsilon\frac{d^2}{dx^2} + V\right)(\psi') - \left(-\epsilon\frac{d^2}{dx^2} + V\right)(x\psi')\right\}$$

$$= (-\iota\hbar)\left\{-\varepsilon x\psi''' + xV\psi' - \left(-2\epsilon\psi'' - \epsilon x\psi''' + xV\psi'\right)\right\}$$

$$= (\iota\hbar)\left\{-2\varepsilon\psi''\right\}.$$

Hence

$$[xp_x, H] = (\iota\hbar)\left\{xV\psi' - Ex\psi' - \varepsilon x\psi''' - 2\varepsilon\psi''\right\},$$

admittedly a not particularly illuminating result.

4.5   Ehrenfest's Theorem

As alluded to in Sect. 4.2, it is possible to show that quantum-mechanical expectation values are in accord with the laws of classical physics if those expectation values are computed according as (4.20). We now prove this assertion, which is known as Ehrenfest's theorem [3]. Our proof will be three-dimensional for sake of generality.

Consider a system with some wavefunction $\Psi(x, t)$; we use an upper-case symbol for the wavefunction as a reminder that it is in general a function of position and time. The expectation value of the linear momentum $\boldsymbol{p}$ is given by

$$\langle \boldsymbol{p} \rangle = -\iota\hbar \int \Psi^* \boldsymbol{\nabla} \Psi dV, \tag{4.32}$$

where dV designates a three-dimensional volume element. Now, Newton's laws relate force to the rate of change of momentum. However, a force can always be expressed as the negative gradient of a potential:

$$F = \frac{d\boldsymbol{p}}{dt} = -\boldsymbol{\nabla}U, \tag{4.33}$$

where we use U for the potential function to avoid confusion with the volume element dV. If we can show that $\langle d\boldsymbol{p}/dt \rangle = -\langle \boldsymbol{\nabla}U \rangle$, then we can conclude that, on average, a quantum system will behave in accord with classical mechanics.

From (4.32) we have

$$\frac{\partial}{\partial t} \langle \boldsymbol{p} \rangle = -\iota \hbar \frac{\partial}{\partial t} \int \Psi^* \boldsymbol{\nabla} \Psi \, dV$$

$$= -\iota \hbar \int \left\{ \frac{\partial \Psi^*}{\partial t} \boldsymbol{\nabla} \Psi + \Psi^* \frac{\partial}{\partial t} (\boldsymbol{\nabla} \Psi) \right\} dV \tag{4.34}$$

$$= -\iota \hbar \int \left\{ \frac{\partial \Psi^*}{\partial t} \boldsymbol{\nabla} \Psi + \Psi^* \boldsymbol{\nabla} \left(\frac{\partial \Psi}{\partial t} \right) \right\} dV.$$

In the last step, the order of differentiation in the second term within the integral was exchanged. Now, (4.34) is an integral over the space coordinates, but derivatives with respect to time appear. We can eliminate these via the time-dependent Schrödinger equation from Chap. 2:

$$\nabla^2 \Psi - \varepsilon U \Psi = \frac{2m}{\iota \hbar} \frac{\partial \Psi}{\partial t}, \tag{4.35}$$

and its complex conjugate,

$$\nabla^2 \Psi^* - \varepsilon U \Psi^* = -\frac{2m}{\iota \hbar} \frac{\partial \Psi^*}{\partial t}, \tag{4.36}$$

where we have defined $\epsilon = 2m/\hbar^2$. Solving (4.35) and (4.36) for the time derivatives gives

$$\frac{\partial \Psi}{\partial t} = \frac{\iota \hbar}{2m} \left\{ \nabla^2 \Psi - \varepsilon U \Psi \right\}, \tag{4.37}$$

and

$$\frac{\partial \Psi^*}{\partial t} = -\frac{\iota \hbar}{2m} \left\{ \nabla^2 \Psi^* - \varepsilon U \Psi^* \right\}. \tag{4.38}$$

Substituting (4.37) and (4.38) into (4.34) gives

$$\frac{\partial < \boldsymbol{p} >}{\partial t} \tag{4.39}$$

$$= -\iota \hbar \int \left\{ -\frac{\iota \hbar}{2m} \left\{ \nabla^2 \Psi^* - \varepsilon U \Psi^* \right\} (\boldsymbol{\nabla} \Psi) + \frac{\iota \hbar}{2m} \Psi^* \boldsymbol{\nabla} \left\{ \nabla^2 \Psi - \varepsilon U \Psi \right\} \right\} dV.$$

On expanding out the gradient operation in the second term within the integral, some cancellations occur and we are left with

$$\frac{\partial <p>}{\partial t} = \frac{\hbar^2}{2m} \int \left\{ -(\nabla\Psi)(\nabla^2\Psi^*) + \Psi^*(\nabla^3\Psi) \right\} dV + \int \Psi(-\nabla U)\Psi^* dV.$$
$$(4.40)$$

Consider the second integral in this expression. It has the physical interpretation

$$\int \Psi(-\nabla U)\Psi^* dV = \langle -\nabla U \rangle = \langle Force \rangle. \qquad (4.41)$$

If we can show that the first integral in (4.40) vanishes, we will have arrived at the desired result. To show this requires knowledge of a result from vector calculus. For convenience, define $f = \Psi^*$ and $g = \nabla\Psi$. Then we have

$$\int \left\{ -(\nabla\Psi)(\nabla^2\Psi^*) + \Psi^*(\nabla^3\Psi) \right\} dV = \int \left\{ f(\nabla^2 g) - g(\nabla^2 f) \right\} dV. \quad (4.42)$$

According to Green's second identity, this integral is equivalent to an integral over the surface that bounds the volume V:

$$\int \left\{ f(\nabla^2 g) - g(\nabla^2 f) \right\} dV = \int \left\{ f(\nabla g) - g(\nabla f) \right\} \cdot dS. \qquad (4.43)$$
$$= \int \left\{ \Psi^*(\nabla^2\Psi) - (\nabla\Psi)(\nabla\Psi^*) \right\} \cdot dS$$

If we make the physically reasonable assumptions that ψ and its gradient both vanish asymptotically, the first integral on the right side of (4.40) vanishes, and what remains proves Ehrenfest's theorem.

Ehrenfest's theorem shows that classical mechanics is the limiting case of quantum mechanics for suitably large averages over time and space. In the words of Max Born, "The motion of particles follows probability laws, but the probability itself propagates according to the law of causality." [4]

4.6 The Orthogonality Theorem

In this section we prove a theorem relating to solutions of Schrödinger's equation that will be of value in the development of perturbation theory in Chap. 9: the orthogonality theorem. Our proof will be one-dimensional; extension to three dimensions should be obvious.

Consider two wavefunctions, $\psi_n(x)$ and $\psi_k(x)$, both of which satisfy Schrödinger's equation for some potential $V(x)$. If their energy eigenvalues are E_n and E_k, respectively, then the orthogonality theorem states that

$$\int \psi_k^*(x)\psi_n(x)dx = 0 \quad (E_k \neq E_n), \tag{4.44}$$

where the integral is over the limits of the system.

The wavefunctions must satisfy

$$-\frac{\hbar^2}{2m}\frac{d^2\psi_n}{dx^2} + V(x)\psi_n = E_n\psi_n \tag{4.45}$$

and

$$-\frac{\hbar^2}{2m}\frac{d^2\psi_k}{dx^2} + V(x)\psi_k = E_k\psi_k. \tag{4.46}$$

Multiply (4.45) and the complex conjugate of (4.46) by ψ_k^* and ψ_n, respectively. Then subtract (4.46) from (4.45) to eliminate $V(x)$. This gives

$$-\frac{\hbar^2}{2m}\left[\psi_k^*\frac{d^2\psi_n}{dx^2} - \psi_n\frac{d^2\psi_k^*}{dx^2}\right] = (E_n - E_k)\psi_k^*\psi_n. \tag{4.47}$$

Integrate this result over the domain relevant to the problem:

$$-\frac{\hbar^2}{2m}\int\left[\psi_k^*\frac{d^2\psi_n}{dx^2} - \psi_n\frac{d^2\psi_k^*}{dx^2}\right]dx = (E_n - E_k)\int\psi_k^*\psi_n dx. \tag{4.48}$$

Consider the integral on the left side. The integrand can be expressed as

$$\left[\psi_k^*\frac{d^2\psi_n}{dx^2} - \psi_n\frac{d^2\psi_k^*}{dx^2}\right] = \frac{d}{dx}\left[\psi_k^*\frac{d\psi_n}{dx} - \psi_n\frac{d\psi_k^*}{dx}\right].$$

With this, the left side of (4.48) becomes

$$-\frac{\hbar^2}{2m}\int\frac{d}{dx}\left[\psi_k^*\frac{d\psi_n}{dx} - \psi_n\frac{d\psi_k^*}{dx}\right]dx = -\frac{\hbar^2}{2m}\left[\psi_k^*\frac{d\psi_n}{dx} - \psi_n\frac{d\psi_k^*}{dx}\right]_{boundaries} = 0.$$

The last step follows if we make the reasonable assumption that the wavefunctions vanish asymptotically. Equation (4.48) then reduces to

$$(E_n - E_k)\int\psi_k^*\psi_n dx = 0.$$

If $E_k \neq E_n$, then (4.44) is proven.

Any set of functions $\psi_j(x)$ such that any two members of the set obey an integral constraint of this form is said to comprise an orthogonal set of functions. If in addition each individual member of the set is normalized, then they are said to comprise an orthonormal set of functions. In this case the mathematical expression of orthogonality can be written succinctly in Dirac notation as

$$\int \psi_k^* \psi_n dx = \langle \psi_k^* | \psi_n \rangle = \delta_k^n,$$ (4.49)

where δ_k^n denotes the *Kronecker delta*, defined as

$$\delta_k^n = \begin{cases} 1 & (k = n) \\ 0 & (k \neq n). \end{cases}$$ (4.50)

4.7 The Superposition Theorem

In this section we examine the superposition theorem, whose mathematical underpinnings will be useful in the development of the perturbation and variational approximation methods taken up in Chap. 9. This theorem is in fact much more generally useful than its application to just these areas; indeed, the entire mathematical development of time-dependent wavefunctions to describe real particles is critically dependent upon it, an issue we will touch upon briefly in Sect. 4.8. If you go on to more advanced study or research of quantum mechanics you will undoubtedly encounter the superposition theorem very frequently.

Imagine that you have some potential $V(x)$ for which Schrödinger's equation has been solved, yielding a (potentially infinite) number of wavefunctions $\psi_i(x)$ and their corresponding eigenenergies E_i. Now imagine constructing a sort of "super wavefunction" Ψ via a linear sum of the individual solutions:

$$\Psi(x) = \sum_i a_i \psi_i(x),$$ (4.51)

where the a_i are expansion coefficients. If we demand in addition that Ψ be normalized over the range over which the $\psi_i(x)$ apply, this leads to a constraint on the a_i:

$$\int \Psi\Psi * dx = 1 \Rightarrow \int \left(\sum_i a_i \psi_i \right) \left(\sum_i a_j^* \psi_j^* \right) dx = 1.$$ (4.52)

You should be able to convince yourself that the order of integration and summation in this expression can be swapped without loss of generality; if you are not sure, try making up a Ψ of a few terms and doing the algebra directly. This means that we can rewrite the normalization condition as

$$\sum_i \sum_j a_i a_j^* \langle \psi_i | \psi_j^* \rangle = 1, \tag{4.53}$$

where we have invoked Dirac notation for the integral. From the orthogonality theorem established in the preceding section, the integral must reduce to δ_i^j. Consequently, in the inner sum over j, only when j is equal to i will nonzero contributions arise. The normalization condition thus reduces to

$$\sum_i a_i^2 = 1, \tag{4.54}$$

where we have written $a_i a_i^* = a_i^2$. This means that if Ψ is to be normalized, the sum of the squares of the expansion coefficients must be unity.

Now suppose we wish to compute the expectation value of the energy of the superposition state represented by Ψ. From Sect. 4.2, we apply the Hamiltonian operator:

$$\langle E \rangle = \int \Psi \left(H \Psi^* \right) dx = \int \left(\sum_i a_i \psi_i \right) \left(H \sum_j a_j^* \psi_j^* \right) dx. \tag{4.55}$$

Taking the operator within the sum and setting $H(\psi_i) = E_i \psi_i$, we have

$$\langle E \rangle = \int \left\{ \left(\sum_i a_i \psi_i \right) \left(\sum_i a_j^* E_j \psi_j^* \right) \right\} dx = \sum_i \sum_j a_i a_j^* E_j \langle \psi_i | \psi_j^* \rangle. \tag{4.56}$$

Applying the orthogonality theorem as in (4.53) reduces this to

$$\langle E \rangle = \sum_i \sum_j a_i a_j^* E_j \delta_i^j = \sum_i a_i^2 E_i. \tag{4.57}$$

The expectation value of the energy for the superposition state is a weighted sum of the energies of the individual contributing states, with the weighting factors being the squares of the expansion coefficients.

It is a fundamental postulate of quantum mechanics that if a system can be in any one of the individual eigenstates ψ_i, then it can also be in a superposition state of the form of (4.51), that is, *the system need not be in an eigenstate*. It is further postulated that when one proceeds to make a measurement to determine what state the system is actually in, then the superposition state Ψ "collapses" into one of the eigenstates ψ_i, with the probability that one will find the system to be in state ψ_i given by a_i^2. These postulates comprise the superposition theorem. Measurements of the energy of a number of identically-prepared superposition-state systems will yield, on average, that predicted by (4.57).

The superposition theorem will be of considerable use in Chap. 9 for exploring the idea of forming a function from a linear sum of eigenfunctions and invoking the orthogonality theorem to simplify integrals such as those appearing in (4.53) and (4.56).

In the following section, we illustrate an important application of the superposition theorem: constructing representations of moving probability distributions.

4.8   Constructing a Time-Dependent Wave Packet

An important application of the superposition theorem is the construction of wavefunctions configured to represent particles that are time-dependently moving in some potential, or even freely. Such wavefunctions can be developed via a superposition of stationary states for the potential at hand multiplied by exponential functions that incorporate the time-dependence. While the emphasis in this book is on time-independent solutions of Schrödinger's equation, we devote this section to an example of how such a "wavepacket" can be constructed and interpreted. If you go on to further study of quantum mechanics you will encounter many such superposition states.

Suppose that some wavefunction $\psi(x)$ with corresponding eigenenergy E is one of the solutions to the time-independent Schrödinger equation for some potential $V(x)$. From (2.33), the corresponding solution $\Psi(x, t)$ to the time-dependent Schrödinger for this wavefunction and energy is given by

$$\Psi(x, t) = \psi(x)e^{-\iota\omega t} = \psi(x)e^{-\iota(E/\hbar)t}. \tag{4.58}$$

While this prescription transforms $\psi(x)$ into a time-dependent form $\Psi(x, t)$, no observable time-dependent consequence arises. This is because the physically observable manifestation of $\Psi(x, t)$, its corresponding probability distribution, retains no time-dependence:

$$\Psi^2 = \Psi^*\Psi = \left[\psi^*(x)e^{+\iota(E/\hbar)t}\right]\left[\psi(x)e^{-\iota(E/\hbar)t}\right] = \psi^*(x)\psi(x). \tag{4.59}$$

This means that the particle's probability distribution would never be observed to change in time: the particle would not move! But knowing that real particles do move, how can we model movement in the context of quantum-mechanical wavefunctions? The answer turns out to lie in formulating a superposition of wavefunctions of the form of (4.58).

We can illustrate such a superposition by using only two wavefunctions. To keep the discussion general for the moment, imagine that $\psi_1(x)$ and $\psi_2(x)$ are two normalized wavefunctions that satisfy the same potential $V(x)$ with energies E_1 and E_2, respectively. For simplicity, we assume that both $\psi_1(x)$ and $\psi_2(x)$ are real functions, although this need not be the case in general. The subscripts should not be interpreted to mean that we are using the $n = 1$ and $n = 2$ states specifically (although such might be the case); they are intended to denote any two solutions of $V(x)$. Now construct a time-dependent superposition state Ψ of the form

$$\Psi = \alpha \psi_1 e^{-\iota \omega_1 t} + \beta \psi_2 e^{-\iota \omega_2 t}, \tag{4.60}$$

where α and β are real-valued constants (again, this need not be the case) and where we drop the space-dependent-only natures of $\psi_1(x)$ and $\psi_2(x)$ for brevity. Now calculate the (observable) probability density of this superposition state:

$$\begin{aligned}
\Psi^2 = \Psi^* \Psi &= \left[\alpha \psi_1 e^{+\iota \omega_1 t} + \beta \psi_2 e^{+\iota \omega_2 t} \right] \left[\alpha \psi_1 e^{-\iota \omega_1 t} + \beta \psi_2 e^{-\iota \omega_2 t} \right] \\
&= \alpha^2 \psi_1^2 + \beta^2 \psi_2^2 + \alpha \beta (\psi_1 \psi_1) \left[e^{+\iota(\omega_1 - \omega_2)t} + e^{-\iota(\omega_1 - \omega_2)t} \right].
\end{aligned} \tag{4.61}$$

Using the Euler identity $e^{\pm \iota \theta} = \cos \theta \pm \iota \sin \theta$, we can cast this result in terms of purely real functions:

$$\Psi^2 = \alpha^2 \psi_1^2 + \beta^2 \psi_2^2 + 2\alpha \beta (\psi_1 \psi_2) \cos \left[(\omega_1 - \omega_2)t \right]. \tag{4.62}$$

Ψ^2 is a real-valued space-and-time dependent function; repeated observations of the system would show a time-dependent probability distribution: The particle moves! Since a function of the form $\cos(\omega t)$ reproduces itself over a time period T such that $\omega T = 2\pi$, Ψ^2 will "oscillate" with a period which depends on the difference in energy of the two states involved:

$$T = \frac{2\pi}{(\omega_1 - \omega_2)} = \frac{2\pi}{(1/\hbar)(E_1 - E_2)} = \frac{h}{(E_1 - E_2)}. \tag{4.63}$$

Even if Ψ^2 oscillates in time, its normalization is time-independent. This follows from

$$\begin{aligned}
\int \Psi^2 dx = 1 \Rightarrow \alpha^2 \int \psi_1^2 dx + \beta^2 \int \psi_2^2 dx \\
+ 2\alpha \beta \cos \left[(\omega_1 - \omega_2)t \right] \int \psi_1 \psi_2 dx = 1.
\end{aligned} \tag{4.64}$$

If E_1 and E_2 are different, then $\psi_1(x)$ and $\psi_2(x)$ must be orthogonal. This means that the last integral will vanish, eliminating any time dependence. Since $\psi_1(x)$ and $\psi_2(x)$ can independently be normalized, (4.64) reduces to

$$\alpha^2 + \beta^2 = 1, \tag{4.65}$$

in analogy to (4.54) for a time-independent summation state. Likewise, the expectation value of the energy of such a superposition system looks like (4.57):

$$\langle E \rangle = \alpha^2 E_1 + \beta^2 E_2. \tag{4.66}$$

It is worth emphasizing that these arguments will hold for any time-dependent superposition of the form of (4.60): no assumptions were made as to the specific potential $V(x)$ that the ψ's satisfy. Equation (4.60) can be extended to a summation of any number of terms; if so, (4.65) and (4.66) will be found to emerge similarly extended as well thanks to the orthogonality theorem.

We can calculate the time-dependence of the average position of a particle that finds itself in such a superposition state by the usual recipe for computing expectation values:

$$\langle x \rangle = \int \left(\Psi^* x \Psi \right) dx. \tag{4.67}$$

This integral proceeds much as (4.64) above; the result emerges as

$$\langle x \rangle = \alpha^2 \int \left(x\psi_1^2 \right) dx + \beta^2 \int \left(x\psi_2^2 \right) dx + 2\alpha\beta \cos\left[(\omega_1 - \omega_2)t \right] \int \left(x\psi_1\,\psi_2 \right) dx. \tag{4.68}$$

Note that even if $\psi_1(x)$ and $\psi_2(x)$ are orthogonal, the last integral in (4.68) will not necessarily vanish.

We now apply these results to a specific case: a superposition state constructed from the sum of two infinite rectangular well wavefunctions:

$$\psi_1(x) = \sqrt{\frac{2}{L}} \, \sin\left(\frac{n_1 \pi x}{L} \right), \quad E_1 = \left(\frac{\pi^2 \hbar^2}{2mL^2} \right) n_1^2,$$

and

$$\psi_2(x) = \sqrt{\frac{2}{L}} \, \sin\left(\frac{n_2 \pi x}{L} \right), \quad E_2 = \left(\frac{\pi^2 \hbar^2}{2mL^2} \right) n_2^2.$$

Figure 4.3 shows plots of Ψ^2 for such a superposition with $n_1 = 2$, $n_2 = 3$, $\alpha = 0.7$ and $\beta = \sqrt{1 - \alpha^2} = \sqrt{0.51}$ at times $t = 0, T/4, T/2$, and $3T/4$. The curves are offset from each other in the vertical direction for clarity; also, the length L has been set to unity. The particle begins at $t = 0$ with a dominant probability-distribution peak at $x \sim 0.2$ and a secondary peak at $x \sim 0.6$. By $t = T/4$ the distribution has spread out to become fairly uniform across the well. At $t = T/2$, one-half a cycle of the overall motion, the original probability peaks have reformed but in locations mirror-imaged from their original positions: the dominant peak has moved rightward and is about to reflect from the wall at $x = L$ while the secondary peak has moved leftward. At $t = 3T/4$ the distribution is the mirror-image of that at $T/4$, although this is not obvious as they are actually symmetric about $x = L/2$. At $t = T$, the pattern at $t = 0$ is recovered. The distribution sloshes back-and-forth in the well every T seconds.

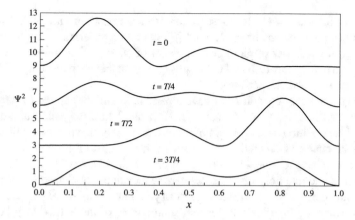

Fig. 4.3 Time-varying probability-density distribution of a superposition state of the $n = 2$ and $n = 3$ infinite rectangular well ($L = 1$) wavefunctions with $\alpha = 0.7$ and $\beta = \sqrt{0.51}$. The curves are vertically offset from each other for clarity; they should all lie at $\Psi^2 = 0$ at $x = 0$ and $x = 1$

Equation (4.68) shows that $\langle x \rangle$ will oscillate with the same period as Ψ^2 if the last integral in that equation does not vanish; if it does, then $\langle x \rangle$ will be independent of time. In the case of the present example, the results of Example 4.1 indicate the first two integrals in (4.67) evaluate to $L/2$. The third integral is not particularly neat; the overall result is

$$\langle x \rangle = \frac{L}{2} + 2\alpha\beta \cos\left[(\omega_1 - \omega_2)t\right] \left\{ \left[\frac{(-1)^{n_1 - n_2} - 1}{(n_1 - n_2)^2} \right] - \left[\frac{(-1)^{n_1 + n_2} - 1}{(n_1 + n_2)^2} \right] \right\},$$

where the normalization $\alpha^2 + \beta^2 = 1$ has been used. If n_1 and n_2 are both even or both odd, then the time-dependent term in this result vanishes, a consequence of the symmetry of the superposition state about $x = L/2$; if they are of opposite parity, then $\langle x \rangle$ oscillates with period T and an amplitude that depends on α, β, n_1, and n_2. For the values of the parameters chosen here, $\langle x \rangle$ oscillates between $0.3055L$ and $0.6945L$. For an electron in this particular superposition state moving in a well with $L = 1$ Å, the period of the motion is $T \sim 2.2 \times 10^{-17}$ seconds. The full back-and-forth displacement of over this time, $2(0.6945 - 0.3055)$Å$= 0.778$ Å, corresponds to a speed of about 3.5×10^6 m/s, or about $0.01c$, comfortably nonrelativistic. A speed computed in this way is known as a "group velocity".

After all of our mathematics and calculations, what does it mean to think of a particle as somehow being in a combination of two energy states at once? It is a fundamental postulate of quantum mechanics that the experimental answer to this question is that if one measured the energies of a large number particles in such systems, any individual measurement would give only either E_1 or E_2, but that the average of the measurements would be as predicted by (4.66). Wavefunctions predict only statistical results of the measurements of positions and energies of particles. It is now in fact experimentally possible to prepare and make measurements on quantum systems in superposition states; observation of a superposition state "collapsing" to a definite single state has been reported [5]. On an engineering level, the unusual metallurgical properties of plutonium, a matter of no small importance to nuclear-weapons designers, are attributed to the ground state of that element being a quantum superposition of two valences [6].

A more practical question is: "I have a particle in an infinite well (or any other potential) at some position "x" at $t = 0$. How can I turn this analysis around to express this condition as such a sum of wavefunctions for that potential? What does quantum mechanics then predict for the subsequent motion of the particle?" To deal with this, you have to model the positional probability distribution of your particle at $t = 0$ as some function, say a tall but narrow rectangularly-shaped function $\phi(x)$ whose width corresponds to the uncertainty in your knowledge of the particle's initial position. Knowing the time-independent solutions ψ_i of the Schrödinger equation for the potential at hand, write $\phi(x)$ as an infinite-sum of the $\psi_i(x)$ with various coefficients a_i: $\phi(x) = \Sigma a_i \psi_i(x)$. Here, $\phi(x)$ is equivalent to $\Psi(x, t)$ at $t = 0$, and the coefficients a_i play the roles of α, β, ... in (4.60). The point now, however, is that rather than being chosen arbitrarily, the a_i need to be chosen to reproduce $\phi(x)$ from the $\psi_i(x)$. An explicit recipe for determining the coefficients is developed in Sect. 9.2. Once the coefficients have been determined, the fully time-dependent superposition state as in (4.60) is then constructed by tacking on the time-dependent exponential factors to each of the $\psi_i(x)$, and from this you can make predictions of observable quantities such as $\langle x \rangle$ and $\langle E \rangle$. In the end, it is only by constructing superposition states in this way that moving probability distributions can be obtained. This analysis can be extended as well to free particles, that is, ones not bound within a potential well.

To close this section, we take up a question which may already have occurred to you: It is clear from Fig. 4.3 that for any wave packet constructed from a superposition of individual wavefunctions that the position uncertainty Δx must be a function of time. This means that even if we begin with a "point" particle represented by a highly localized wave packet (one with a very small Δx), the packet must eventually "smear out." Why do we not see objects such as baseballs, planets, or people exhibiting such behavior?

A rigorous analysis of this question is beyond the scope of this book, but we can summarize the results of such research fairly easily. (If you are familiar with Sect. 9.2 of this book and are comfortable with Fourier analysis, there are a number of very readable treatments available. What follows is adopted from Shankar, Bransden and Joachim, Liboff, and French and Taylor [7–10]).

Suppose that the wavefunction representing a "free" particle, that is, one not subject to a potential $V(x)$, initially has a Gaussian shape at $t = 0$:

$$\psi(x, 0) = \frac{1}{(\pi \Delta^2)^{1/4}} e^{-x^2/2\Delta^2} \quad (-\infty \leq x \leq \infty), \tag{4.69}$$

where Δ is a parameter we can choose to control the width of ψ as we please; the factor of $(\pi \Delta^2)^{1/4}$ ensures proper normalization. The position uncertainty of this function at $t = 0$ is (prove it!)

$$\Delta x_0 = \frac{\Delta}{\sqrt{2}}. \tag{4.70}$$

As shown in Shankar [7], this wave packet will retain its normalization and Gaussian shape as a function of time, but it will widen. At any later time t, the position uncertainty is given by

$$\Delta x(t) = \Delta x_0 \sqrt{1 + \frac{\hbar^2}{4 m^2 (\Delta x_0)^4} t^2}, \tag{4.71}$$

where m is the mass of the particle. To get a sense of how rapidly this spreading occurs, consider the time required for Δx to double from its initial value:

$$t_{double} = \sqrt{12} \left(\frac{m}{\hbar}\right) (\Delta x_0)^2. \tag{4.72}$$

For an electron with $\Delta x_0 = 10^{-15}$ m, $t_{double} \sim 3 \times 10^{-26}$ s, an incredibly brief interval. On the other hand, for a 145-gram baseball with $\Delta x_0 = 1$ millimeter, $t_{double} \sim 1.5 \times 10^{20}$ years, some 10 billion times the estimated age of the Universe. The trajectory of a macroscopic particle can therefore be teated classically for any sensible length of time. In the case of the electron, however, one must be cautious in interpreting this calculation. The electron will become spread out in the sense that its probability distribution will be spatially extensive, but as soon as one detectes it (with a Geiger counter, say), it will be recorded as being at a particular place at a particular time; so far as we know, electrons are indivisible fundamental particles. Also, as soon as the detection takes place, the electron will no longer be a free particle and would have to be described by some different wavefunction. Further, we cannot really model a baseball as a point mass; it is composed of an enormous number of interacting atoms which also interact with the outside environment. These calculations have to be taken with a grain of salt, but the idea should be clear.

4.9 The Virial Theorem

In this section we take up the virial theorem, an expression which relates the expectation values of the kinetic and potential energy operators for any potential. For practical purposes, the virial theorem is not particularly helpful for establishing or even estimating solutions to the Schrödinger equation, but we examine it as there is a corresponding theorem in classical dynamics and it provides an interesting general-ization of the relationship between kinetic and potential energies in central potentials.

To set the stage for this, suppose that we have some operator A which is known to be time-independent. Then operate the commutator comprising A and the Hamil-tonian operator H on some wavefunction ψ:

$$[A, H]\psi = A(H\psi) - H(A\psi). \tag{4.73}$$

In general, it is assumed here that ψ may be both space and time-dependent. Now, from (2.28) we can write the Hamiltonian operator as

$$H \equiv \iota\hbar\frac{\partial}{\partial t}, \tag{4.74}$$

so that (4.73) becomes

$$[A, H]\psi = (\iota\hbar)\left\{A\left(\frac{\partial\psi}{\partial t}\right) - \frac{\partial}{\partial t}(A\psi)\right\}. \tag{4.75}$$

If A is independent of time (and only if it is so), it can be factored out of the brackets in the last term in this expression:

$$[A, H]\psi = (\iota\hbar)\left\{A\left(\frac{\partial\psi}{\partial t}\right) - A\left(\frac{\partial\psi}{\partial t}\right)\right\} = 0, \tag{4.76}$$

that is, we must have $[A, H]\psi = 0$ whenever A is time-independent.

The virial theorem involves using this identity in the case where A is defined as the vector dot product of the position and momentum operators:

$$A \equiv \boldsymbol{r}_{op} \cdot \boldsymbol{p}_{op} = -(\iota\hbar)\boldsymbol{r} \cdot \nabla, \tag{4.77}$$

where ∇ is the usual gradient operator. Since we are assuming that $A \equiv \boldsymbol{r}_{op} \cdot \boldsymbol{p}_{op}$ is time-independent, the following derivation applies only for stationary states, that is, ones for which $\langle \boldsymbol{r}\boldsymbol{p} \rangle = 0$; it does *not* apply for superposition states such as those considered in Sect. 4.8.

Applying the commutator of A and H to some wavefunction ψ that is a solution of the Schrödinger equation for some potential V gives:

$$[A, H]\psi = \iota\left(\frac{\hbar^3}{2m}\right)[\boldsymbol{r} \cdot \nabla(\nabla^2\psi) - \nabla^2(\boldsymbol{r} \cdot \nabla\psi)] \tag{4.78}$$
$$+ (\iota\hbar)[V(\boldsymbol{r} \cdot \nabla\psi) - \boldsymbol{r} \cdot \nabla(V\psi)],$$

where we have used the "spatial" form of the Hamiltonian in the usual Schrödinger equation, $H\psi = -(\hbar^2/2m)\nabla^2\psi + V\psi$.

Call the first square bracket appearing on the right side of (4.78) "part 1", and the second one "part 2". We analyze each individually. Consider first the first term within part 1. The gradient operations can be written as

$$\nabla(\nabla^2\psi) = \frac{\partial}{\partial x}(\nabla^2\psi)\boldsymbol{x} + \frac{\partial}{\partial y}(\nabla^2\psi)\boldsymbol{y} + \frac{\partial}{\partial z}(\nabla^2\psi)\boldsymbol{z},$$

where $(\boldsymbol{x}, \boldsymbol{y}, \boldsymbol{z})$ denote the usual Cartesian unit vectors. The Laplacian operator appears as

$$\nabla^2\psi = \frac{\partial^2\psi}{\partial x^2} + \frac{\partial^2\psi}{\partial y^2} + \frac{\partial^2\psi}{\partial z^2}.$$

In expanding out this term, it is helpful to write derivatives in the more compact notation

$$\psi_{abc} \equiv \frac{\partial^3\psi}{\partial a\,\partial b\,\partial c}.$$

Hence we have

$$\nabla(\nabla^2\psi) = (\psi_{xxx} + \psi_{xyy} + \psi_{xzz})\boldsymbol{x} + (\psi_{yxx} + \psi_{yyy} + \psi_{yzz})\boldsymbol{y} + (\psi_{zxx} + \psi_{zyy} + \psi_{zzz})\boldsymbol{z}.$$

From the usual procedure for taking a dot product, this gives the first term in part 1 as

$$\boldsymbol{r}\cdot\nabla(\nabla^2\psi) = x(\psi_{xxx} + \psi_{xyy} + \psi_{xzz}) + y(\psi_{yxx} + \psi_{yyy} + \psi_{yzz}) + z(\psi_{zxx} + \psi_{zyy} + \psi_{zzz}). \tag{4.79}$$

Now look at the second term of part 1. In similar notation for the derivatives, it evaluates to

$$\begin{aligned} \nabla^2(\boldsymbol{r}\cdot\nabla\psi) = {}&2(\psi_{xx} + \psi_{yy} + \psi_{zz}) \\ &+ x(\psi_{xxx} + \psi_{yyx} + \psi_{zzx}) + y(\psi_{xxy} + \psi_{yyy} + \psi_{zzy}) \\ &+ z(\psi_{xxz} + \psi_{yyz} + \psi_{zzz}). \end{aligned} \tag{4.80}$$

In combining (4.79) and (4.80) to formulate part 1, all of the third-derivative terms cancel because the order of the derivatives is irrelevant. Also, $\psi_{xx} + \psi_{yy} + \psi_{zz} \equiv \nabla^2\psi$, so part 1 reduces to

$$\iota\left(\frac{\hbar^3}{2m}\right)[\boldsymbol{r}\cdot\nabla(\nabla^2\psi) - \nabla^2(\boldsymbol{r}\cdot\nabla\psi)] = 2\iota\hbar\left(-\frac{\hbar^2}{2m}\nabla^2\psi\right) = 2\iota\hbar(KE_{op}\psi). \tag{4.81}$$

Part 2 is much more straightforward. Recalling that we can write $\nabla(V\psi) = \psi(\nabla V) + V(\nabla\psi)$, we have

$$V(\mathbf{r} \cdot \nabla\psi) - \mathbf{r} \cdot \nabla(V\psi) = V(\mathbf{r} \cdot \nabla\psi) - \mathbf{r} \cdot (\psi\nabla V) - \mathbf{r} \cdot (V\nabla\psi).$$

In the last term, V can be brought outside the brackets because it is a purely scalar function; the first and last terms on the right side then cancel. In the middle term on the right side, the scalar function ψ can likewise be brought outside the brackets, leaving

$$(\iota\hbar)[V(\mathbf{r} \cdot \nabla\psi) - \mathbf{r} \cdot \nabla(V\psi)] = -\iota\hbar(\mathbf{r} \cdot \nabla V)\psi. \tag{4.82}$$

Equations (4.78), (4.81), and (4.82) then give

$$[A, \mathbf{H}]\psi = 2\iota\hbar(KE_{op}\psi) - \iota\hbar(\mathbf{r} \cdot \nabla V)\psi. \tag{4.83}$$

Multiplying through by ψ^* and integrating over the domain of the wavefunction gives, in terms of expectation values,

$$\langle[A, \mathbf{H}]\rangle = \iota\hbar\left(2\langle KE\rangle - \langle \mathbf{r} \cdot \nabla V\rangle\right). \tag{4.84}$$

However, because $A \equiv \mathbf{r}_{op} \cdot \mathbf{p}_{op}$ is a time-independent operator, (4.76) tells us that $\langle[A, \mathbf{H}]\rangle = 0$, so we can conclude that, for any potential V,

$$2\langle KE\rangle = \langle \mathbf{r} \cdot \nabla V\rangle. \tag{4.85}$$

This is the virial theorem (VT).

One of the more interesting conclusions resulting from the VT concerns a purely radial potential of the form

$$V(r) = kr^n, \tag{4.86}$$

where n is any power. In this case

$$\nabla V = \left(\frac{\partial V}{\partial r}\right)\mathbf{r}' = nkr^{n-1}\mathbf{r}',$$

where $\mathbf{r}'$ is used temporarily to denote the spherical-coordinate unit vector. Hence

$$\mathbf{r} \cdot \nabla V = (r\mathbf{r}') \cdot (nkr^{n-1}\mathbf{r}') = nkr^n = nV.$$

The VT then predicts

$$2\langle KE\rangle = n\langle V\rangle. \tag{4.87}$$

The Coulomb potential is of this form, with $n = -1$. This means that the VT predicts that, on average, the potential energy of a hydrogen atom will be negative twice the kinetic energy, a result consistent with the Bohr model. Because $\langle E\rangle = \langle KE\rangle + \langle V\rangle$ in general, this result also means that, for the Coulomb potential, $\langle E\rangle = -\langle KE\rangle$, and hence that the expectation value of the momentum, $\langle p^2\rangle = 2m_e\langle KE\rangle$, can be expressed as $\langle p^2\rangle = -2m_e\langle E_n\rangle = \hbar^2/(a_o^2 n^2)$.

Example 4.6 The harmonic oscillator potential $V(x) = kx^2/2$, $(-\infty \le x \le \infty)$ leads to a ground-state wavefunction $\psi(x) = Ae^{-\alpha^2 x^2/2}$ where A is a normalization constant and $\alpha = (mk/\hbar^2)^{1/4}$. This potential is analyzed in detail in Chap. 5. Verify that the virial theorem is satisfied by this potential and wavefunction.

We need to show that $2\langle KE \rangle = \langle r \cdot \nabla V \rangle$. Deal with the left side first. The KE operator is $2KE_{op}\psi = -\hbar^2/2m(\partial^2 \psi/\partial x^2)$, hence

$$2KE_{op}\psi = -\frac{\hbar^2}{m}\frac{\partial^2 \psi}{\partial x^2} = -\frac{A\hbar^2}{m}\left(-\alpha^2 e^{-\alpha^2 x^2/2} + \alpha^4 x^2 e^{-\alpha^2 x^2/2}\right).$$

The expectation value is

$$\langle 2KE_{op} \rangle = \int\limits_{-\infty}^{\infty} \psi^*(KE_{op}\psi)dx = -\frac{A^2\hbar^2}{m}\left(-\alpha^2 \int\limits_{-\infty}^{\infty} e^{-\alpha^2 x^2}dx + \alpha^4 \int\limits_{-\infty}^{\infty} x^2 e^{-\alpha^2 x^2}dx\right)$$

The integrals are standard; the result is

$$\langle 2KE_{op} \rangle = \frac{A^2\hbar^2\alpha\sqrt{\pi}}{2m}.$$

On the right side of the VT we have

$$r \cdot \nabla V = (x\boldsymbol{x}) \cdot (kx\boldsymbol{x}) = kx^2.$$

Hence

$$\langle r \cdot \nabla V \rangle = A^2 k \int\limits_{-\infty}^{\infty} x^2 e^{-\alpha^2 x^2}dx = \frac{A^2 k\sqrt{\pi}}{2\alpha^3}.$$

Setting $\alpha = (mk/\hbar^2)^{1/4}$ shows that the VT is satisfied:

$$\langle 2KE_{op} \rangle = \frac{A^2\hbar^2\alpha\sqrt{\pi}}{2m} = \frac{A^2\hbar^2\sqrt{\pi}}{2m}\left(\frac{mk}{\hbar^2}\right)^{1/4} = \frac{A^2\sqrt{\pi}\,\hbar^{3/2}k^{1/4}}{2m^{3/4}}$$

and

$$\langle r \cdot \nabla V \rangle = \frac{A^2 k\sqrt{\pi}}{2\alpha^3} = \frac{A^2 k\sqrt{\pi}}{2}\left(\frac{\hbar^2}{mk}\right)^{3/4} = \frac{A^2\sqrt{\pi}\,\hbar^{3/2}k^{1/4}}{2m^{3/4}}.$$

Problem 5–19 investigates this question more fully.

Example 4.7 Suppose that you are given the potential $V(x) = kx^n/2$, $(-\infty \leq x \leq \infty)$, with n being a positive, even-valued number. Hypothesizing that a solution of Schrödinger's equation for this potential is $\psi(x) = Ae^{-\alpha^2 x^2/2}$ for a particle of mass m, how must α depend on n, m, $\hbar$, and k such that the virial theorem is satisfied? Can this wavefunction be made to be a solution of Schrödinger's equation for this potential for any n?

This is the same wavefunction as appeared in Example 4.6 above, so we have

$$\langle 2KE_{op} \rangle = \frac{A^2 \hbar^2 \alpha \sqrt{\pi}}{2m}.$$

Now

$$\mathbf{r} \cdot \nabla V = (x\mathbf{x}) \cdot \left(\frac{nk}{2} x^{n-1} \mathbf{x} \right) = \frac{nk}{2} x^n.$$

The right side of the VT is then

$$\langle \mathbf{r} \cdot \nabla V \rangle = \frac{A^2 nk}{2} \int_{-\infty}^{\infty} x^n e^{-\alpha^2 x^2} dx = \frac{A^2 nk}{2\alpha^{n+1}} \Gamma \left(\frac{n+1}{2} \right).$$

The gamma function $\Gamma(x)$ is defined in Appendix C. Equating $\langle 2KE_{op} \rangle$ to $\langle \mathbf{r} \cdot \nabla V \rangle$ and solving for α gives

$$\alpha^{n+2} = \frac{n m k}{\sqrt{\pi} \hbar^2} \Gamma \left(\frac{n+1}{2} \right).$$

We can check this result by setting $n = 2$ and invoking the identity $\Gamma(3/2) = \sqrt{\pi}/2$; these give $\alpha^4 = mk/\hbar^2$, as claimed in Example 4.6.

While we can make this trial wavefunction satisfy the VT for this potential, it does *not* satisfy Schrödinger's equation for this potential. To see this, substitute $\psi(x)$ and $V(x)$ into Schrödinger's equation; the result is

$$\frac{\hbar^2 \alpha^2}{2m} - \frac{\hbar^2 \alpha^4}{2m} x^2 + \frac{k}{2} x^n = E.$$

For this to be meaningful, the two terms with x-dependences must cancel each other; this demands $n = 2$ and $\alpha^4 = mk/\hbar^2$, as above. The remaining terms give $E = \hbar^2 \alpha^2/2m$, which is known to be the ground state energy for this potential when $n = 2$. However, no solution is possible for $n \neq 2$.

This result seems to lead to a contradiction: we can apparently make a hypothetical wavefunction to satisfy the virial theorem for some potential just by including an adjustable parameter α in the wavefunction. However, the wavefunction is not guaranteed to be a solution of Schrödinger's equation for that potential. How can this be? *The answer is that the Schrödinger equation was never used in the derivation of the virial theorem!* Our derivation of the virial theorem depended on the definition of the Hamiltonian operator H and the fact that any time-independent operator A must satisfy $[A, H] = 0$, but it was never demanded that ψ satisfy $H\psi = E\psi$. This what limits the practicality of the virial theorem.

4.10 Momentum-Space Wavefunctions

This section can be considered optional; no subsequent material will refer to it.

We have seen that a fundamental postulate of quantum mechanics is that the wavefunction $\psi_n(x)$ of some quantum state gives the probability density $\psi_n^*(x)\psi_n(x)$ of finding the particle involved between x and $x + dx$. From the discussion in Sect. 4.3 concerning how a wave pattern with a finite extent will involve contributions from many different values of the momentum p, you might surmise that an alternate description of the state of a particle along the lines of a "momentum probability density" could be conceived. This is indeed so. This function is usually designated as $\phi_n(p)$, with $\phi_n^*(p)\phi_n(p)dp$ giving the probability that the particle has momentum between p and $p + dp$ if it is in state n of the system involved. This section explores this concept.

In terms of $\psi_n(x)$, $\phi_n(p)$ is given by

$$\phi_n(p) = \frac{1}{\sqrt{2\pi\hbar}} \int_{-\infty}^{\infty} \psi_n(x)e^{-\iota px/\hbar}dx. \tag{4.88}$$

The limits on the integral are formal; in practice, they go over the domain of the $\psi_n(x)$. With the factor of $\iota = \sqrt{-1}$ in the exponential, this is an explicitly complex integral; after integrating over x, what remains is a function of p. A proof of this relationship can be found in [11].

These integrals can get complicated, but we can explore a simple example by appealing to our old friend, the infinite rectangular well of Sect. 3.2; see also [12]. From (3.11) and (3.15), the wavefunctions and energies for this system are

$$\psi_n(x) = \sqrt{\frac{2}{L}} \sin\left(\frac{n\pi}{L}x\right), \quad (0 \leq x \leq L), \quad n = 1, 2, 3, \ldots. \tag{4.89}$$

and

$$E_n = \left(\frac{\pi^2 \hbar^2}{2mL^2}\right) n^2.   \tag{4.90}$$

In what follows we drop any factors of ι which end up as only overall multiplicative constants when evaluating $\phi_n(p)$ because they will reduce to unity when computing $\phi_n^*(p)\phi_n(p)$. Factors of ι *within* the integral for $\phi_n(p)$ must be carefully maintained, however, as they could affect the final results.

The integral for $\phi_n(p)$ is then

$$\phi_n(p) = \frac{1}{\sqrt{L\pi\hbar}} \int\limits_0^L \sin\left(\frac{n\pi}{L}x\right) e^{-\iota px/\hbar} dx.   \tag{4.91}$$

To simplify this, it is handy to invoke Euler's identity and write

$$\sin\theta = \frac{1}{2\iota}(e^{\iota\theta} - e^{-\iota\theta}),$$

where we have $\theta = n\pi x/L$. Hence we have (dropping external factors of ι as described above)

$$\phi_n(p) = \frac{1}{2\sqrt{L\pi\hbar}} \int\limits_0^L [e^{\iota(n\pi x/L)} - e^{-\iota(n\pi x/L)}] e^{-\iota px/\hbar} dx$$

$$\tag{4.92}$$

$$= \frac{1}{2\sqrt{L\pi\hbar}} \left[\int\limits_0^L e^{\iota(n\pi/L - p/\hbar)x} dx - \int\limits_0^L e^{-\iota(n\pi/L + p/\hbar)x} dx \right].$$

Both integrands here are of the form e^{ax} where $a = \pm\iota(n\pi/L \mp p/\hbar)$; both will consequently integrate to e^{ax}/a. Be careful with signs; integrating gives

$$\phi_n(p) = \frac{1}{2\sqrt{L\pi\hbar}} \left[\frac{e^{\iota(n\pi/L - p/\hbar)x}}{(n\pi/L - p/\hbar)} + \frac{e^{-\iota(n\pi/L + p/\hbar)x}}{(n\pi/L + p/\hbar)} \right]_0^L.   \tag{4.93}$$

We work out the first term explicitly. Evaluating the limits gives

$$\left[\frac{e^{\iota(n\pi/L - p/\hbar)x}}{(n\pi/L - p/\hbar)} \right]_0^L = \frac{\left[e^{\iota(n\pi - pL/\hbar)} - e^0\right]}{(n\pi/L - p/\hbar)} = \frac{\left[e^{\iota(n\pi)}e^{-\iota pL/\hbar} - 1\right]}{(n\pi/L - p/\hbar)}.$$

In the $e^{i(n\pi)}$ term in the numerator we can invoke Euler's identity and write $e^{i(n\pi)} = \cos(n\pi) + i\sin(n\pi)$. But $\sin(n\pi) = 0$ for all n, and $\cos(n\pi) = (-1)^n$ for all n, so this expression reduces to

$$\frac{\left[(-1)^n e^{-ipL/\hbar} - 1\right]}{(n\pi/L - p/\hbar)}.$$

In the second term in (4.93), the numerator turns out to be identical to that in the first term; the denominator is $(n\pi/L + p/\hbar)$. Gathering the two terms and bringing things to a common denominator gives

$$\phi_n(p) = \frac{n\sqrt{\pi}}{L^{3/2}\sqrt{\hbar}}\left[\frac{1 - (-1)^n e^{-ipL/\hbar}}{(n\pi/L)^2 - (p/\hbar)^2}\right]. \tag{4.94}$$

We now invoke a notational simplification that has a physical rationale. For a particle in an infinite well the potential energy is always zero, so *all* of the energy E_n for some state can be considered to be kinetic. Classically, kinetic energy and momentum are related as $K = p^2/2m$, so we can write $p_n^2 = 2mK = 2mE_n$, or $p_n = \pm\sqrt{2mE_n}$. From (4.90) this gives

$$p_n = \pm\sqrt{2m\left(\frac{\pi^2\hbar^2}{2mL^2}\right)n^2} = \pm\left(\frac{\pi\hbar}{L}\right)n = \pm\left(\frac{h}{2L}\right)n. \tag{4.95}$$

Classically, our particle can have either only forward or backward momentum of magnitude $p_n = n\pi\hbar/L$, which we can write as $p_n/\hbar = n\pi/L$. But $n\pi/L$ is exactly what appears in the denominator of (4.94). Writing the n in the prefactor as $n = Lp_n/\pi\hbar$ and extracting a factor of $\hbar^2$ from the term in the denominator of the square brackets then lets us write $\phi_n(p)$ more compactly as

$$\phi_n(p) = \sqrt{\frac{\hbar}{L\pi}}\left(\frac{p_n}{p_n^2 - p^2}\right)\left[1 - (-1)^n e^{-ipL/\hbar}\right]. \tag{4.96}$$

Be careful here to distinguish between p_n and p: The former is the "classical" momentum corresponding to a given energy state according as (4.95); the latter is momentum as a general variable.

To facilitate plotting $\phi_n(p)$ it is convenient to cast the exponential function in terms of familiar sine and cosine functions. To this end, first put $\hbar = h/2\pi$ and write the term in square brackets as $\left[1 - (-1)^n e^{-2i\pi pL/h}\right]$. Then extract a factor of $e^{-i\pi pL/h}$ to put this is the form

$$\left[1 - (-1)^n e^{-2i\pi pL/h}\right] = e^{-i\pi pL/h}\left[e^{i\pi pL/h} - (-1)^n e^{-i\pi pL/h}\right].$$

Temporarily abbreviate $\pi pL/h$ as θ. If n is even, then the terms in the square brackets become $e^{i\theta} - e^{-i\theta} = 2i\sin\theta$. Similarly, if n is odd, they become $e^{i\theta} + e^{-i\theta} = 2\cos\theta$. Hence we can cast (4.96) as

$$\phi_n(p) = 2\sqrt{\frac{\hbar}{L\pi}} \left(\frac{p_n}{p_n^2 - p^2}\right) e^{-\iota\pi pL/h} \times \begin{cases} \sin\left(\frac{\pi pL}{h}\right) & n \text{ even} \\ \\ \cos\left(\frac{\pi pL}{h}\right) & n \text{ odd.} \end{cases} \tag{4.97}$$

We can now write an expression for the momentum probability density $\phi_n^*(p)\phi_n(p)$:

$$\phi_n^2(p) = \frac{2h}{L\pi^2} \left(\frac{p_n}{p_n^2 - p^2}\right)^2 \times \begin{cases} \sin^2\left(\frac{\pi pL}{h}\right) & n \text{ even} \\ \\ \cos^2\left(\frac{\pi pL}{h}\right) & n \text{ odd,} \end{cases} \tag{4.98}$$

where we set $\hbar = h/2\pi$ in the prefactor.

To plot this, it is convenient to cast p into dimensionless units. Notice from (4.95) that $h/2L$ serves as a natural unit of momentum for this problem. We then define dimensionless momentum w as $p = w(h/2L)$, and write $\phi_n^2(p)$ as

$$\phi_n^2(p) = \frac{8L}{\pi^2 h} \left(\frac{n}{n^2 - w^2}\right)^2 \times \begin{cases} \sin^2\left(\frac{\pi}{2}w\right) & n \text{ even} \\ \\ \cos^2\left(\frac{\pi}{2}w\right) & n \text{ odd.} \end{cases} \tag{4.99}$$

The prefactor emerges from a factor of $1/(h/2L)^2$ arising from the large round brackets; you should confirm that $8L/\pi^2 h$ has units of inverse momentum, as it should if $\phi_n^2(p)dp$ is to be the probability of our particle having momentum between p and $p + dp$.

Figure 4.4 shows a plot of $\phi_n^2(p)$ for $n = 4$. In creating this plot, the factor of $8L/\pi^2 h$ has been neglected as it does not affect the shape of the curve or the locations of the maxima and minima. The plot runs over the domain $-6 \le w \le 6$, but should extend over $-\infty \le w \le \infty$; this is a plot in *momentum space*, not *physical* space, where the well is restricted to $0 \le x \le L$. However, little action happens for $|w| > 6$.

The striking feature of this plot is that while it indicates that the most probable momentum of the particle is close to the classical prediction of $\pm 4(h/2L)$, it is possible that we will observe the momentum to be any other value except for those where $\sin(w\pi/2) = \sin(\pi pL/h) = 0$. In particular, there are secondary maxima at $\pm \sim 1(h/2L)$. The price of confining a particle-wave in physical space is is that it bears the fingerprints of an infinitude of momentum contributions. For this case of $n = 4$, the large maxima occur at $w \sim \pm 3.85$.

A further observation to made here is that if n is even, we will never observe the particle to have zero momentum; conversely, if n is odd, there is always some non-zero probability of the particle having zero momentum. Curiously, if $n = 1$, the most probable momentum is zero!

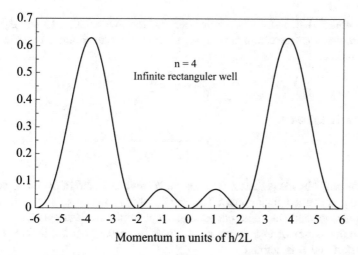

Fig. 4.4 Momentun-space probability distribution for n = 4 infinite rectangular well state

The symmetry of the curve in Fig. 4.4 indicates that $\langle p \rangle = 0$, as we deduced in Example 4.1. For a test of your integration skills, try computing $\langle p^2 \rangle$:

$$\langle p^2 \rangle = \int_{-\infty}^{-\infty} \phi_n^*(p) \, p^2 \phi_n(p) dp.$$

In p-space the operator for p is just p itself, as it is for x in x-space. There are no derivatives to compute, but the integral is not trivial. Also, you should be able to convince yourself that the expression for $\phi_n^2(p)$, (4.99), does not diverge for $w = n$ despite the denominator.

Summary

The expectation value of some observable whose operator is Op is given by

$$\langle Op \rangle = \int \psi^*(x)[Op\psi(x)] \, dx.$$

It is important to compute $[Op\psi(x)]$ first, then multiply by $\psi^*(x)$ before carrying out the integral. A notational simplification can be had by using "Dirac brackets":

$$\langle \psi^*(x) \, | f(x) | \, \psi(x) \rangle = \int \psi^*(x)\{Op[f(x)]\}\psi(x) dx.$$

Heisenberg's Uncertainty Principle holds that the uncertainties Δx and Δp inherent in simultaneous measurements of the position and momentum of a particle obey the restriction

$$\Delta x \, \Delta p \geq \hbar/2,$$

where Δx is defined as

$$\Delta x = \sqrt{\langle x^2 \rangle - \langle x \rangle^2},$$

where the angle brackets denote expectation values as defined in the Dirac notation above; Δp is defined analogously. This relationship is most useful for establishing approximate ground-state energies for a given potential; see Example 4.1.

The commutator of two operators A_{op} and B_{op} corresponding to two physical quantities A and B is defined as

$$[A_{op}, B_{op}] = A_{op} B_{op} - B_{op} A_{op}.$$

If $[A, B]$ operates on some wavefunction ψ and yields a nonzero result, then there will be an uncertainty relation linking A and B. This has the form

$$(\Delta A)(\Delta B) \geq \frac{1}{2}\sqrt{-[A, B]^2}.$$

If $[A, B] = 0$, then the eigenvalues of A and B can be measured simultaneously.

The orthogonality theorem states that for two wavefunctions $\psi_n(x)$ and $\psi_k(x)$ which satisfy Schrödinger's equation for some potential,

$$\langle \psi_k^* | \psi_n \rangle = \delta_k^n,$$

where δ_k^n is the Kronecker delta symbol, which is defined to be equal to 1 if $k = n$, but is zero otherwise.

The superposition theorem states that if one forms a wavefunction from a linear sum of eigenstates,

$$\Psi(x) = \sum_i a_i \psi_i(x),$$

then the expectation value of the energy of such a superposition system is given by

$$\langle E \rangle = \sum_i a_i^2 E_i.$$

The virial theorem states that the expectation values of kinetic energy and the potential function V are related according to

$$2 \langle KE \rangle = \langle r \cdot \nabla V \rangle,$$

where r is the position vector.

Problems

4.1(E) For the data of Fig. 4.1, compute $\langle \text{height} \rangle$, $\langle \text{height} \rangle^2$, $\langle \text{height}^2 \rangle$, and $\Delta(\text{height})$. The bins are 0.05 m wide; take the effective height of each to be its midpoint, that is, there are three people of height 1.525 m, seven of height 1.575 m, and so on.

4.2(I) Prove that for the infinite potential well wavefunctions, $\langle xp \rangle = -\langle px \rangle = \iota\hbar/2$.

4.3(I) A particle of mass m has the hypothetical wavefunction $\psi(x) = Ae^{-\alpha x}$, $(0 \leq x \leq \infty)$; $\psi = 0$ otherwise). Determine the normalization constant A in terms of α. Determine $\langle x \rangle$ and $\langle KE \rangle$ in terms of α and m.

4.4(E) The uncertainty principle tells us that $\Delta x \Delta p \geq \hbar/2$. Suppose that in some experiment, the position of a proton at some initial time is known to an accuracy of $\Delta x = 10^{-15}$ meters. Using the classical definition of momentum, determine the uncertainty in the speed of the proton. Given that distance $= vt$, how much time must elapse before the uncertainty in the position of the proton has grown to 1 m?

4.5(E) A brief radio-wave pulse of photons is of duration $\tau = 0.001$ s. The pulse must then have a length of $\tau c = 3 \times 10^5$ meters, and, since an individual photon might be anywhere in the pulse, the uncertainty in the position of that photon will be $\Delta x = 3 \times 10^5$ meters. What is the corresponding uncertainty in the momentum of the photon? If the momentum and frequency of a photon are related by $p = h\nu/c$, what is the uncertainty in the frequency of the photon?

4.6(I) The ground state wavefunction of the harmonic oscillator potential (see Chap. 5) is given by

$$\psi_o(x) = \frac{\sqrt{\alpha}}{\pi^{1/4}} e^{-\alpha^2 x^2/2},$$

where α is a constant. Verify that $\Delta x \Delta p = \hbar/2$ for this state.

4.7(I) Suppose that the hypothetical wavefunction $\psi(x) = A/(a^2 + x^2)$ $(-\infty \leq x \leq \infty)$ is a solution of Schrödinger's equation for some potential $V(x)$ with energy E. (a) What must be the value of the normalization constant A in terms of a? (b) What is the potential $V(x)$? (c) Evaluate $\Delta x \Delta p$ for this function; is the uncertainty principle satisfied?

4.8(I) A hypothetical wavefunction is given by, $\psi(x) = Ax^2 e^{-\alpha x}$, $(0 \leq x \leq \infty)$. Normalize this wavefunction and compute $\Delta x \Delta p$ for it.

4.9(A) Generalize your calculations in the previous problem to a wavefunction of the form $\psi(x) = Ax^{n/2}\exp(-\beta x^q/2)$, $(0 \leq x \leq \infty)$; $n, q > 0$. Plot $\Delta x \Delta p$ as a function of n and q. Can you discover any trends?

4.10(I) Use the uncertainty principle to estimate the ground state energy of a particle of mass m moving in the linear potential

$$V(x) = \begin{cases} \infty & (x \leq 0), \\ \alpha x & (x \geq 0). \end{cases}$$

HINT: Imagine some energy level E that cuts the potential at $x = 0$ and $x = E/\alpha$. Take $x \sim \Delta x$ and $p \sim \Delta p \sim \hbar/2(\Delta x)$.

4.11(E) Consider a non-relativistic "free" particle (that is, one not subject to a potential) of mass m moving with speed v. The energy of this particle will be $mv^2/2 = p^2/2m$, where $p = mv$. Show that the uncertainty principle $\Delta x \Delta p \geq \hbar/2$ becomes $\Delta E \Delta t \geq \hbar/2$ in this case. According to Einstein, this mass has an energy equivalent given by $E = mc^2$; show also that the uncertainty principle can then be written in the form $\Delta m \Delta t \geq \hbar/2c^2$. A free neutron (one not bound within a nucleus) has a half-life against beta-decay of 10.25 min. Taking this time as an estimate of Δt, what is the corresponding uncertainty in the mass of the neutron?

4.12(I) Show that (a) $[x^n, p_x] \equiv \iota\hbar n x^{n-1}$ and (b) $[x, p_x^n] \equiv \iota\hbar n p_x^{n-1}$. In the latter case, it is helpful to know that the n-th derivative of a product of two functions $u(x)$ and $v(x)$ is given by

$$\frac{d^n}{dx^n}(uv) = \sum_{j=0}^{n} \binom{n}{j} \frac{d^j v}{dx^j} \frac{d^{n-j} u}{dx^{n-j}}.$$

4.13(I) Show that $\langle [x, H] \rangle = \frac{\iota\hbar}{m} \langle p \rangle$. H is the usual Hamiltonian operator and p is momentum.

4.14(E) Show that for any function $k(x)$, $[k(x), p_x] \equiv \iota\hbar \frac{dk}{dx}$.

4.15(E) Show that for four operators A, B, C, and D,

$$[(A + D), (B + C)] = [A, B] + [A, C] + [D, B] + [D, C].$$

4.16(E) Verify that the infinite potential well wavefunctions form an orthonormal set.

4.17(E) In Sect. 4.3 it was shown that if a particle of mass m is trapped within a linear distance $\Delta x \sim L$, then its energy must satisfy $E \geq \hbar^2/8mL^2$. Early in the history of nuclear physics (before the discovery of the neutron), it was speculated that the neutral mass that was known must reside within nuclei might be neutral particles composed of combinations of protons and electrons. However, the discovery of the uncertainty principle refuted this theory. What would be the minimum energy of an electron trapped in a nuclear potential of $L \sim 10^{-15}$ meters? Compare to the typical depth of nuclear wells for heavy elements computed in Sect. 3.9.

4.18(E) Current versions of string theory hold that the most fundamental structures in the Universe may be "strings" of length $L \sim 10^{-35}$ meters vibrating in various energy states. The mass of these strings is taken to be the Planck mass, $m_P = \sqrt{\hbar c/G}$, where G is the Newtonian gravitational constant. Using the argument presented at the end of Sect. 4.3, estimate the lowest possible energy for a vibrating string. Express your answer in giga-electron-volts (GeV = 10^9 eV). Are such energies obtainable with current particle accelerators?

4.19(I) Consider a superposition wavefunction formed from the lowest three states of an infinite potential well:

$$\Psi = \frac{1}{\sqrt{3}}\psi_1 + \frac{1}{\sqrt{4}}\psi_2 + \sqrt{\frac{5}{12}}\psi_3.$$

Verify that ψ is normalized. What is $\langle E \rangle$ for such a system? Put your answer in terms of the energy E_1 of the lowest infinite-well state.

4.20(E) Squares of wavefunctions must be normalized to unity. Express this in Dirac notation.

4.21(I) In Chap. 5 we will explore the simple harmonic oscillator potential. The wavefunctions for the two lowest energy states of this potential are of the forms $\psi_0 = A_0 e^{-\gamma^2 x^2/2}$ and $\psi_1 = A_1 x e^{-\gamma^2 x^2/2}$, $(-\infty \leq x \leq \infty)$, where A_0 and A_1 are normalization constants given by $A_0 = \sqrt{\gamma/\sqrt{\pi}}$ and $A_1 = \sqrt{2\gamma^3/\sqrt{\pi}}$, and where γ is a constant that depends upon the mass involved and the characteristics of the potential. (As explained in Chap. 5, there are good reasons why the quantum numbering for the states of this potential conventionally start at $n = 0$). Construct a time-dependent superposition state from these two functions following the approach of Sect. 4.8. Show that the amplitude of the oscillations in $\langle x \rangle$ for your superposition state is given by $\langle x \rangle = \alpha\beta\sqrt{2}/\gamma$, where α and β are as in (4.60). Do the units of this result make sense?

4.22(E) Consider an electron in the two-state wavepacket developed in Sect. 4.8 to be trapped in an infinite rectangular well with $L = 1$ Å. What would be the expectation value of its energy, $\langle E \rangle$? Express your result as a multiple of the ground-state energy E_1 of an electron in such a well.

4.23(I) Suppose it is claimed that the function $\psi(x) = Axe^{-\beta x/2}$ is a solution of Schrödinger's equation for a particle of mass m moving in the linear potential $V(x) = kx$ with $\beta, k > 0, 0 \leq x \leq \infty$. What is β in terms of k if the virial theorem is to be satisfied? Is this function actually a solution of Schrödinger's equation for this potential?

References

1. S. Gasiorowicz, *Quantum Physics* (John Wiley and Sons, New York, 1974) App. B
2. A. Pais A, *Inward Bound. Of Matter and Forces in the Physical World* (Oxford University Press, New York, 1986), p. 262
3. P. Ehrenfest, Zeitschrift fur Physik **45**, 455 (1927)
4. A.A. Pais, *Inward Bound. Of Matter and Forces in the Physical World* (Oxford University Press, New York, 1986), p. 258
5. P.F. Schewe, B.P. Stein, Phys. Today **50**(1), 9 (1997)
6. J.H. Shim, K. Haule, G. Kotliar, Nature **446**(7135), 513 (2007)
7. R. Shankar, *Principles of Quantum Mechanics*, 2nd edn. (Kluwer Academic/Plenum Publishers, New York, 1994). (Ch. 5)
8. B. H. Bransden, C. J. Joachain, *Introduction to Quantum Mechanics* (Longman Scientific & Technical, Essex, England, 1989) (Sects. 2.3 and 2.4)
9. R.L. Liboff, *Introductory Quantum Mechanics*, 2nd edn. (Addison-Wesley, Reading, Massachusetts, 1992) (Sect. 6.1)
10. A.P. French, E.F. Taylor, *An Introduction to Quantum Physics* (W. W. Norton, New York, 1978) (Sect. 8.9)
11. D.J. Griffiths, *Introduction to Quantum Mechanics* (Prentice Hall, Upper Saddle River, NJ, 1995), Sect. 3.3
12. Y.Q. Liang, H. Zhang, Y.X, Dardenne, J. Chem Educ. 72(2), 148 (1995)

Chapter 5
The Harmonic Oscillator

Summary This chapter is devoted to solving Schrödinger's equation for the important *harmonic oscillator potential*, $V(x) = kx^2/2$. In classical physics this potential
describes the motion of a mass attached to a spring of force constant k; at the atomic
level, it serves as a model for the attractive force between the atoms in a diatomic
molecule. Ultimately, we will see how the behavior of energy levels for such systems
relates to the values of the effective spring constants of molecular bonds. Further, this
system serves as a platform for introducing some very powerful and general techniques for solving differential equations, namely the idea of expressing a general
solution as a product of an *asymptotic solution* and a *series solution*; it turns out that
solving these separately is easier than trying to secure an overall solution in one step.
Because the classical solution for the harmonic potential is well-known, this system
is also well-suited for comparing classical and quantum predictions with the idea of
getting a sense of seeing how they merge into each other at macroscopic energies.
We also extend some of the operator concepts developed in Chap. 4.

In this chapter we take up the solution of Schrödinger's equation for the *harmonic
oscillator* potential. This potential corresponds to any situation where a mass is
subject to a *linear restoring force*:

$$F = -kx \quad (-\infty \leq x \leq \infty).$$

The classical model for such a system is a mass m attached to a spring of force
constant k; if the spring is stretched (or compressed) by a displacement x from its
equilibrium position, it will pull (or push) back with a force proportional to x, and the
result is sinusoidal simple harmonic motion that repeats with period $T = 2\pi\sqrt{m/k}$
seconds. At the atomic level, diatomic molecules behave to a good approximation as
two nuclei joined by a spring.

The potential corresponding to the linear restoring force is obtained by integrating
the force equation:

© The Author(s), under exclusive license to Springer Nature Switzerland AG 2022
B. C. Reed, *Quantum Mechanics*, https://doi.org/10.1007/978-3-031-14020-4_5

$$V(x) = -\int F\,dx = \int kx\,dx = kx^2/2 + c,$$

where c is a constant of integration. This is the harmonic oscillator potential. Normally, the constant c is dropped from further consideration as only differences in potential energy are physically meaningful; the effect of a non-zero value for c would be to raise or lower the permitted energy levels by that amount. Schrödinger's equation for the harmonic potential then appears as

$$-\frac{\hbar^2}{2m}\frac{d^2\psi(x)}{dx^2} + \frac{1}{2}kx^2\,\psi(x) = E\,\psi(x).$$

The solution of Schrödinger's equation for this potential is detailed in Sects. 5.1–5.4. Since the simple harmonic oscillator is such a well-known system from Newtonian mechanics, it is instructive to compare results for the classical and quantum oscillators, an issue taken up in Sect. 5.5. An alternate solution to the harmonic oscillator potential is developed in Sect. 5.6, which can be considered optional.

5.1 A Lesson in Dimensional Analysis

The harmonic potential is sketched in Fig. 5.1. Classically, a system of total energy E moving in this potential cannot find itself beyond the turning points $\pm x_o$ given by $E = kx_o^2/2$. Quantum-mechanically, we can anticipate some finite probability of tunneling beyond these limits because ψ will not in general be zero for $|x| > x_o$.

Because $V(x)$ here is a function of x, the situation is more complex than most of the potential well/barrier problems previously encountered. However, because the expression for $V(x)$ is valid for all x, this complexity is mitigated somewhat in that there are no "inside" or "outside" regions to link via the continuity of ψ and $d\psi/dx$, that is, there will be but a single expression for $\psi(x)$ valid for all x. At the same time, with $V(x)$ extending to $\pm\infty$, we must ultimately force ψ to vanish as $x \to \pm\infty$. When $E > V$ (that is, within the well: $|x| \leq x_o$), we can expect some type of potential-well-sinusoidal behavior for ψ, while for $|x| > x_o$ we can expect to find an exponential-decay form for ψ. Also, since the harmonic potential is symmetric about $x = 0$, we can expect solutions of alternating parity.

Before plunging into the solution of Schrödinger's equation for the harmonic potential, we cast it into a dimensionless form. This is a valuable lesson in its own right, and will make subsequent algebra easier.

Almost any incarnation of Schrödinger's equation can be made dimensionless by finding two combinations of the mass of the particle involved (m), $\hbar$, and any constants that appear in the potential such that one combination has dimensions of reciprocal length and the other dimensions of reciprocal energy. Call these constants α and ϵ, respectively. Dimensionless variables $\xi = \alpha x$ and $\lambda = \epsilon E$ are then defined, and x and E in Schrödinger's equation are cast in terms of ξ and λ. The result is a

Fig. 5.1 Harmonic potential $V(x) = kx^2/2$ for $k = 0.7$. For a particle moving in this potential with the hypothetical energy shown (dashed line), the classical limits of the motion are restricted to lying within $\pm x_0$

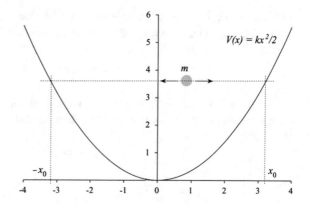

dimensionless differential equation that is free of constants and consequently easier to manipulate than the original version. In particular, dimensionless formulations are particularly amenable to numerical computation because one is freed from concerns with units and quantities of exceptionally large or small magnitude. Once solutions to the dimensionless differential equation have been established, they can be put back in terms of the original variables by simple substitutions.

The harmonic oscillator potential involves only one constant, k. We begin by finding a combination of m, $\hbar$, and k that has units of reciprocal meters. This means that we seek values of the powers a, b, and c such that $m^a k^b \hbar^c$ has units of $(m)^{-1}$. Recalling that mass has units of kg, k has units of N/m ($= \text{kg/s}^2$), and $\hbar$ units of J-s ($= \text{kg m}^2/\text{s}$), we have

$$m^{-1} \equiv (kg)^a \left(\frac{kg}{s^2}\right)^b \left(\frac{kg\, m^2}{s}\right)^c \equiv kg^{a+b+c}\ m^{2c}\ s^{-(2b+c)}.$$

To balance units, we must have $c = -1/2$. Further, because $(2b + c)$ must equal zero, then $b = -c/2 = +1/4$. Finally, because $(a + b + c)$ must also equal zero, $a = +1/4$. Hence we have

$$\alpha = m^a\, k^b\, \hbar^c = m^{1/4}\, k^{1/4}\, \hbar^{-1/2} = \left(\frac{m\,k}{\hbar^2}\right)^{1/4}. \tag{5.1}$$

Second, we desire a combination $m^a k^b \hbar^c$ whose units are inverse Joules, that is

$$\frac{s^2}{kg\, m^2} \equiv (kg)^a \left(\frac{kg}{s^2}\right)^b \left(\frac{kg\, m^2}{s}\right)^c \equiv kg^{a+b+c}\ m^{2c}\ s^{-(2b+c)}.$$

Here we must have $c = -1$, $b = -1/2$, and $a = +1/2$, so

$$\varepsilon = m^a\, k^b\, \hbar^c = m^{1/2}\, k^{-1/2}\, \hbar^{-1} = \frac{1}{\hbar}\sqrt{\frac{m}{k}} = \frac{1}{\hbar\,\omega},$$

where $\omega = \sqrt{k/m}$ is the angular frequency of a classical mass/spring system. For further algebraic convenience, ϵ is defined with an additional (dimensionless) factor of 2:

$$\varepsilon = \frac{2}{\hbar\omega}. \tag{5.2}$$

The reader should verify that α and ϵ do have units of reciprocal length and energy, respectively. The next step is to define dimensionless variables for position and energy

$$\xi = \alpha x \tag{5.3}$$

and

$$\lambda = \varepsilon E. \tag{5.4}$$

On substituting (5.3) and (5.4) into Schrödinger's equation for the harmonic oscillator, remember to make the substitution in the derivative as well, that is,

$$\frac{d^2}{dx^2} \equiv \alpha^2 \frac{d^2}{d\xi^2}.$$

These substitutions reduce Schrödinger's equation to

$$\frac{d^2\psi}{d\xi^2} + (\lambda - \xi^2)\psi = 0, \tag{5.5}$$

a considerably cleaner expression to work with than its ancestor. Had ϵ not been defined with an additional factor of two in (5.2), such a factor would remain in the λ term in (5.5). As λ involves the energy of the system, we can expect to see some boundary condition arise that restricts it to certain values.

Example 5.1 In some cosmological theories, string theories, and "grand-unified" theories, the Planck time plays the role of a "quantum of time", that is, a smallest possible time interval. This quantity is a combination of the constants G, c, and h that has units of time. Use dimensional analysis to derive an expression for the Planck time. What is it's value?

We need to find powers (a, b, d) such that

$$s^1 \equiv c^a \, h^b \, G^d.$$

Note that we use d as the power of G as opposed to c to avoid confusion with the speed of light. Substituting the appropriate units gives

$$s^1 \equiv \left(\frac{m}{s}\right)^a \left(\frac{kg\,m^2}{s}\right)^b \left(\frac{m^3}{kg\,s^2}\right)^d \equiv m^{a+2b+3d}\,kg^{b-d}\,s^{-a-b-2d}.$$

Matching powers shows that we have three constraints:

$$a + 2b + 3d = 0,$$

$$b - d = 0,$$

and

$$-a - b - 2d = 1.$$

The middle one of these shows that $d = b$. Substituting this into the first constraint gives $a = -5b$. The last constraint then gives $b = 1/2$. Hence we have $b = d = 1/2$ and $a = -5/2$. The Planck time is often abbreviated t_P:

$$t_P = c^{-5/2}\,h^{1/2}\,G^{1/2} = \sqrt{\frac{hG}{c^5}}.$$

Putting in the numbers gives

$$t_P = \sqrt{\frac{(6.626 \times 10^{-34})(6.674 \times 10^{-11})}{(2.998 \times 10^8)^5}} = 1.351 \times 10^{-43}\ s,$$

a tiny value indeed. Try comparing the Planck time to the orbital period of the electron in the Bohr atom.

5.2 The Asymptotic Solution

Our task is to solve (5.5) for $\psi(\xi)$, then revert the result back to x-space via the substitution $\xi = \alpha x$. Because of the form of the potential, the usual linear-equation-with-constant-coefficient recipes cannot be applied. One might then surmise that a brute-force series solution is called for, and indeed a series solution will figure prominently in what follows. There is, however, a more elegant approach available which exploits knowledge of the physics of the situation.

Recall the finite rectangular well of Chap. 3. That situation was also characterized by turning points (the walls of the well) beyond which the wavefunction behaved in an asymptotically exponential fashion. Given the similarity of that situation to the present problem, we might infer that $\psi(\xi)$ approaches zero as ξ (or, equivalently, x) $\rightarrow \infty$.

This inference proves to be correct, and leads to a technique of considerable power and generality: that of first establishing the *asymptotic form* of ψ. The problem then reduces to finding a solution valid for "small" values of ξ to be grafted onto the asymptotic solution in order to establish a solution valid for all ξ. The goal of the present section is to establish the asymptotic solution for the harmonic oscillator potential. The detailed general solution is then developed in Sect. 5.3, and some characteristics of the resulting wavefunctions are examined in Sect. 5.4.

To establish the asymptotic behavior of ψ, consider (5.5) when ξ is large. In this case the ξ^2 term will dominate, leaving

$$\frac{d^2\psi}{d\xi^2} \sim \xi^2\psi \qquad (\xi \to \pm\infty). \tag{5.6}$$

Guided by the finite-well solution, we posit a trial solution for this behavior of the form

$$\psi(\xi) \sim A\exp(\beta\xi^n), \qquad (\xi \to \pm\infty), \tag{5.7}$$

where A, β, and n are to be determined. Differentiating this trial function gives

$$\frac{d^2\psi(\xi)}{d\xi^2} = [n^2\beta^2\xi^{2n-2} + n(n-1)\beta\xi^{n-2}]\psi(\xi), \tag{5.8}$$

which, upon back-substitution into (5.6) gives

$$n^2\beta^2\xi^{2n-2} + n(n-1)\beta\xi^{n-2} \sim \xi^2. \tag{5.9}$$

Equation (5.9) is meaningless unless the various powers of ξ can be brought into accord. From the form of the trial solution, we can infer that n must be positive (why?). For large ξ, the first term will dominate the left side, leaving

$$n^2\beta^2\xi^{2n-2} \sim \xi^2. \tag{5.10}$$

Equality will hold for the powers of ξ only if $2n - 2 = 2$, or $n = 2$. For the multiplicative factors of $(n^2\beta^2)$ on the left and unity on the right to agree, we must then have $4\beta^2 = 1$, or $\beta = \pm 1/2$. Choosing $\beta = 1/2$ would result in an asymptotic solution of the form $\psi(\xi) \propto \exp(+\xi^2/2)$, which would be impossible to normalize. On the other hand, choosing $\beta = -1/2$ leads to a normalizable solution of the form

$$\psi(\xi) \sim A\exp(-\xi^2/2) \qquad (\xi \to \pm\infty). \tag{5.11}$$

This result tells us that the postulate of an exponential form for the asymptotic solution is valid if we pick the particular form $\psi(\xi) \propto \exp(-\xi^2/2)$: $n = 2$ is the only possibility if our postulated solution, (5.7), is to satisfy (5.6) as $\xi \to \infty$. At this point there is nothing to be gained in attempting to establish the normalization constant A as we have only a partial solution as yet; we can take care of normalization later. The important thing is that we have established the general form of ψ for large ξ.

5.3 The Series Solution

What of the general solution of Schrödinger's equation for the harmonic potential? In the absence of any recipe solution, we are reduced to a process of trial and error: postulate a trial solution, which is usually taken to be a polynomial series in the independent variable, and tinker with it until the differential equation is satisfied. This may seem a circuitous way of solving a problem, but it is legitimate: A solution is, after all, a solution, and can always be checked by back-substitution into the differential equation.

By whatever means we solve Schrödinger's equation for this problem, we wish to explicitly build in knowledge of the exponential asymptotic behavior of ψ established in the preceding section. One way to do this is by assuming that ψ can be expressed as the product of two functions, one of which embodies the asymptotic behavior and another, which is an as-yet unknown function $H(\xi)$,

$$\psi(\xi) = H(\xi)e^{-\xi^2/2}. \tag{5.12}$$

How do we know if this is a valid assumption? The answer is that we don't, at least not yet. As in any trial-and-error circumstance, this can be known only after the fact by seeing if it can be made to work.

What of $H(\xi)$? From elementary calculus we know that any function of x (or ξ) can be represented as an infinite polynomial, that is, as a Taylor or MacLaurin series expansion. We will assume that $H(\xi)$ can be expressed as such a series:

$$H(\xi) = a_o + a_1\xi + a_2\xi^2 + \cdots = \sum_{n=0}^{\infty} a_n \xi^n, \tag{5.13}$$

where the a_n are (numerical) expansion coefficients to be manipulated until the series satisfies the differential equation at hand. Again, this is an assumption, the validity of which can be established only if we can force it to satisfy the differential equation.

Note that the index of summation n begins at zero; we cannot admit negative powers of ξ, because $\xi = 0$ is a legitimate point in the domain of the potential, and negative powers would lead to infinities. (Actually, infinities do not automatically disqualify a ψ as being legitimate: so long as ψ is normalizable, it is fine. But the form of ψ being dealt with here does not admit this possibility.)

Now, our task is to solve Schrödinger's equation in the form,

$$\frac{d^2\psi}{d\xi^2} + (\lambda - \xi^2)\psi = 0. \tag{5.14}$$

The next step is to transform this into a differential equation for $H(\xi)$ by direct substitution of (5.12). Differentiating (5.12) twice gives

$$\frac{d^2\psi}{d\xi^2} = \left(\frac{d^2 H}{d\xi^2} - 2\xi \frac{dH}{d\xi} + (\xi^2 - 1)H \right) e^{-\xi^2/2}, \tag{5.15}$$

where we have dropped the explicit dependence of H on ξ for sake of clarity. Substituting this result and (5.12) into (5.14) and canceling the common exponential factors gives the desired differential equation for H:

$$\frac{d^2 H}{d\xi^2} - 2\xi \frac{dH}{d\xi} + (\lambda - 1)H = 0. \tag{5.16}$$

We now attempt a series solution of the form

$$H(\xi) = \sum_{n=0}^{\infty} a_n \xi^n. \tag{5.17}$$

Differentiating this gives

$$\frac{dH}{d\xi} = \sum_{n=0}^{\infty} n a_n \xi^{n-1} \tag{5.18}$$

and

$$\frac{d^2 H}{d\xi^2} = \sum_{n=0}^{\infty} n(n-1) a_n \xi^{n-2}. \tag{5.19}$$

Substituting (5.17), (5.18), and (5.19) into (5.16) and gathering terms with the same power of ξ gives

$$\sum_{n=0}^{\infty} n(n-1) a_n \xi^{n-2} + \sum_{n=0}^{\infty} [\lambda - 1 - 2n] a_n \xi^n = 0. \tag{5.20}$$

As it stands, (5.20) is awkward: expanding out the sums would give contributions of the form ξ^j from both terms for $j > 2$. It would be convenient to have all terms involving the same power in ξ collected together. We can do this with a trick of notation. Consider the first sum in (5.20). When $n = 0$ or 1, this sum makes no contribution due to the factor of $n(n-1)$:

$$\sum_{n=0}^{\infty} n(n-1) a_n \xi^{n-2} = 0 + 0 + \sum_{n=2}^{\infty} n(n-1) a_n \xi^{n-2}. \tag{5.21}$$

Now, the index of summation n is known as a *dummy* index. This means that because the sum runs over all possible integral values (> 0) up to infinity, it is irrelevant what label we assign the index of summation. This opens up the possibility of performing a transformation of index in exactly the same way as one does a transformation of variable to simplify an integral. To this end, define a new index j such that $j = n - 2$. We can then write (5.21) as

$$\sum_{n=0}^{\infty} n(n-1)\,a_n\xi^{n-2} = \sum_{n=2}^{\infty} n(n-1)a_n\xi^{n-2} = \sum_{j=0}^{\infty}(j+2)(j+1)\,a_{j+2}\xi^j. \quad (5.22)$$

Note that in effecting this change, the substitution is made everywhere the original index appears. Since the lowest value of n that makes a non-zero contribution to the sum is $n = 2$, then the lowest value of j that makes a non-zero contribution is $j = 0$ since $j = n - 2$.

Now, in (5.22), both j and n are dummy indexes. We are at liberty to re-label j as n (in effect, define a new index $n = j$), and write

$$\sum_{n=0}^{\infty} n(n-1)a_n\xi^{n-2} = \sum_{n=0}^{\infty}(n+2)\,(n+1)a_{n+2}\xi^n. \quad (5.23)$$

If you have doubts as to the veracity of (5.23), simply expand out the sums on both sides: they are absolutely identical. The value of this manipulation is that it transforms the ξ^{n-2} term in (5.20) into one involving ξ^n, which is exactly the form in which ξ appears in the other term. We can then combine the terms in (5.20) to give

$$\sum_{n=0}^{\infty}[(n+1)(n+2)a_{n+2} + (\lambda - 1 - 2n)a_n]\xi^n = 0. \quad (5.24)$$

This transformation-of-index procedure is a useful artifice often employed in series solutions of differential equations.

Examine (5.24) carefully. The terms within the square brackets are pure numbers; they have no dependence on ξ. In general, ξ is not equal to zero (indeed, $-\infty \leq \xi \leq \infty$), yet the sum evaluates to zero. The only way this can happen for all possible values of ξ is if the quantity in square brackets vanishes in general. This demands

$$a_{n+2} = \frac{(2n+1-\lambda)}{(n+1)(n+2)}a_n. \quad (5.25)$$

Equation (5.25) is a *recursion relation*: it specifies a given expansion coefficient in terms of a preceding coefficient in the series. To complete the solution, it is necessary to supply values for the first two coefficients, a_0 and a_1. All subsequent even-indexed coefficients can then be expressed in terms of a_0, and all subsequent odd-indexed coefficients in terms of a_1. The need to supply two coefficients can be understood on the basis that second-order differential equations always yield two constants of integration to be set by boundary conditions.

In view of the fact that our recursion relation links pairs of successive coefficients, we can imagine splitting the solution for $H(\xi)$ into a sum of two sub-series:

$$H(\xi) \sim \left(a_0 + a_2\xi^2 + a_4\xi^4 + \cdots\right) + \left(a_1\xi + a_3\xi^3 + a_5\xi^5 + \cdots\right). \qquad (5.26)$$

with the coefficients in both sub-series related by (5.25).

In the present case there is only one boundary condition to satisfy: that $\psi(\xi)$ remain finite as $\xi \to \pm\infty$ (that is, as $x \to \pm\infty$). Now, from (5.12), $\psi(\xi)$ is the product of two functions, $H(\xi)$ and $\exp(-\xi^2/2)$. The latter of these is clearly convergent; indeed, we deliberately chose this form based on the argument in Sect. 5.2. Thus, as long as $H(\xi)$ is convergent, or at worst if it diverges no more strongly than $\exp(+\xi^2/2)$, then the overall solution will remain convergent. The issue is then the behavior of $H(\xi)$. Unfortunately, $H(\xi)$ proves to behave asymptotically as $\exp(+\xi^2)$. To see this, it is simplest to work with one of the subseries in (5.26). We work with the even-indexed one, but the same conclusion emerges for the odd-indexed one as well. Write the subseries as

$$a_0 + a_2\xi^2 + a_4\xi^4 + \cdots = \sum_{j=0,1,2,\ldots}^{\infty} a_{2j}\,\xi^{2j}.$$

The ratio of two successive terms in this sub-series is

$$\left[\frac{\text{Term }(j+1)}{\text{Term }(j)}\right]_{sub-series} = \frac{a_{2j+2}}{a_{2j}}\xi^2. \qquad (5.27)$$

To express the ratio of the a coefficients here we can use (5.25), but we must be careful to transform from index n to index j; this is because (5.25) concerns every other term of the original series for $H(\xi)$, whereas we are now dealing with successive terms of one of the subseries that comprise $H(\xi)$. By comparing Equations (5.25) and (5.27), it should be clear that the transformation is $n = 2j$. Hence (5.27) can be written as

$$\left[\frac{\text{Term }(j+1)}{\text{Term }(j)}\right]_{sub-series} = \frac{(4j+1-\lambda)}{(2j+1)(2j+2)}\xi^2. \qquad (5.28)$$

Now consider a series expansion of the exponential function $\exp(\eta\xi^p)$:

$$\exp\left(\eta\xi^p\right) = 1 + \eta\xi^p + \frac{\eta^2\xi^{2p}}{2!} + \frac{\eta^3\xi^{3p}}{3!} + \cdots$$

$$\qquad (5.29)$$

$$+ \frac{\eta^j\,\xi^{jp}}{j!} + \frac{\eta^{j+1}\xi^{(j+1)p}}{(j+1)!} + \frac{\eta^{j+2}\,\xi^{(j+2)p}}{(j+2)!} + \cdots.$$

In this case the ratio of successive terms behaves as

$$\left[\frac{\text{Term }(j+1)}{\text{Term }(j)}\right]_{exp} = \frac{\eta\,\xi^p}{(j+1)}.$$

(5.30)

Now compare (5.28) and (5.30) as $j \to \infty$:

$$\left[\frac{\text{Term }(j+1)}{\text{Term }(j)}\right]_{\substack{sub-series \\ j\to\infty}} \sim \frac{\xi^2}{j}$$

(5.31)

and

$$\left[\frac{\text{Term }(j+1)}{\text{Term }(j)}\right]_{\substack{exp \\ j\to\infty}} \sim \frac{\eta\,\xi^p}{j}.$$

(5.32)

The denominators of (5.31) and (5.32) are identical; this means that for high-order terms, either subseries behaves as an exponential function. Specifically, comparing (5.31) and (5.32) shows that $p = 2$ and $\eta = 1$, that is, that the high-order terms of $H(\xi)$ behave as $H(\xi) \sim \exp(\xi^2)$. This has the consequence that the overall solution $\psi(\xi)$ behaves as

$$\psi(\xi) = H(\xi)e^{-\xi^2/2} \sim e^{\xi^2}e^{-\xi^2/2} \sim e^{\xi^2/2}.$$

(5.33)

This result is consistent with our conclusion in Sect. 5.2 that $\psi(\xi)$ must behave asymptotically as $\exp(\pm\xi^2/2)$. However, (5.33) is the divergent, non-normalizable one of these two possibilities. The only way to prevent this catastrophe is to somehow prevent the series for $H(\xi)$ from extending out to an infinite number of terms in order to allow the convergent factor of $\exp(-\xi^2/2)$ in (5.12) to dominate as $\xi \to \infty$. This means that the series for $H(\xi)$ must be forced to terminate after a finite number of terms; a polynomial function with a finite number of terms multiplied by a convergent exponential function will always give a function whose overall behavior is convergent. (Why?)

The demand that $H(\xi)$ so terminate has the physical consequence of quantizing the energy levels of the system. If the highest power of ξ appearing in $H(\xi)$ is n, that is, if all $a_j = 0$ for $j > n$, then the recursion relation demands

$$\lambda = 2n + 1.$$

(5.34)

One problem remains: since the recursion relation (5.25) links two coefficients that straddle another coefficient, we cannot simultaneously terminate both the even $(a_0, a_2, \ldots)$ and odd $(a_1, a_3, \ldots)$ sets of coefficients with this expedient. We must in addition arbitrarily demand that either a_0 or a_1 be zero.

The general result obtained here is a family of functions $H_n(\xi)$, with each member being a polynomial involving only even or odd powers of ξ (but not both) up to order ξ^n. The index n serves to label the different solutions.

From the termination condition given above and the definition of λ [see (5.2) and (5.4)], we can determine the energy corresponding to quantum number n:

$$\lambda = 2n + 1 = \frac{2E}{\hbar\omega},$$

or,

$$E = \left(n + \frac{1}{2}\right)\hbar\omega = \left(n + \frac{1}{2}\right)\hbar\sqrt{\frac{k}{m}}, \qquad n = 0, 1, 2, 3, \ldots \qquad (5.35)$$

Notice that E is proportional to n, that is, harmonic-potential energy levels are equally spaced. The *zero-point energy* of the harmonic oscillator is $\hbar\omega/2$; from Example 4.4 we can regard this as an aspect of Heisenberg's uncertainty principle.

These equally-spaced energy levels are known as vibrational levels in recognition of the fact that the harmonic potential mimics the attractive forces within molecules. For real molecules, the force constants involved are on the order of a few hundred Newtons per meter, as shown in what follows.

For a diatomic molecule comprising atoms of masses m_1 and m_2, the analysis of the harmonic potential goes through as has been done here but for one change: the mass m in the angular frequency $\sqrt{k/m}$ must be replaced by the reduced mass $\mu = m_1 m_2/(m_1 + m_2)$. Spectroscopists tabulate information on vibrational energy levels via species-specific constants designated with the symbol ω_e, defined such that the energy-level expression, (5.35), takes the form

$$E = \left(n + \frac{1}{2}\right)h c \omega_e, \qquad (5.36)$$

where c is the speed of light. ω_e is usually quoted in units of reciprocal centimeters; this designates the number of wavelengths which fit in one centimeter. Since $1\,\mathrm{cm} = 10^{-2}\,\mathrm{m}$, $1\,\mathrm{cm}^{-1} = 100\,\mathrm{m}^{-1}$, that is, any value of ω_e expressed in reciprocal centimeters can be converted to an equivalent number of reciprocal meters by multiplying by 100.

By comparing (5.35) and (5.36), we can extract the effective spring constant of the molecular bond from knowledge of the reduced mass and ω_e:

$$\hbar\sqrt{\frac{k}{\mu}} = h c \omega_e \quad \Rightarrow \quad k = 4\pi^2 c^2 \omega_e^2 \mu. \qquad (5.37)$$

The following example illustrates an application of this result.

Example 5.2 The *CRC Handbook of Chemistry and Physics* lists the ω_e value of hydrogen chloride, $^1\mathrm{H}^{35}\mathrm{Cl}$, as $2990.95\,\mathrm{cm}^{-1} = 2.99095 \times 10^5\,\mathrm{m}^{-1}$. Chlorine has two main isotopes, of masses 35 and 37 atomic mass units; the isotope here is that of mass 35. In atomic mass units, the reduced mass is (ignoring the fact that the true masses are very slightly different than their integerized values)

$$\mu = \frac{m_H\, m_{Cl}}{m_H + m_{Cl}} = \frac{1 \times 35}{1 + 35} = \frac{35}{36}.$$

With one mass unit being 1.6605×10^{-27} kg, this gives $\mu = 1.6144 \times 10^{-27}$ kg. Hence

$$\omega_e = 4\pi^2 c^2 \omega_e^2 \mu$$
$$= 4\pi^2 (2.9979 \times 10^8 \text{m s}^{-1})^2 (2.99095 \times 10^5 \text{m}^{-1})^2 (1.6144 \times 10^{-27} \text{kg})$$
$$= 512.4 \quad \text{N/m}.$$

Check that the units do reduce to N/m. A force constant of 500 N/m is surprisingly strong, corresponding to that of a spring of strength ~ 3.5 pounds per inch. This would be a robust laboratory spring.

Force constants of this magnitude can be understood with a simple estimate. If you compute the Coulomb force between two electrons separated by 1 Å, you will find a magnitude $F \sim 2.3 \times 10^8$ Newtons. The force per unit distance is then $\sim (2.3 \times 10^8 \text{N})/(10^{-10} \text{m}) \sim 230$ N/m, of the same order of magnitude determined here for k. The details will depend on the molecule at hand, but we can expect atomic-scale situations to be characterized by charges and physical sizes of these magnitudes.

The advantage of tabulating data as ω_e values is that they can be quickly converted to transition energies. For this example, 2.991×10^5 m^{-1} corresponds to a wavelength of $\lambda = 1/\omega_e = 3.343 \times 10^{-6}$ m, or 33,400 Å, which is in the far infrared. From Problem 1-8, this corresponds to an energy of $E = 12,398/34,000 \sim 0.37$ eV.

In reality, molecules are three-dimensional entities, so we cannot expect our one-dimensional analysis to reveal the entire story. Molecules also exhibit *rotational* energy levels due to the quantization of the rotational angular momentum of the molecule about its center of mass. This is discussed in more detail in Chaps. 6 and 8. The situation is further complicated by the fact that diatomic-bond potentials are more complex than $V = kx^2/2$. In more advanced treatments, the Morse potential [1]

$$V(r) = D\{1 - \exp[-a(x - x_o)]\}^2$$

is found to give a better fit to observed spectra; the result of this is that (5.36) is but the first term in an infinite series in powers of $(n + 1/2)$.

5.4   Hermite Polynomials and Harmonic Oscillator Wavefunctions

The family of solutions $H_n(\xi)$ to (5.16) are known as Hermite polynomials. Full details on their properties and normalization can be found in Chap. 13 of Weber and Arfken. The first few Hermite polynomials are listed in Table 5.1. Note that in each case, the highest power of ξ appearing in the series for $H_n(\xi)$ is ξ^n. To get the polynomials in terms of real (x) space, substitute $\xi = \alpha x$.

The normalized harmonic oscillator wavefunctions are given by

$$\psi_n(x) = A_n H_n(\xi) e^{-\xi^2/2} = A_n H_n(\alpha x) e^{-\alpha^2 x^2/2}, \tag{5.38}$$

where

$$A_n = \sqrt{\frac{\alpha}{\sqrt{\pi} \, 2^n n!}}, \tag{5.39}$$

and where

$$\alpha = \left(mk/\hbar^2\right)^{1/4}. \tag{5.40}$$

Hermite polynomials of order $n \geq 2$ can be generated from $H_0(x)$ and $H_1(x)$ from the recursion relation

$$H_{n+1}(x) = 2x H_n(x) - 2n H_{n-1}(x).$$

Figure 5.2 shows the $n = 0$ and $n = 5$ harmonic-oscillator wavefunctions (assuming $\alpha = 1$). $\psi_n(\xi)$ is symmetric for even values of n and antisymmetric for odd values of n; for a given n, there are $n + 1$ maxima. As n increases, the value(s) of ξ at which ψ reaches its maximum values move closer to the classical turning points, which are indicated in the figure by vertical lines. The probability of finding the oscillator outside the well is greatest for $n = 0$.

Table 5.1 Hermite polynomials

n	$H_n(\xi)$
0	1
1	2ξ
2	$4\xi^2 - 2$
3	$8\xi^3 - 12\xi$
4	$16\xi^4 - 48\xi^2 + 12$
5	$32\xi^5 - 160\xi^3 + 120\xi$
6	$64\xi^6 - 480\xi^4 + 720\xi^2 - 120$
7	$128\xi^7 - 1344\xi^5 + 3360\xi^3 - 1680\xi$

Fig. 5.2 Harmonic oscillator wavefunctions for $n = 0$ (solid curve) and $n = 5$ (thick dashed curve). The vertical lines designate the classical turning points for each curve; $\xi_{turn} = \pm 1$ and $\pm\sqrt{11}$ for $n = 0$ and $n = 5$, respectively. See Problem 5.9

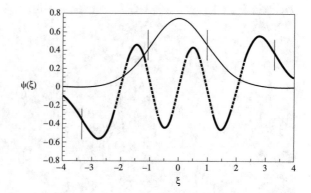

Example 5.3 Determine the probability that a harmonic oscillator in the ground state ($n = 0$) will be found beyond the classical turning points.

The normalized wavefunction for the ground-state harmonic oscillator is

$$\psi_0(x) = \frac{\sqrt{\alpha}}{\pi^{1/4}} e^{-\alpha^2 x^2/2},$$

with

$$E_o = \hbar\omega/2.$$

The classically forbidden region corresponds to $V(x) > E_0$, or

$$kx^2/2 > \hbar^2\alpha^2/2m,$$

that is,

$$|x| > \beta, \quad \text{where} \quad \beta = \sqrt{\frac{\hbar^2\alpha^2}{km}} = \sqrt{\frac{\hbar\omega}{k}}.$$

It is easier to calculate the desired probability by determining the probability that the oscillator will be found inside the well and subtracting the result from unity:

$$P(\text{outside}) = 1 - P(\text{inside}).$$

Hence

$$P(\text{inside}) = 2 \int_0^\beta \psi_0^* \psi_0 dx = \frac{2\alpha}{\sqrt{\pi}} \int_0^\beta e^{-\alpha^2 x^2} dx.$$

The factor of 2 comes from the symmetry of $\psi^*\psi$ about $x = 0$. Setting $\xi = \alpha x$ gives

$$P(\text{inside}) = \frac{2}{\sqrt{\pi}} \int_0^{\alpha\beta} e^{-\xi^2} d\xi.$$

This integral cannot be solved analytically; it has to be evaluated numerically. Integrals of this form turn up in a number of problems in mathematical physics (in particular, statistics), and are known as *error functions*:

$$erf(z) = \frac{2}{\sqrt{\pi}} \int_0^z e^{-\xi^2} d\xi.$$

Values of $erf(z)$ are tabulated in a number of mathematical handbooks. In the present case we have $erf(1)$ since

$$z = \alpha\beta = \left(\frac{mk}{\hbar^2}\right)^{1/4}\left(\frac{\hbar\omega}{k}\right)^{1/2} = \left(\frac{m\omega^2}{k}\right)^{1/4} = 1,$$

given that $\omega^2 = k/m$. Since $erf(1) = 0.843$, we find $P(\text{outside}) = 0.157$, a distinctly non-negligible result. This is another manifestation of the phenomena of quantum tunneling explored in Chap. 3.

Example 5.4 Determine the root-mean-square position $\sqrt{\langle x^2 \rangle}$ for a ground-state harmonic oscillator, and compare the result to the classical turning point.
Here we have

$$\langle x^2 \rangle = \int_{-\infty}^{\infty} \psi_0^* x^2 \psi_0 dx = \frac{\alpha}{\sqrt{\pi}} \int_{-\infty}^{\infty} x^2 e^{-\alpha^2 x^2} dx = \frac{2\alpha}{\sqrt{\pi}} \int_0^{\infty} x^2 e^{-\alpha^2 x^2} dx$$

$$= \frac{2\alpha}{\sqrt{\pi}}\left(\frac{\sqrt{\pi}}{4\alpha^3}\right) = \frac{1}{2\alpha^2} = \frac{\hbar}{2\sqrt{mk}}.$$

Because $x_{turn} = \sqrt{\hbar\omega/k}$,

$$\frac{\langle x^2 \rangle}{x_{turn}^2} = \frac{\hbar}{2\sqrt{mk}}\frac{k}{\hbar\omega} = \frac{k}{2\omega\sqrt{mk}} = \frac{k}{2\sqrt{k/m}\sqrt{mk}} = \frac{1}{2}.$$

Hence

$$\sqrt{\langle x^2 \rangle} = \frac{x_{turn}}{\sqrt{2}}.$$

The RMS displacement of the oscillator from the origin is less that the turning-point position, an intuitively sensible result.

5.5 Comparing the Classical and Quantum Harmonic Oscillators

Our analysis of the quantum harmonic oscillator has resulted in wavefunctions which, for a given state n, are products of an n-th order polynomial and an exponential function. These wavefunctions look nothing like what one sees as the time-dependent functions used to describe a mass-spring system in Newtonian physics. In the latter analysis, if a mass m is attached to a spring of force constant k, displaced a distance D from the spring's equilibrium position $(x = 0)$ and then released, the resulting motion is described by

$$x = D\cos(\omega t), \tag{5.41}$$

where the angular frequency ω is identical to that appearing in the quantum analysis, $\omega = \sqrt{k/m}$. The total energy of the classical oscillator is given by $E = kD^2/2$, a continuous function of total displacement.

Because the simple harmonic oscillator is such a well-known Newtonian system, it is instructive to compare the quantum and classical solutions for it and to try to reconcile their apparently very different descriptions. Quantum-mechanical predictions are invariably given as probabilities, so we effect our comparison via probabilistic predictions.

Quantum-mechanically, the probability of finding the oscillator between x and $x + dx$ is given by $P_{quant}(x, x + dx) = |\psi|^2 dx$; our task is to derive an analogous expression for a classical oscillator and then to compare them for oscillators of the same energy. Now, Equation (5.41) tells us where the classical oscillator will be as a function of time. We can turn this into an expression for the velocity of the oscillator at any time by taking a derivative:

$$v = \frac{dx}{dt} = -\omega D\sin(\omega t). \tag{5.42}$$

For a reason that will become clear momentarily, it is helpful to rewrite this expression so that the right side is in terms of $x(t)$. From the identity $\sin^2 x + \cos^2 x = 1$, we can write $D\sin(\omega t) = \sqrt{D^2 - D^2\cos^2(\omega t)}$, hence

$$v = -\omega D\sin(\omega t) = -\omega\sqrt{D^2 - D^2\cos^2(\omega t)} = -\omega\sqrt{D^2 - x^2}. \tag{5.43}$$

Now, imagine that you observe a classical mass-spring system by taking a large number of superimposed stop-action photographs of it at random times over many periods of the motion. As you examine the resulting photograph, you could tabulate a census of how many exposures out of the total number taken show the mass to be between positions x and $x + dx$. Since all oscillations of the system are identical, this number would be given by the amount of time that the oscillator spends between those limits over the course of each full oscillation divided by the period $T = 2\pi/\omega$ for a single oscillation. For a small spatial slice of width dx, the passage time would be $dx/v(x)$ where $v(x)$ is the velocity of the oscillator at position x; this is why we cast the velocity of (5.42) into the position form of (5.43). On recalling that a full back-and-forth motion of the oscillator will bring it through the relevant region of space twice during each cycle, we can write

$$
P_{class}(x, x + dx) = 2\frac{[dx/v(x)]}{T}
$$
$$
= \left[\frac{2}{T\omega\sqrt{D^2 - x^2}}\right] dx = \left[\frac{1}{\pi\sqrt{D^2 - x^2}}\right] dx, \tag{5.44}
$$

where we have dropped the negative sign in (5.43) because we are interested only in the magnitude of the time that the oscillator spends in the relevant region.

For a meaningful comparison, we need to have our quantum and classical oscillators be of the same energy:

$$
\left(n + \frac{1}{2}\right)\hbar\omega = \frac{1}{2}kD^2 \implies D^2 = (2n + 1)\frac{\hbar\omega}{k}. \tag{5.45}
$$

It is convenient to cast (5.44) into a form involving the dimensionless variable $\xi = \alpha x$ that was introduced in Sect. 5.1; see (5.1) and (5.3). Using (5.45) to eliminate D, the reader should verify that this transformation brings (5.44) to the form

$$
P_{class}(\xi, \xi + d\xi) = \left[\frac{1}{\pi\sqrt{2n + 1 - \xi^2}}\right] d\xi. \tag{5.46}
$$

The quantity within square brackets is the classical probability density.

For the quantum oscillator we have

$$
P_{quant}(x, x + dx) = \psi^2(x)dx = \left[A_n H_n(\alpha x)e^{-\alpha^2 x^2/2}\right]^2 dx.
$$

On converting to ξ-space, this becomes

$$P_{quant}(\xi, \xi + d\xi) = \left[\frac{1}{\sqrt{\pi}\, 2^n n!} H_n^2(\xi) e^{-\xi^2} \right] d\xi. \qquad (5.47)$$

For quantum state n, the turning points of the motion are given by $\xi = \pm\sqrt{2n + 1}$ (see Problem 5.9). Note that P_{class} diverges at these points; this corresponds to the fact that the oscillator is momentarily at rest at those points and hence has a high probability of being found at one or the other of them in a series of short-exposure photographs of the system.

Equations (5.46) and (5.47) are difficult to compare analytically, so we content ourselves with a numerical example. If we desired a macroscopic-scale energy, we might set $m = 1\,\text{kg}$, $k = 500\,\text{N/m}$, and $D = 1\,\text{m}$, in which case $n \sim 10^{35}$. Given the oscillatory nature of the Hermite polynomials, however, it would be impossible to plot P_{quant} and still resolve the individual probability peaks. We will have to settle for a much smaller value of n.

Figure 5.3 shows the probability densities [square-bracketed quantities in (5.46) and (5.47)] for the case of $n = 15$. The wiggly line shows P_{quant}; the upward-opening smooth curve is P_{class}. Eyeball inspection of the graph shows that P_{class} tracks closely to the average run of P_{quant}. Indeed, on considering that this energy would be microscopic, the agreement is striking: For a 1 kg oscillator with $k = 500\,\text{N/m}$ and $n = 15$, (5.45) gives $D \sim 10^{-17}$ meters; the energy would be about 10^{-13} eV. If n is increased, the wiggles in the P_{quant} curve would come closer and closer together since the spatial density of peaks is $n/\xi_{turn} \propto \sqrt{n}$, which grows as n increases. (P_{class} always has the upward-opening form shown in the Figure.) For practical purposes, measurements of the position of the oscillator would necessarily sample a finite range of distance incorporating many such wiggles; the derived quantum probability distribution would then agree with the classical one to an incredible degree of accuracy: if the agreement is good for $n = 15$, try to imagine what it would be like for $n = 10^{35}$! It is in this way that we say that the predictions of quantum mechanics merge into their classical counterparts when the quantum number of thesystem is

Fig. 5.3 Probability distribution for $n = 15$ classical (smooth, upward-opening curve) and quantum (oscillating curve) oscillators. The classical turning points of the motion are $\xi = \pm 5.57$

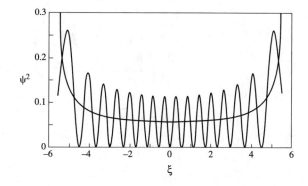

large. Real mass and spring systems have quantized energies, but the energy levels and spatial probability distributions are, for all practical purposes, continuous.

5.6 Raising and Lowering Operators

In this section we explore an alternate operator-based solution to the harmonic oscillator potential, a method originally developed by Dirac. The advantage of this approach is that it allows one to completely circumvent the lengthy solution of Schrödinger's equation as laid out in Sects. 5.1–5.3. While this technique can be extended to other potentials, the disadvantage is that is can be difficult to establish the appropriate form for the relevant operators.

Dirac's approach is predicated on defining two operators:

$$A^+ = \frac{\iota}{\alpha\sqrt{2}}\left(-\frac{d}{dx} + \alpha^2 x\right) \tag{5.48}$$

and

$$A^- = \frac{\iota}{\alpha\sqrt{2}}\left(-\frac{d}{dx} - \alpha^2 x\right), \tag{5.49}$$

where α is as defined in (5.1),

$$\alpha = \left(\frac{mk}{\hbar^2}\right)^{1/4}.$$

For reasons that will soon become clear, these operators are respectively known as the raising and lowering operators.

In order to see how defining A^+ and A^- lead to solutions for the harmonic-oscillator potential, it is necessary to first establish some of their properties. Let ψ be some wavefunction on which these operators can act. Although α here is defined as that for the harmonic oscillator states, this dummy wavefunction can be any wavefunction; it need not be one of the harmonic oscillator wavefunctions. First consider what happens if A^- is allowed to operate on the result of $A^+\psi$:

$$\begin{aligned}
A^-(A^+\psi) &= \frac{\iota}{\alpha\sqrt{2}}\left(-\frac{d}{dx} - \alpha^2 x\right)\left[\frac{\iota}{\alpha\sqrt{2}}\left(-\frac{d\psi}{dx} + \alpha^2 x\psi\right)\right] \\
&= \frac{\iota^2}{2\alpha^2}\left(\frac{d^2\psi}{dx^2} - \alpha^2\psi - \alpha^2 x\frac{d\psi}{dx} + \alpha^2 x\frac{d\psi}{dx} - \alpha^4 x^2\psi\right) \\
&= -\frac{1}{2\alpha^2}\left(\frac{d^2\psi}{dx^2} - \alpha^2\psi - \alpha^4 x^2\psi\right).
\end{aligned} \tag{5.50}$$

Similarly,

$$A^+(A^-\psi) = -\frac{1}{2\alpha^2}\left(\frac{d^2\psi}{dx^2} + \alpha^2\psi - \alpha^4 x^2\psi\right). \tag{5.51}$$

The commutator of A^+ and A^- operating on ψ is then

$$\left[A^-, A^+\right]\psi = (A^-A^+ - A^+A^-)\psi$$

$$= -\frac{1}{2\alpha^2}\left(\frac{d^2\psi}{dx^2} - \alpha^2\psi - \alpha^4 x^2\psi\right) + \frac{1}{2\alpha^2}\left(\frac{d^2\psi}{dx^2} + \alpha^2\psi - \alpha^4 x^2\psi\right)$$

$$= \psi.$$

$$\tag{5.52}$$

This result indicates that we can regard $[A^-, A^+]$ as being a sort of "unity" operator:

$$\left[A^-, A^+\right] \equiv (A^-A^+ - A^+A^-) \equiv 1. \tag{5.53}$$

This result is entirely independent of the particular wavefunction that $[A^-, A^+]$ should happen to be operating on. Now consider the result of operating $(A^-A^+ + A^+A^-)$ on ψ:

$$(A^-A^+ + A^+A^-)\psi = -\frac{1}{2\alpha^2}\left(\frac{d^2\psi}{dx^2} - \alpha^2\psi - \alpha^4 x^2\psi\right) - \frac{1}{2\alpha^2}\left(\frac{d^2\psi}{dx^2} + \alpha^2\psi - \alpha^4 x^2\psi\right)$$

$$= \left(-\frac{1}{\alpha^2}\frac{d^2}{dx^2} + \alpha^2 x^2\right)\psi.$$

$$\tag{5.54}$$

But for a multiplicative factor, the bracketed quantity on the right side of (5.54) is precisely the Hamiltonian operator for the harmonic-oscillator potential:

$$-\frac{1}{\alpha^2}\frac{d^2\psi}{dx^2} + \alpha^2 x^2\psi = -\frac{\hbar}{\sqrt{mk}}\frac{d^2\psi}{dx^2} + \frac{\sqrt{mk}}{\hbar}x^2\psi$$

$$\tag{5.55}$$

$$= \frac{2\sqrt{m/k}}{\hbar}\left(-\frac{\hbar^2}{2m}\frac{d^2}{dx^2} + \frac{k}{2}x^2\right)\psi.$$

Defining $\omega = \sqrt{k/m}$ as previously, (5.54) and (5.55) give

$$\boldsymbol{H} \equiv \frac{\hbar\omega}{2}(A^-A^+ + A^+A^-). \tag{5.56}$$

Now, from (5.53) we can write

$$(A^- A^+ - A^+ A^-) \equiv 1 \Rightarrow A^- A^+ \equiv 1 + A^+ A^-. \tag{5.57}$$

This result, when substituted into (5.56) gives

$$\boldsymbol{H} \equiv \hbar\omega \, (A^+ A^- + 1/2). \tag{5.58}$$

With this result, form the commutator of $\boldsymbol{H}$ and A^+ acting on ψ:

$$
\begin{aligned}
\left[\boldsymbol{H}, A^+\right] &= (\boldsymbol{H} A^+ - A^+ \boldsymbol{H})\psi \\
&= \hbar\omega \, (A^+ A^- + 1/2)(A^+ \psi) - \hbar\omega \, A^+ (A^+ A^- + 1/2)\,\psi \\
&= \hbar\omega \, (A^+ A^- A^+ - A^+ A^+ A^-)\,\psi \\
&= \hbar\omega \, A^+ (A^- A^+ - A^+ A^-)\,\psi \\
&= \left(\hbar\omega \, A^+\right)\psi,
\end{aligned}
\tag{5.59}
$$

where (5.53) was used to simplify $(A^- A^+ - A^+ A^-)$ in the second-last line. This result gives an identity:

$$[\boldsymbol{H}, A^+] \equiv \hbar\omega \, A^+. \tag{5.60}$$

Similarly, it is easy to show that

$$[\boldsymbol{H}, A^-] \equiv -\hbar\omega \, A^-. \tag{5.61}$$

Circular as it may seem, now redo the calculation of (5.59), but using (5.60). This will reveal a remarkable conclusion.

$$
\begin{aligned}
(\boldsymbol{H} A^+ - A^+ \boldsymbol{H})\psi &= \hbar\omega \, (A^+ \psi) \\
\Rightarrow \boldsymbol{H}(A^+ \psi) - A^+ (\boldsymbol{H}\psi) &= \hbar\omega \, (A^+ \psi).
\end{aligned}
$$

Now, the action of $\boldsymbol{H}$ on ψ must be to return $E\psi$, where E is the energy corresponding to ψ, that is, we must have

$$\boldsymbol{H}(A^+ \psi) - A^+ (E\psi) = \hbar\omega(A^+)\psi,$$

or

$$\boldsymbol{H}(A^+ \psi) = (E + \hbar\omega)(A^+ \psi). \tag{5.62}$$

Look at the structure of (5.62): It is an eigenvalue equation. The remarkable conclusion here is that when A^+ acts on ψ, it gives rise to a new function, $(A^+\psi)$, whose energy eigenvalue is just that of ψ, plus $\hbar\omega$. By a similar approach, it is easy to show that

$$H(A^-\psi) = (E - \hbar\omega)(A^-\psi), \tag{5.63}$$

that is, when A^- acts on ψ, it gives rise to a new function, $(A^-\psi)$, whose energy eigenvalue is just that of ψ, less $\hbar\omega$. It is worth reiterating that these results are quite independent of the details of the wavefunctions that A^+ and A^- are operating on; nowhere in the derivation have we assumed that the ψ's are harmonic oscillator wavefunctions.

The reason why A^+ and A^- are known as raising and lowering operators should now be clear: when applied to any wavefunction, they give rise to new wavefunctions whose energies are raised (in the case of A^+) or lowered (in the case of A^-) from that of the original ones by an amount $\hbar\omega$. A^+ and A^- are also known as ladder operators.

To establish the harmonic-oscillator wavefunctions and energies by this approach requires an inference. The potential is

$$V(x) = \frac{k}{2}x^2. \tag{5.64}$$

Since $V > 0$ everywhere, we can infer that the energy eigenvalues must be positively-valued. Suppose, then, that ψ is some solution of this potential with energy $E > 0$. Successive applications of the lowering operator A^- would yield functions whose energies are decremented by $\hbar\omega$ at each application. If this process continued indefinitely, we would arrive at negative energies, in contradiction to the conclusion that all eigenenergies of this potential must be positive. We infer, then, that there must exist some state of lowest positive energy, say (ψ_0, E_0), which cannot be allowed to generate a yet lower state when acted upon by A^-, that is, ψ_0 must be such that

$$A^-\psi_0 = 0. \tag{5.65}$$

ψ_0 can be explicitly determined by demanding the veracity of (5.65):

$$\frac{\iota}{\alpha\sqrt{2}}\left(-\frac{d\psi_0}{dx} - \alpha^2 x\psi_0\right) = 0. \tag{5.66}$$

The solution of this simple differential equation is $\psi_0 = Ke^{-\alpha^2 x^2/2}$, where K is a constant of integration; normalizing yields the ground-state harmonic-oscillator wavefunction. E_0 can be determined by substituting ψ_0 back into Schrödinger's equation or by applying (5.58):

$$\boldsymbol{H}\psi_0 = E_0\psi_0 \Rightarrow \hbar\omega\,(A^+A^- + 1/2)\psi_0 = \left(\frac{1}{2}\hbar\omega\right)\psi_0 = E_0\psi_0, \qquad (5.67)$$

where we have used the fact that $A^-\psi_0 = 0$.

Do not think that because A^+ and A^- act to raise and lower the energy of a wavefunction on which they operate on by $\pm\hbar\omega$ that it follows that they must give rise exactly to other harmonic oscillator wavefunctions. In fact, they act as

$$A^+\psi_n = \iota\sqrt{n+1}\,\psi_{n+1} \qquad (5.68)$$

and

$$A^-\psi_n = -\iota\sqrt{n}\,\psi_{n-1}. \qquad (5.69)$$

To prove these, it is crucial to have on hand two recursion relationships satisfied by Hermite polynomials:

$$-\frac{dH_n}{d\xi} + 2\xi\,H_n = H_{n+1} \qquad (5.70)$$

and

$$\frac{dH_n}{d\xi} = 2n\,H_{n-1}. \qquad (5.71)$$

Equation (5.68), for example, can be proven as follows. In terms of $\xi = \alpha x$, A^+ appears as

$$A^+ = \frac{\iota}{\sqrt{2}}\left(-\frac{d}{d\xi} + \xi\right).$$

Hence, from (5.38),

$$A^+\psi_n = \frac{\iota}{\sqrt{2}}\left(-\frac{d}{d\xi} + \xi\right)\left(A_n H_n e^{-\xi^2/2}\right)$$

$$\qquad\qquad\qquad\qquad (5.72)$$

$$= \frac{\iota A_n}{\sqrt{2}}\left(-\frac{dH_n}{d\xi} + 2\xi\,H_n\right)e^{-\xi^2/2}.$$

From (5.39), the normalization factors satisfy

$$A_n = \sqrt{2(n+1)}\,A_{n+1}. \qquad (5.73)$$

Equations (5.70) and (5.73) in (5.72) give

$$A^+\psi_n = \frac{\iota\sqrt{2(n+1)}A_{n+1}}{\sqrt{2}}(H_{n+1})\,e^{-\xi^2/2}$$

(5.74)

$$= \iota\sqrt{n+1}A_{n+1}H_{n+1}e^{-\xi^2/2} = \iota\sqrt{n+1}\psi_{n+1}.$$

Equation (5.69) follows similarly, using (5.71).

This operator approach to the harmonic potential has other advantages. Since the momentum operator is $p_{op} = -\iota\hbar(d/dx)$, (5.48) and (5.49) can be rearranged to give

$$p_{op} = \frac{\alpha\hbar}{\sqrt{2}}(A^+ + A^-)$$

(5.75)

and

$$x_{op} = -\frac{\iota}{\sqrt{2}\alpha}(A^+ - A^-).$$

(5.76)

Suppose we wish to compute the expectation value of x^2 for any harmonic-oscillator state, say ψ_n, with $E_n = (n+1/2)\hbar\omega$. This demands calculating

$$\langle x^2 \rangle = \int \psi_n(x_{op}^2 \psi_n)dx$$

$$= \frac{\iota^2}{2\alpha^2}\int \psi_n[(A^+ - A^-)(A^+\psi_n - A^-\psi_n)]dx$$

$$= -\frac{1}{2\alpha^2}\int \psi_n(A^+A^+\psi_n - A^+A^-\psi_n - A^-A^+\psi_n + A^-A^-\psi_n)dx.$$

Look at the first term within the bracket in the integrand. $A^+A^+\psi_n$ will give rise to ψ_{n+2}, which will be orthogonal to ψ_n; the integral of the product of this result and ψ_n must consequently be zero. A similar argument applies for the last term within the brackets. Only the second and third terms will make any contribution to the integral:

$$\langle x^2 \rangle = \frac{1}{2\alpha^2}\int \psi_n(A^+A^- + A^-A^+)\psi_n dx.$$

The bracketed term in the integrand is just the Hamiltonian operator of (5.56):

$$\langle x^2 \rangle = \frac{1}{2\alpha^2}\frac{2}{\hbar\omega}\int \psi_n(H\psi_n)dx = \frac{1}{\alpha^2\hbar\omega}\int \psi_n(E_n\psi_n)dx$$

(5.77)

$$= \frac{E_n}{\alpha^2\hbar\omega} = \frac{1}{\alpha^2}(n+1/2).$$

$\langle p^2 \rangle$ follows similarly. Doing these calculations directly from the wavefunctions and properties of Hermite polynomials is considerably more involved; see Problem 5.11.

The power of the operator approach derives from but is limited by the fact that the harmonic-oscillator Hamiltonian has a sort of "sum of squares" form, [see (5.55)] and can be "factored" as in (5.56). While this sort of manipulation will not in general be possible for any arbitrary potential, there are situations where defining similar raising and lowering operators can be valuable in more advanced work; we will explore such a situation concerning angular momentum in Chap. 8.

Summary

The solution of the harmonic oscillator potential $V(x) = kx^2/2$ is one of the classic triumphs of quantum mechanics, and can be found in virtually any text on the subject. Variants of this potential are central to our understanding of molecular spectra.

Almost as important as obtaining the wavefunctions and energies of the harmonic-oscillator potential is that its solution illustrates three techniques of considerable generality and power: (i) casting Schrödinger's equation into dimensionless form; (ii) establishing and factoring out an asymptotic solution valid for extreme values of x; and (iii) establishing a series solution to be grafted onto the asymptotic solution to yield an overall solution valid for all x. Forcing the convergence of the series solution leads to the energy eigenvalues of the system.

The normalized harmonic oscillator wavefunctions are given by

$$\psi_n(x) = A_n \, H_n(\xi) e^{-\xi^2/2} = A_n \, H_n(\alpha x) e^{-\alpha^2 x^2/2},$$

where A_n is a normalization factor:

$$A_n = \sqrt{\frac{\alpha}{\sqrt{\pi} \, 2^n n!}},$$

with

$$\alpha = \left(mk/\hbar^2 \right)^{1/4}.$$

$H_n(x)$ denotes a Hermite polynomial of order n, with n being the quantum number ($n = 0, 1, 2, 3, \ldots$). Explicit expressions for the first few $H_n(x)$ are given in Table 5.1; expressions for any others can be produced from the recursion relation

$$H_{n+1}(x) = 2x \, H_n(x) - 2n \, H_{n-1}(x),$$

or, if you are ambitious, directly from the generating function

$$H_n(x) = (-1)^n e^{x^2} \frac{d^n}{dx^n} (e^{-x^2}).$$

The energy of a harmonic oscillator in state n is given by

$$E = \left(n + \frac{1}{2}\right) \hbar \sqrt{\frac{k}{m}}, \qquad n = 0, 1, 2, 3 \ldots.$$

The raising and lowering operators are given by

$$A^+ = \frac{\iota}{\alpha\sqrt{2}} \left(-\frac{d}{dx} + \alpha^2 x\right)$$

and

$$A^- = \frac{\iota}{\alpha\sqrt{2}} \left(-\frac{d}{dx} - \alpha^2 x\right).$$

In terms of these, the harmonic oscillator Hamiltonian can be written as

$$H \equiv \hbar\omega (A^+ A^- + 1/2).$$

These operators act on the harmonic oscillator states according as

$$A^+ \psi_n = \iota\sqrt{n+1}\psi_{n+1}$$

and

$$A^- \psi_n = -\iota\sqrt{n}\,\psi_{n-1}.$$

In terms of these operators, the position and momentum operators appear as

$$x_{op} = -\frac{\iota}{\sqrt{2}\alpha}(A^+ - A^-)$$

and

$$p_{op} = \frac{\alpha\hbar}{\sqrt{2}}(A^+ + A^-).$$

Problems

5.1(E) Use dimensional analysis to construct a quantity whose units are that of force from the physical constants c, h, and G. See Example 5.1.

5.2(I) Verify that (5.5) results upon substitution of Eqs. (5.2), (5.3) and (5.4) into Schrödinger's equation for the harmonic oscillator potential.

5.3(E) Convince yourself that (5.23) is valid by expanding out the first few terms on both sides.

5.4(A) Working from (5.12), (5.17), (5.25), and the normalization condition, develop explicitly the expression for the $n = 2$ harmonic oscillator wavefunction. Compare to Table 5.1.

5.5(I) From Table 5.1 and Equations (5.38), (5.39), and (5.40), write out an explicit expression for the wavefunction for the third excited state ($n = 3$) of the harmonic oscillator. Verify that it satisfies Schrödinger's equation for the harmonic oscillator potential.

5.6(I) Verify by explicit calculation that the $n = 0$ and $n = 2$ harmonic oscillator wavefunctions are orthonormal.

5.7(I) Consider a particle of mass m in the first excited state ($n = 1$) of a harmonic oscillator potential. Compute the probability of finding the particle outside the classically allowed region.

5.8(E) The *Handbook of Chemistry and Physics* gives the vibrational energy constant for diatomic hydrogen ($^1H_2$) as $\omega_e = 4401$ cm^{-1}. Determine the effective force constant k for H_2.

5.9(E) Show that the classical turning points for a harmonic oscillator in energy state n are given by $\xi_{turn} = \pm\sqrt{2(n + 1/2)}$.

5.10(A) In Chap. 9 we will study the so-called linear potential

$$V(x) = \begin{cases} \infty & x \leq 0 \\ \alpha x & x > 0. \end{cases}$$

Following the development in Sects. 5.1 and 5.2, cast Schrödinger's equation for this potential into a dimensionless form and derive an asymptotically valid solution to it. Then hypothesize a series solution and attempt to establish a recursion relation. Can this potential be solved analytically?

5.11(A) Prove that $\Delta x \Delta p = (n + 1/2)\hbar$ for the harmonic oscillator wavefunctions. HINTS:

(i) What can you say of $\langle x \rangle$ and $\langle p \rangle$?

(ii) It is helpful to know that Hermite polynomials satisfy

$$\int_{-\infty}^{\infty} \xi^2 H_n^2(\xi) e^{-\xi^2} d\xi = \sqrt{\pi} 2^n n! (n + 1/2).$$

(iii) When computing $\langle p^2 \rangle$, (5.5) may be helpful; convert from ξ to x.

5.12(I) A form of the Morse potential is

$$V(x) = V_o \left\{ e^{-2x/b} - 2e^{-x/b} \right\},$$

where b is a constant. Show by appropriate expansions that for small values of (x/b), this potential is equivalent to a harmonic-oscillator potential with $k = 2V_o/b^2$ and which is offset by an additive energy $-V_o$, that is,

$$V(x) \sim -V_o + \left(\frac{V_o}{b^2} \right) x^2.$$

5.13(A) In Sect. 5.1 we rendered Schrödinger's equation for the harmonic oscillator dimensionless by establishing quantities α and ϵ with units of inverse meters and inverse Joules, respectively, and then replacing x and E with the dimensionless proxies ξ and λ. Consider a more general potential of the form $V(x) = A|x|^p$. Generalize the expressions for α and ϵ to be in terms of the power p. Include with each a purely numerical multiplicative constant (K_α, K_ϵ) and determine these as functions of p such that Schrödinger's equation can be reduced to the form

$$\left(\frac{d^2\psi}{d\xi^2} \right) + (\lambda - |\xi|^p)\psi = 0.$$

Check your results against the harmonic oscillator case. Are there any values of p for which this scheme will not work?

5.14(I) As Problem 5.13, consider a potential of the form $V(x) = Ax^p$ ($-\infty \le x \le \infty$); presume that p is even so that no absolute values are involved. Generalize the asymptotic solution of Sect. 5.2 to this case; put $\psi \sim \exp(-\beta\xi^n)$. How do n and β depend on p? Then generalize the series solution of Sect. 5.3 to this case by setting $\psi = F(\xi)\exp(-\beta\xi^n)$. Show that the differential equation for $F(\xi)$ analogous to (5.16) is

$$\frac{d^2 F}{d\xi^2} - 2\xi^{p/2}\frac{dF}{d\xi} + \left[\lambda - (p/2)\xi^{p/2-1} \right] F = 0.$$

5.15(I) In setting up the series solution for $H(\xi)$, we assumed a summation of the form $H(\xi) = \Sigma a_n \xi^n$. To accommodate the possibility that the solution might not be a series in integral powers of ξ, we should be more careful and include an "indicial" power c in ξ: $H(\xi) = \Sigma a_n \xi^{n+c}$. Rework the solution with this assumption, and show that we must have $c = 0$.

5.16(E) Show that (5.75) and (5.76) follow from (5.48) and (5.49).

5.17(E) Show that application of the lowering operator A^- to the $n = 3$ harmonic-oscillator wavefunction leads to the result predicted by (5.69).

5.18(E) Following the approach that led to (5.77), use the raising and lowering operators to show that $\langle p^2 \rangle = \alpha^2 \hbar^2 (n + 1/2)$ for the nth harmonic-oscillator state, and hence verify your result for $\Delta x \Delta p$ obtained in Problem 5.11.

5.19(I) Show that the virial theorem holds for all harmonic-oscillator states. The identity given in Problem 5.11 is helpful.

5.20(I) Use dimensional analysis to construct a quantity whose units are that of time from the physical constants ϵ_o, m_e, e, and c; the latter is the speed of light.

Reference

1. P. Morse, Phys. Rev. **34**, 57 (1934)

Chapter 6
Schrödinger's Equation in Three Dimensions and the Quantum Theory of Angular Momentum

Summary Real atoms and molecules live in three dimensions. This chapter is devoted to analyzing some general features of solutions to Schrödinger's equation in three dimensions, with particular emphasis on the role of angular momentum operators expressed in spherical coordinates. This leads to a set of mathematical functions known as *spherical harmonics*, which play an important role in the solution of the hydrogen atom in Chap. 7.

In Chaps. 3 and 5 we examined solutions of Schrödinger's equation for various one-dimensional potentials. Those analyses yielded a number of uniquely quantum-mechanical effects such as discrete energy states, alternating parity wavefunctions, barrier penetration, and resonance scattering. While these examples are instructive in that they make clear the differences between classical and quantum mechanics, it is evident that if we seek to examine truly realistic problems it will be necessary to consider three-dimensional situations. This is the purpose of both this chapter and Chap. 7.

Examining three-dimensional solutions to Schrödinger's equation for even relatively simple potentials will prove to be an involved task. Culminating as it does with a study of the hydrogen-atom Coulomb potential in Chap. 7, however, this work very much comprises the heart of this book. Here we lay out the plan of attack for these two chapters. Ultimately, we will see that the presence of extra dimensions leads to two physical phenomena not present in purely one-dimensional problems: *energy degeneracy* and *quantization of angular momentum*.

The three-dimensional Schrödinger equation was given in Chap. 2:

$$-\frac{\hbar^2}{2m} \nabla^2 \psi(r) + V(r)\psi(r) = E\psi(\mathbf{r}),$$

where r denotes a three-dimensional vector position. In Sect. 6.1 we examine a quite powerful method, known as separation of variables, for solving this equation in multidimensional Cartesian coordinates. Application of this technique to a three-dimensional analog of the infinite square well will allow us to develop an approximate expression for the number of photon states lying within a narrow band of frequencies, (1.7), one of the great problems in the history of quantum mechanics. Since our ultimate goal is to study solutions of Schrödinger's equation for the Coulomb potential,

B. C. Reed, *Quantum Mechanics*, https://doi.org/10.1007/978-3-031-14020-4_6

Sect. 6.2 is devoted to a review of some of the properties of spherical coordinates, the coordinate system in which that potential proves most easily separable. An important aspect of solutions of Schrödinger's equation for potentials like the Coulomb potential is quantization of angular momentum, so Sect. 6.3 is devoted to establishing some properties of angular momentum operators which will prove extremely useful throughout the remainder of this chapter and in Chap. 7.

The Coulomb potential is an example of a *central potential* or *radial potential*, one that depends only on the magnitude of the separation between two particles and not on their orientation in space. Any central potential is thus normally symbolized in the form $V(r)$, as opposed to $V(\boldsymbol{r})$. In the case of the Coulomb potential, r is the electron/nucleus separation as in the treatment of the Bohr model in Chap. 1. Section 6.4 is devoted to separating Schrödinger's equation for central potentials in spherical coordinates. We will see that an important result of this separation is that Schrödinger's equation naturally breaks into a "radial equation" and an "angular equation", with the potential $V(r)$ being involved only in the radial part. As a consequence, the solution to the angular part is quite general for any central potential, and turns out to involve the angular momentum operators developed in Sect. 6.3. Section 6.5 is devoted to the general solution of this angular part of Schrödinger's equation; this leads to some rather complicated functions known as *spherical harmonics*. Anticipating things somewhat, we might expect to see no suggestion of energy levels in the angular solution inasmuch as the energy eigenstates of a system depend exclusively on the potential and the boundary conditions. This proves to be the case: for central potentials, energy levels are the purview of the radial equation. Chap. 7 opens with a general introduction to the radial equation for central potentials (Sect. 7.1), and then takes up its solution for three specific central potentials: the infinite and finite spherical potential wells (Sects. 7.2 and 7.3; these are analogs of the one-dimensional wells of Chap. 3), and then, in Sections 7.4 and 7.5, the Coulomb potential. Chapter 8, which should be considered optional, explores some more advanced aspects of the angular momentum operators introduced in the present chapter.

This is a lengthy, somewhat abstract chapter. The best advice that can be offered is to go through it step by step, convincing yourself of the manipulations and arguments at each point, and then do the end-of-chapter problems.

6.1 Separation of Variables: Cartesian Coordinates

If the potential energy function $V(\boldsymbol{r})$ is a function of all three coordinates, solving Schrödinger's equation could be a daunting task. Fortunately, there is a technique available that often simplifies the job considerably. We illustrate this technique, known as separation of variables, by example.

Our example is the three-dimensional analog of a particle of mass m in an infinite potential well, a so-called infinite potential box. In analogy to the one-dimensional potential well, this is a box of dimensions (a, b, c) with impenetrable walls, as sketched in Fig. 6.1.

Fig. 6.1 Infinite potential box

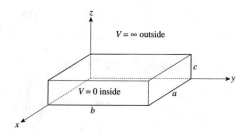

Formally, the infinite potential box is defined by

$$V(x, y, z) = \begin{cases} 0, & (0 \le x \le a,\ 0 \le y \le b,\ 0 \le z \le c) \\ \infty, & \text{otherwise}. \end{cases} \tag{6.1}$$

Outside the box we must have $\psi = 0$ for the same reasons as were elucidated in the one-dimensional case. Inside the box, Schrödinger's equation becomes

$$-\frac{\hbar^2}{2m}\left(\frac{\partial^2 \psi}{\partial x^2} + \frac{\partial^2 \psi}{\partial y^2} + \frac{\partial^2 \psi}{\partial z^2}\right) = E\psi. \tag{6.2}$$

Equation (6.2) is a partial differential equation, which is much more difficult to solve, in general, than the ordinary differential equations we have heretofore encountered.

The essence of the separation of variables technique is to assume that ψ can be written as the product of three separate functions, each depending on only one of the coordinates. That is, we assume that we can write

$$\psi(x,\ y,\ z) = X(x)\,Y(y)\,Z(z). \tag{6.3}$$

Note that upper-case letters denote functions of the coordinates represented by lower-case letters. We have no rigorous justification for this assumption; its validity can only be known after attempting to solve the differential equation. If it is a valid assumption, we will be able to obtain a solution; if not, we will have to try some other approach.

On substituting (6.3) into (6.2), two of the three functions X, Y, and Z come through each of the partial derivatives as would constants since they do not depend on the coordinate involved in the derivative:

$$-\frac{\hbar^2}{2m}\left(YZ\frac{\partial^2 X}{\partial x^2} + XZ\frac{\partial^2 Y}{\partial y^2} + XY\frac{\partial^2 Z}{\partial z^2}\right) = E(XYZ). \tag{6.4}$$

Since each partial derivative operates on a function of only one variable, they can be replaced with ordinary derivatives:

$$-\frac{\hbar^2}{2m}\left(YZ\frac{d^2X}{dx^2}+XZ\frac{d^2Y}{dy^2}+XY\frac{d^2Z}{dz^2}\right)=E\,(X\,Y\,Z).\qquad(6.5)$$

Dividing through by (XYZ) gives

$$-\frac{\hbar^2}{2m}\left(\frac{1}{X}\frac{d^2X}{dx^2}+\frac{1}{Y}\frac{d^2Y}{dy^2}+\frac{1}{Z}\frac{d^2Z}{dz^2}\right)=E.\qquad(6.6)$$

Examine (6.6) carefully. Each term within the brackets is a function of only one coordinate. Inasmuch as the coordinates are independent, these terms can be varied independently of one another. The right side, however, is a constant. The result of our assumption is that three independently variable functions add up to a constant. This can only be so if each function is itself equal to a constant. Labeling these constants E_x, E_y, and E_z, we can split (6.6) into three simultaneous ordinary differential equations:

$$-\frac{\hbar^2}{2m}\left(\frac{1}{X}\frac{d^2X}{dx^2}\right)=E_x,\qquad(6.7)$$

$$-\frac{\hbar^2}{2m}\left(\frac{1}{Y}\frac{d^2Y}{dy^2}\right)=E_y,\qquad(6.8)$$

$$-\frac{\hbar^2}{2m}\left(\frac{1}{Z}\frac{d^2Z}{dz^2}\right)=E_z,\qquad(6.9)$$

with

$$E=E_x+E_y+E_z.\qquad(6.10)$$

E_x, E_y, and E_z are known as *separation constants*; in any successful application of this method, one will always end up with as many separation constants as there are dimensions.

The essential result here is that we have reduced the original partial differential equation to three ordinary differential equations, one for each dimension. Equations (6.7)–(6.9) are of exactly the same form as Schrödinger's equation for the one-dimensional infinite well. From Chap. 3, we can write the normalized solutions for X, Y, and Z as

$$X(x)=\sqrt{\frac{2}{a}}\sin\left(\frac{n_x\pi x}{a}\right),\qquad E_x=\frac{h^2n_x^2}{8ma^2},\qquad(6.11)$$

$$Y(y)=\sqrt{\frac{2}{b}}\sin\left(\frac{n_y\pi y}{b}\right),\qquad E_y=\frac{h^2n_y^2}{8mb^2},\qquad(6.12)$$

$$Z(z) = \sqrt{\frac{2}{c}} \sin\left(\frac{n_z \pi z}{c}\right), \qquad E_z = \frac{h^2 n_z^2}{8mc^2}. \tag{6.13}$$

Note that each dimension acquires its own quantum number and that the (x, y, z) dimensions of the box appear in their corresponding wavefunctions. This is so because (6.7)–(6.9) must be solved independently of each other.

Recognizing that the volume of the box is given by $V = abc$, we can write the overall wavefunction as

$$\psi(x, y, z) = \sqrt{\frac{8}{V}} \sin\left(\frac{n_x \pi x}{a}\right) \sin\left(\frac{n_y \pi y}{b}\right) \sin\left(\frac{n_z \pi z}{c}\right), \tag{6.14}$$

and the total energy as

$$E = \frac{h^2}{8m}\left(\frac{n_x^2}{a^2} + \frac{n_y^2}{b^2} + \frac{n_z^2}{c^2}\right), \quad (n_x, n_y, n_z) = 1, 2, 3, \ldots. \tag{6.15}$$

The infinite potential box wavefunction can thus be regarded as the product of three one-dimensional infinite potential well wavefunctions, with a total energy equal to that given by summing contributions from each dimension. Since each of X, Y, and Z are independently normalized, ψ will be normalized as well. That we have found a solution to the problem means that our initial assumption of separation of variables was valid.

To plot such three-dimensional wavefunctions on a two-dimensional piece of paper is a challenge: (6.14) associates a value (ψ) with each point in (x, y, z) space, so we have the problem of trying to represent four numbers on a two-dimensional surface. Instead, wavefunctions for a two-dimensional infinite well are easier to deal with. Figure 6.2 shows $\psi(x, y)$ for the case $(a, b) = (1, 1)$ and $(n_x, n_y) = (5, 2)$; in creating the plot, the two-dimensional amplitude factor $\sqrt{4/ab}$ has been reduced to unity for convenience. Notice that there are five and two maxima in the x and y directions, respectively. It is left as a mental exercise for the reader to extrapolate to a three-dimensional situation!

An important facet of such multi-dimensional solutions of Schrödinger's equation is *energy degeneracy*. For simplicity, consider an infinite potential with sides all of the same length L, that is, $a = b = c = L$. The possible energy states are then

$$E = \frac{h^2}{8mL^2}\left(n_x^2 + n_y^2 + n_z^2\right), \quad (n_x, n_y, n_z) = 1, 2, 3, \ldots. \tag{6.16}$$

Since each of the quantum numbers can independently take on the values 1, 2, 3, …, it follows that different combinations of (n_x, n_y, n_z) can yield the same total energy. For example, $E = 6h^2/8ml^2$ can be realized in three different ways: $(n_x, n_y, n_z) = (2, 1, 1)$ or $(1, 2, 1)$ or $(1, 1, 2)$. This energy level is said to be triply degenerate. Degeneracy means a condition where different individual quantum states

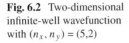

Fig. 6.2 Two-dimensional infinite-well wavefunction with $(n_x, n_y) = (5,2)$

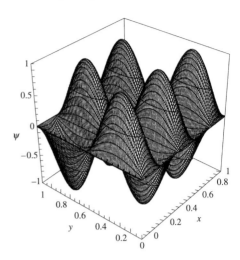

(n_x, n_y, n_z) result in the same "level" of total energy E. We can expect the degree of degeneracy of a given energy level to rise with the energy, as it can be realized with more and more combinations of (n_x, n_y, n_z). But this does not mean that *any* infinite potential box will exhibit degeneracy: From (6.15) it can be seen that only when the ratio of side lengths $a : b : c$ is as a ratio of squares of integers will degeneracy arise. The possibility of degeneracy is inherent in multi-dimensional problems; its presence in any particular case depends on the boundary conditions.

The concept of degeneracy can be tricky to get hold of if it is new to you. A simple analogy might help. Consider rolling two regular dice. There is one way, and only one way, in which a total of 12 can appear: both dice must show 6. On the other hand, there are four distinct ways ("states") in which a total of 5 can arise: (1,4), (2,3), (3,2), and (4,1). We can say that the level wherein the total is 12 has a degeneracy of one, while that in which the total is 5 has a degeneracy of 4.

How can one know, a priori, if the separation of variables technique will work in any particular case? The answer to this depends on what coordinate system is being used and on the form of the potential function. Eisenhart has shown that Schrödinger's equation is separable in no less than 11 coordinate systems, including the familiar Cartesian, cylindrical, and spherical ones, provided that the potential is of the form

$$V = AX_1(x_1) + BX_2(x_2) + CX_3(x_3),$$

where the upper-case letters X_i designate functions of the coordinates x_i [1].

To close this section we take up a problem with connections to the development of Planck's theory of blackbody radiation discussed in Chap. 1: A derivation of the number of individual quantum states available to a particle of mass m trapped within an infinite three-dimensional potential box. This problem is analogous to the derivation in Sect. 3.5 of an expression for the number of energy states available to a particle of mass m moving in a finite one-dimensional rectangular well.

Fig. 6.3 Positive octant of
an ellipsoid of semi-axes
(α, β, γ)

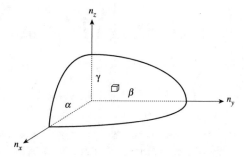

From (6.15), the energy corresponding to quantum state (n_x, n_y, n_z) is given by

$$E = \frac{h^2}{8m}\left(\frac{n_x^2}{a^2} + \frac{n_y^2}{b^2} + \frac{n_z^2}{c^2}\right), \quad (n_x, n_y, n_z) = 1, 2, 3, \ldots. \quad (6.17)$$

The question of the number of available quantum states can be more precisely phrased as "How many individual quantum states with energies $< E$ are available to the particle?" A derivation of this is an interesting exercise in geometry.

First, rearrange (6.17) by dividing through by E:

$$1 = \left(\frac{n_x^2}{\alpha^2}\right) + \left(\frac{n_y^2}{\beta^2}\right) + \left(\frac{n_z^2}{\gamma^2}\right), \quad (6.18)$$

where we have defined

$$\{\alpha, \beta, \gamma\} = \left\{\sqrt{\frac{8mEa^2}{h^2}}, \sqrt{\frac{8mEb^2}{h^2}}, \sqrt{\frac{8mEc^2}{h^2}}\right\}. \quad (6.19)$$

Consult a textbook on three-dimensional geometry: You will find that (6.18) describes the surface of a three-dimensional ellipsoid of semi-axes (α, β, γ) along the (n_x, n_y, n_z) axes. This is sketched in Fig. 6.3, where only the positive octant of the ellipsoid is shown. The volume of the entire ellipsoid (all eight octants) is $(4\pi/3)\alpha\beta\gamma$.

In our case, the dimensionless axes (n_x, n_y, n_z) do not correspond to continuous variables, but rather to discrete integer quantum numbers. Imagine the space represented in Fig. 6.3 to be divided up into a large number of tiny cubes of volume $\Delta n_x \Delta n_y \Delta n_z$ centered at integral values of (n_x, n_y, n_z), each of side lengths $\Delta n_x = \Delta n_y = \Delta n_z = 1$. Each such tiny volume (one is shown in the figure) would then correspond to a single quantum state (n_x, n_y, n_z). The total volume within the ellipsoid then represents the number of individual quantum states lying below energy E, a quantity usually designated as $N(E)$. This is given by the volume of only the positive octant of the ellipsoid, since (n_x, n_y, n_z) are restricted to positive integers:

$$N(E) = \frac{1}{8}\left(\frac{4\pi}{3}\right)\alpha\beta\gamma = \frac{\pi}{6}\sqrt{\frac{8mEa^2}{h^2}}\sqrt{\frac{8mEb^2}{h^2}}\sqrt{\frac{8mEc^2}{h^2}},$$

or

$$N(E) = \frac{8^{3/2}\pi}{6}\frac{Vm^{3/2}}{h^3}E^{3/2}, \tag{6.20}$$

where we have again put $V = abc$ for the volume of the box.

Equation (6.20) represents the number of individual quantum states of energy $< E$ available to a particle of mass m trapped in a box of volume V with impenetrable walls. If it is true that if $N(E)$ is small (that is, if E is small), then $N(E)$ will not be accurately predicted by (6.20) because the ellipsoidal surface in Fig. 6.3 will cut through a relatively large number of the individual unit volumes and we will not get an accurate count. But if E is large, the approximation made by representing the volume enclosed within the ellipsoid as little cubes will be quite accurate.

The main quantity of interest to Planck was the number of quantum states lying between energies E and $E + dE$. We can get an expression for this by differentiating (6.20):

$$\frac{dN}{dE} = \frac{8^{3/2}\pi}{4}\frac{V\,m^{3/2}}{h^3}E^{1/2} \Rightarrow dN = \frac{8^{3/2}\pi}{4}\frac{V\,m^{3/2}}{h^3}E^{1/2}dE. \tag{6.21}$$

This is often designated as $N(E)dE$.

The connection between this result and Planck's radiation formula can now be made. First, for a reason that will become clear momentarily, we transform (6.21) to an equivalent momentum form $N(p)dp$ by writing $E = p^2/2m$ and $dE = (p/m)dp$. This gives

$$N(p)dp \sim \frac{4\pi V}{h^3}p^2dp. \tag{6.22}$$

This expression gives the number of quantum states where a particle's momentum would lie between p and $p + dp$. Now suppose that the particles are photons. In Chap. 1 we saw how Einstein's energy-momentum-rest mass relationship $E^2 = p^2c^2 + m_o^2c^4$ can be used to derive the wavelength-momentum relationship for photons, $\lambda = h/p$, by setting the rest mass m_o to be zero. Setting $\lambda = c/\nu$, we can put this into the form $p = h\nu/c$, hence $dp = (h/c)d\nu$. In frequency form, (6.22) becomes

$$N(\nu)d\nu \sim \frac{4\pi V}{c^3}\nu^2d\nu. \tag{6.23}$$

But for a factor of two, this is Planck's expression (1.7) for the number of photon "normal modes" in a chamber of volume V; note that Planck's constant cancelled out. The reason for first transforming to a momentum distribution is that putting $m = 0$ in (6.21) directly would give a zero result. The missing factor of two can be attributed to different photon polarizations. It is in fact somewhat surprising that this derivation gives a result that even looks correct: Schrödinger's equation is not relativistic and was not designed to deal with massless particles! A rigorous derivation involves computing the number of normal modes directly from Maxwell's equations [2]. The present derivation should consequently not be regarded as definitive; it is included here to give a sense of the situation.

Example 6.1 Estimate the volume of space necessary to support a single quantum state of energy $E < 1$ eV to be occupied by an electron.

Rearranging (6.20) to solve for the volume gives

$$V = \frac{6 N h^3}{8^{3/2} \pi m^{3/2} E^{3/2}}.$$

Setting $N = 1$ and with $E = 1$ eV $= 1.602 \times 10^{-19}$ Joules, we have, in MKS units,

$$V = \frac{6 N h^3}{8^{3/2} \pi m^{3/2} E^{3/2}} = \frac{(1.745 \times 10^{-99})}{8^{3/2} \pi (8.694 \times 10^{-46}) (6.412 \times 10^{-29})}$$

$$= 4.403 \times 10^{-28} \text{m}^3.$$

If this volume were in the shape of a sphere, the radius would be about 4.7 Ångstroms: atomic-scale.

6.2 Spherical Coordinates

Problems wherein the quantity of interest is a function only of the distance from some point (the origin, say) and not of direction are said to possess spherical symmetry. For example, the magnitude of the electric field due to a point charge Q is given by

$$E = \frac{Q}{4\pi \varepsilon_o r^2}, \tag{6.24}$$

where r is the distance between the charge and the field point. As long as one remains at distance r from the charge, the magnitude of E will not change, irrespective of how you are oriented with respect to the charge. To describe such situations it is convenient to develop a coordinate system wherein one of the coordinates is simply r itself. Specifying r indicates only that you are somewhere on the surface of a sphere; to completely specify your location in space requires two other coordinates,

Fig. 6.4 Spherical
coordinates

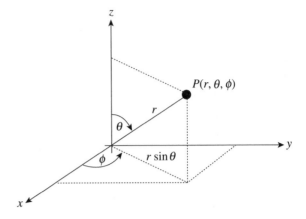

analogous to latitude and longitude. The spherical polar system, with coordinates
designated by (r, θ, ϕ), has been defined with such situations in mind.

The relationship between spherical and Cartesian coordinates is illustrated in
Fig. 6.4. Cartesian coordinates corresponding to a point $P(r, \theta, \phi)$ are defined by

$$\left. \begin{array}{l} x = r \sin \theta \cos \phi \\ y = r \sin \theta \sin \phi \\ z = r \cos \theta \end{array} \right\}, \tag{6.25}$$

or equivalently by the inverse relations

$$\left. \begin{array}{l} r = \sqrt{x^2 + y^2 + z^2} \\ \theta = \cos^{-1}(z/r) \\ \phi = \tan^{-1}(y/x) \end{array} \right\}. \tag{6.26}$$

Note that θ is measured from the z-axis, not in the "equatorial" xy plane. The
volume of space enclosed between (r, θ, ϕ) and $(r + dr, \theta + d\theta, \phi + d\phi)$ is given
by

$$dV = r^2 \sin \theta \, dr \, d\theta \, d\phi. \tag{6.27}$$

The drawback of spherical coordinates is that the unit vectors are functions of
the coordinates themselves. This is due to the fact that the coordinates are defined
in terms of combinations of Cartesian coordinates; see (6.26). In terms of the usual
Cartesian units vectors (x, y, z), the spherical units vectors (r, θ, ϕ) are

$$\left. \begin{array}{l} r = \sin \theta \cos \phi x + \sin \theta \sin \phi y + \cos \theta z \\ \theta = \cos \theta \cos \phi x + \cos \theta \sin \phi y - \sin \theta z \\ \phi = -\sin \phi x + \cos \phi y \end{array} \right\}. \tag{6.28}$$

Conversely, Cartesian unit vectors can be expressed in terms of the spherical unit vectors as

$$
\left.\begin{aligned}
x &= \sin\theta\cos\phi\, r + \cos\theta\cos\phi\, \theta - \sin\phi\, \phi \\
y &= \sin\theta\sin\phi\, r + \cos\theta\sin\phi\, \theta + \cos\phi\, \phi \\
z &= \cos\theta\, r - \sin\theta\, \theta
\end{aligned}\right\} . \tag{6.29}
$$

In the following section, it will prove useful to have available the derivatives of the spherical unit vectors with respect to θ and ϕ, written in terms of each other:

$$
\left.\begin{aligned}
\frac{\partial r}{\partial \theta} &= \theta & \frac{\partial r}{\partial \phi} &= \sin\theta\, \phi \\[2mm]
\frac{\partial \theta}{\partial \theta} &= -r & \frac{\partial \theta}{\partial \phi} &= \cos\theta\, \phi \\[2mm]
\frac{\partial \phi}{\partial \theta} &= 0 & \frac{\partial \phi}{\partial \phi} &= -\sin\theta\, r - \cos\theta\, \theta
\end{aligned}\right\} . \tag{6.30}
$$

These expressions were derived by differentiating (6.28) (remembering that (x, y, z) act like constants) and then substituting (6.29) into the results. It will also prove helpful to have expressions for gradient and Laplacian operators in spherical coordinates:

$$
\nabla \equiv r\frac{\partial}{\partial r} + \theta\frac{1}{r}\frac{\partial}{\partial \theta} + \phi\frac{1}{r\sin\theta}\frac{\partial}{\partial \phi}, \tag{6.31}
$$

and

$$
\nabla^2 \equiv \frac{1}{r^2}\frac{\partial}{\partial r}\left(r^2\frac{\partial}{\partial r}\right) + \frac{1}{r^2\sin\theta}\frac{\partial}{\partial \theta}\left(\sin\theta\frac{\partial}{\partial \theta}\right) + \frac{1}{r^2\sin^2\theta}\frac{\partial^2}{\partial \phi^2}. \tag{6.32}
$$

Complicated as these expressions are, we will find that they greatly facilitate application of Schrödinger's equation to naturally spherical systems such as the hydrogen atom.

One can imagine an "infinite spherical potential" analogous to the infinite box potential taken up in Sect. 6.1. Such a construct is considered to be a central potential and is examined in Chap. 7.

6.3 Angular Momentum Operators

The concept of angular momentum will prove central to the analysis of Schrödinger's equation for the hydrogen atom. In classical mechanics, the angular momentum of a particle is a vector quantity given by $L = r \times p$, where r and p are respectively the position and linear momentum of the particle and where $\times$ denotes a vector cross-product. The purpose of this section is to develop general expressions for quantum-mechanical operators for the components of angular momentum and for the square of the magnitude of angular momentum. Our derivation of the latter follows that developed by P. T. Leung [3].

We saw in Chap. 4 that the operator for linear momentum is given by

$$\boldsymbol{p}_{op} = -\iota\hbar\nabla, \tag{6.33}$$

where ∇ is the gradient operator. The quantum-mechanical operator for $\boldsymbol{L}$ is then defined as:

$$\boldsymbol{L}_{op} = \boldsymbol{r}_{op} \times \boldsymbol{p}_{op} = \iota\hbar(\boldsymbol{r} \times \nabla). \tag{6.34}$$

We reintroduce temporarily here the "op" notation of Chap. 4 to make clear what quantities are operators.

In Cartesian coordinates, $\boldsymbol{L}_{op}$ appears as

$$
\begin{aligned}
\boldsymbol{L}_{op} &\equiv \boldsymbol{r} \times \boldsymbol{p}_{op} \equiv (x\,\boldsymbol{x} + y\,\boldsymbol{y} + z\,\boldsymbol{z}) \times (-\iota\hbar) \left(\boldsymbol{x}\frac{\partial}{\partial x} + \boldsymbol{y}\frac{\partial}{\partial y} + \boldsymbol{z}\frac{\partial}{\partial z} \right) \\
&\equiv -\iota\hbar \left[xz\frac{\partial}{\partial y} - xy\frac{\partial}{\partial z} - yz\frac{\partial}{\partial x} + yx\frac{\partial}{\partial z} + zy\frac{\partial}{\partial x} - zx\frac{\partial}{\partial y} \right] \\
&\equiv -\iota\hbar \left[\boldsymbol{x}\left(y\frac{\partial}{\partial z} - z\frac{\partial}{\partial y} \right) + \boldsymbol{y}\left(z\frac{\partial}{\partial x} - x\frac{\partial}{\partial z} \right) + \boldsymbol{z}\left(x\frac{\partial}{\partial y} - y\frac{\partial}{\partial x} \right) \right],
\end{aligned}
$$

where $(\boldsymbol{x}, \boldsymbol{y}, \boldsymbol{z})$ again represent the usual Cartesian-coordinate units vectors. We can then write operators for the components of $\boldsymbol{L}$ as

$$L_x \equiv -\iota\hbar \left(y\frac{\partial}{\partial z} - z\frac{\partial}{\partial y} \right), \tag{6.35}$$

$$L_y \equiv -\iota\hbar \left(z\frac{\partial}{\partial x} - x\frac{\partial}{\partial z} \right), \tag{6.36}$$

and

$$L_z \equiv -\iota\hbar\left(x\frac{\partial}{\partial y} - y\frac{\partial}{\partial x}\right). \tag{6.37}$$

Notice the way in which the various coordinates enter these operators: no "x" appears in the operator for L_x; similarly no "y" appears in L_y, nor a "z" in L_z. In fact, L_y and L_z can be derived from L_x by performing what is called a "cyclical permutation" of the coordinates: everywhere a "y" appears, replace it by "z"; similarly, replace "z" with "x" (think of the order $x \to y \to z \to x \to$ etc.) L_x and L_z can similarly be derived from L_y, or L_x and L_y from L_z.

Example 6.2 Prove that $[L_x, L_y]\psi = (\iota\hbar L_z)\psi$.
The operators are

$$L_x \equiv -\iota\hbar\left(y\frac{\partial}{\partial z} - z\frac{\partial}{\partial y}\right),$$

$$L_y \equiv -\iota\hbar\left(z\frac{\partial}{\partial x} - x\frac{\partial}{\partial z}\right),$$

and

$$L_z \equiv -\iota\hbar\left(x\frac{\partial}{\partial y} - y\frac{\partial}{\partial x}\right).$$

Apply $[L_x, L_y]$ to some dummy wavefunction ψ:

$$\psi = (L_x L_y - L_y L_x)\psi$$

$$= -\hbar^2\left[\left(y\frac{\partial}{\partial z} - z\frac{\partial}{\partial y}\right)\left(z\frac{\partial\psi}{\partial x} - x\frac{\partial\psi}{\partial z}\right) - \left(z\frac{\partial}{\partial x} - x\frac{\partial}{\partial z}\right)\left(y\frac{\partial\psi}{\partial z} - z\frac{\partial\psi}{\partial y}\right)\right].$$

For convenience, denote derivatives by subscripts; e.g., $\psi_{xy} = \partial^2\psi/\partial x\partial y$. Expanding out the commutator gives

$$[L_x, L_y]\psi = -\hbar^2[y\psi_x + yz\psi_{zx} - xy\psi_{zz} - z^2\psi_{yx} + xz\psi_{yz}]$$

$$-[zy\psi_{xz} - z^2\psi_{xy} - xy\psi_{zz} + x\psi_y + xz\psi_{zy}].$$

On recalling the order of derivatives is irrelevant, that is, that $\psi_{xy} = \psi_{yx}$, a number of terms cancel; what remains is

$$\psi = -\hbar^2[y\psi_x - x\psi_y] = -\hbar^2\left[y\frac{\partial}{\partial x} - x\frac{\partial}{\partial y}\right]\psi$$

$$= -\hbar^2\left(\frac{L_z}{\iota\hbar}\right)\psi = (\iota\hbar L_z)\psi.$$

Problem 6.10 elaborates on this example to further show that $[L_y, L_z]\psi = (\iota\hbar L_x)\psi$ and $[L_z, L_x]\psi = (\iota\hbar L_y)\psi$. The physical meaning of these results is that it is not possible to measure simultaneous eigenvalues for the various components of L; they will be connected by uncertainty relations.

Since our ultimate purpose is to apply L_{op} to the Coulomb potential, it is convenient to have some of these results expressed purely in terms of spherical coordinates. Working from the expression for the gradient in spherical coordinates, (6.31), we have

$$L_{op} = -\iota\hbar(\boldsymbol{r} \times \nabla) = -\iota\hbar\left[r\boldsymbol{r} \times \left(\boldsymbol{r}\frac{\partial}{\partial r} + \boldsymbol{\theta}\frac{1}{r}\frac{\partial}{\partial\theta} + \boldsymbol{\phi}\frac{1}{r\sin\theta}\frac{\partial}{\partial\phi}\right)\right]$$

$$\tag{6.38}$$

$$= -\iota\hbar\left[\boldsymbol{\phi}\frac{\partial}{\partial\theta} - \boldsymbol{\theta}\frac{1}{\sin\theta}\frac{\partial}{\partial\phi}\right],$$

where we have used the identities $\boldsymbol{r} \times \boldsymbol{r} = 0$, $\boldsymbol{r} \times \boldsymbol{\theta} = \boldsymbol{\phi}$, and $\boldsymbol{r} \times \boldsymbol{\phi} = -\boldsymbol{\theta}$.

A physically important quantity is the square of the magnitude of L. An operator for computing this quantity, usually designated L_{op}^2 or just L^2, can be derived by operating L_{op} on itself. Since the vector equivalent of multiplication is to take a dot product, we have

$$L_{op} \cdot L_{op} = -\iota\hbar\left[\boldsymbol{\phi}\frac{\partial}{\partial\theta} - \boldsymbol{\theta}\frac{1}{\sin\theta}\frac{\partial}{\partial\phi}\right] \cdot -\iota\hbar\left[\boldsymbol{\phi}\frac{\partial}{\partial\theta} - \boldsymbol{\theta}\frac{1}{\sin\theta}\frac{\partial}{\partial\phi}\right]. \tag{6.39}$$

In expanding out this expression, we must be careful to respect the order of dot products and derivatives. Expanding from left to right gives

$$L_{op}^2 = -\hbar^2\left[\left(\boldsymbol{\phi}\frac{\partial}{\partial\theta}\right) \cdot \left(\boldsymbol{\phi}\frac{\partial}{\partial\theta}\right) - \left(\boldsymbol{\phi}\frac{\partial}{\partial\theta}\right) \cdot \left(\boldsymbol{\theta}\frac{1}{\sin\theta}\frac{\partial}{\partial\phi}\right)\right.$$

$$\tag{6.40}$$

$$\left. - \left(\boldsymbol{\theta}\frac{1}{\sin\theta}\frac{\partial}{\partial\phi}\right) \cdot \left(\boldsymbol{\phi}\frac{\partial}{\partial\theta}\right) + \left(\boldsymbol{\theta}\frac{1}{\sin\theta}\frac{\partial}{\partial\phi}\right) \cdot \left(\boldsymbol{\theta}\frac{1}{\sin\theta}\frac{\partial}{\partial\phi}\right)\right].$$

When evaluating the four terms in this expression, care must be taken to first compute the partial derivatives, including those of the unit vectors, and then to do the dot products. The third term is worked through as an example:

$$\left(\boldsymbol{\theta}\frac{1}{\sin\theta}\frac{\partial}{\partial\phi}\right)\cdot\left(\boldsymbol{\phi}\frac{\partial}{\partial\theta}\right) = \boldsymbol{\theta}\frac{1}{\sin\theta}\cdot\left[\frac{\partial\boldsymbol{\phi}}{\partial\phi}\frac{\partial}{\partial\theta} + \boldsymbol{\phi}\frac{\partial^2}{\partial\phi\partial\theta}\right]$$

$$= \frac{1}{\sin\theta}\boldsymbol{\theta}\cdot(-\sin\theta\boldsymbol{r} - \cos\theta\boldsymbol{\theta})\frac{\partial}{\partial\theta} = -\cot\theta\frac{\partial}{\partial\theta}. \tag{6.41}$$

In the first line of (6.41), the second term inside the square brackets vanished because $\boldsymbol{\theta}\cdot\boldsymbol{\phi} = 0$; in the second line, (6.30) in the preceding section was used to simplify $\partial\boldsymbol{\phi}/\partial\phi$, and in the last line we have written $\cos\theta/\sin\theta = \cot\theta$. The other three terms in (6.40) similarly simplify, leaving

$$L_{op}^2 = -\hbar^2\left[\frac{\partial^2}{\partial\theta^2} + \cot\theta\frac{\partial}{\partial\theta} + \frac{1}{\sin^2\theta}\frac{\partial^2}{\partial\phi^2}\right], \tag{6.42}$$

which, upon combining the first two terms, can be expressed more compactly as

$$L_{op}^2 = -\hbar^2\left[\frac{1}{\sin\theta}\frac{\partial}{\partial\theta}\left(\sin\theta\frac{\partial}{\partial\theta}\right) + \frac{1}{\sin^2\theta}\frac{\partial^2}{\partial\phi^2}\right]. \tag{6.43}$$

This is our desired expression for the operator for the squared magnitude of angular momentum expressed in spherical coordinates. Now compare (6.43) with the angular part of the expression for the the Laplacian operator in spherical coordinates, (6.32). But for a factor of $-\hbar^2/r^2$, they are identical. We can thus write the Laplacian in terms of L_{op}^2 as

$$\nabla^2 = \frac{1}{r^2}\frac{\partial}{\partial r}\left(r^2\frac{\partial}{\partial r}\right) - \frac{L_{op}^2}{\hbar^2 r^2}. \tag{6.44}$$

The Hamiltonian operator $H = -(\hbar^2/2m)\nabla^2 + V(r, \theta, \phi)$ then appears in spherical coordinates as

$$H \equiv -\frac{\hbar^2}{2mr^2}\frac{\partial}{\partial r}\left(r^2\frac{\partial}{\partial r}\right) + \frac{L_{op}^2}{2mr^2} + V(r, \theta, \phi), \tag{6.45}$$

where $V(r, \theta, \phi)$ is the potential function.

One further expression that will prove useful in the following section is an operator for the magnitude of the z-component of angular momentum, $(L_z)_{op}$. The component of any vector in any desired direction can always be found by taking the dot product of the unit vector representing the desired direction with the vector at hand. From (6.38) and (6.29),

$$(L_z)_{op} = \boldsymbol{z}\cdot\boldsymbol{L}_{op} = (\cos\theta\boldsymbol{r} - \sin\theta\boldsymbol{\theta})\cdot -\iota\hbar\left[\boldsymbol{\phi}\frac{\partial}{\partial\theta} - \boldsymbol{\theta}\frac{1}{\sin\theta}\frac{\partial}{\partial\phi}\right],$$

that is,

$$
(L_z)_{op} = -\iota\hbar\left[(\cos\theta r)\cdot\left(\boldsymbol{\phi}\frac{\partial}{\partial\theta}\right) - (\cos\theta r)\cdot\left(\boldsymbol{\theta}\frac{1}{\sin\theta}\frac{\partial}{\partial\phi}\right)\right.
$$
$$
\left. - (\sin\theta\boldsymbol{\theta})\cdot\left(\boldsymbol{\phi}\frac{\partial}{\partial\theta}\right) + (\sin\theta\boldsymbol{\theta})\cdot\left(\boldsymbol{\theta}\frac{1}{\sin\theta}\frac{\partial}{\partial\phi}\right)\right].
$$

In this expression, the first three terms vanish because the dot product of orthogonal unit vectors always vanishes, no matter what coordinate system one is working in. What remains from the last term is

$$
(L_z)_{op} \equiv -\iota\hbar\frac{\partial}{\partial\phi}. \tag{6.46}
$$

This result can also be derived by substituting the expressions for the spherical-coordinate unit vectors in terms of their Cartesian counterparts, (6.28), into (6.38) and then taking the dot product with z.

A number of important points regarding angular momentum are worth emphasizing here. In Example 6.2 we concluded that since the operators for L_x, L_y, and L_z do not commute with each other, any pair of them will be linked by an uncertainty relation, that is, a measurement of any one component of L will introduce uncertainty into our knowledge of the other two. A measurement of L_y, for example, will force a system into an eigenstate of L_y which cannot simultaneously be an eigenstate of L_x or L_z. In Problem 6.18 and Chap. 8, however, we will see that $[L_{op}^2, (L_x)_{op}]\psi = [L_{op}^2, (L_y)_{op}]\psi = [L_{op}^2, (L_z)_{op}]\psi = 0$ for any wavefunction that is a product of separable functions of θ and ϕ. This means that eigenvalues for L^2 and any one component of L can in fact be found, but, once a measurement has been made, knowledge of the other two components is irretrievably lost. The measured component is conventionally taken to be L_z; the corresponding eigenfunctions of L^2 and L_z are the spherical harmonics developed in Sect. 6.5 below. Further, we will see in Sect. 6.4 that for a potential which is a function of r only, the wavefunction can be separated into the form $\psi(r, \theta, \phi) = R(r)g(\theta, \phi)$, and that it will satisfy $[H, L_{op}^2] = 0$ (Problem 6.19). This means that both energy and L_{op}^2 will possess simultaneous eigenvalues; the hydrogen-atom Coulomb potential satisfies this criterion. In such cases we can ultimately expect three quantum numbers to arise: one for energy, one for L_{op}^2, and one for one of the components of L, conventionally taken to be L_z. Solutions for the L_{op}^2 and L_z pieces are developed in Sect. 6.5.

Finally, we remark that it is possible to define raising and lowering operators for angular momentum in analogy to the harmonic oscillator raising and lowering operators introduced in Chap. 5 (see Problem 6.8). These operators do not play a role in our solution of the Coulomb potential, however, so we do not elaborate on them here; readers interested in learning more about these operators are directed to Chap. 8.

6.4 Separation of Variables in Spherical Coordinates: Central Potentials

In this section we apply the separation of variables technique to Schrödinger's equation written in spherical coordinates for central potentials, that is, ones which can be written in the form $V(r)$. As emphasized in the introduction to this chapter, these are potentials which depend only on the magnitude of the separation between two particles and not on their orientation in space. Spherical coordinates are the natural system within which to examine this type of potential; the usual xyz Cartesian coordinates are not used because any function of $r = \sqrt{x^2 + y^2 + z^2}$ is not suitable for separation.

It is worth emphasizing that the results developed in the present section are very general for any central potential, irrespective of the precise dependence of $V(r)$ on r. In anticipation of applying these results to the Coulomb potential, one might wonder how the fact that the hydrogen atom is really a two-body system with both the proton and electron "orbiting" their mutual center of mass is built into Schrödinger's equation. For the moment, suffice it to say that the situation can be reduced to an equivalent one-body problem in a way that has no effect on the radial/angular separation that is the concern of the present section; the two-body issue is taken up in Sect. 4 of Chap. 7.

Using the expression for ∇^2 in spherical coordinates, Schrödinger's equation for a particle of mass μ moving in a central potential $V(r)$ is

$$
-\frac{\hbar^2}{2\mu}\left[\frac{1}{r^2}\frac{\partial}{\partial r}\left(r^2\frac{\partial}{\partial r}\right) + \frac{1}{r^2\sin\theta}\frac{\partial}{\partial\theta}\left(\sin\theta\frac{\partial}{\partial\theta}\right)\right.
$$
$$
\left. +\frac{1}{r^2\sin^2\theta}\frac{\partial^2}{\partial\phi^2}\right]\psi(r,\theta,\phi) + V(r)\psi(r,\theta,\phi) = E\psi(r,\theta,\phi). \tag{6.47}
$$

We use μ to designate mass in this chapter and in Chap. 7 as the symbol "m" will soon be adopted to designate a quantity known as the magnetic quantum number.

We now attempt a separation of variables into radial and angular terms by assuming that $\psi(r, \theta, \phi)$ can be expressed as the product of a "radial wavefunction" $R(r)$ and an "angular wavefunction" $Y(\theta, \phi)$:

$$
\psi(r,\theta,\phi) = R(r)Y(\theta,\phi). \tag{6.48}
$$

As in Sect. 6.1, upper-case letters represent functions of variables represented by lower-case letters. For convenience we shall often refer to these functions simply as R and Y, dropping the implicitly understood arguments. Substituting (6.48) into (6.47) gives

$$-\frac{\hbar^2}{2\mu}\left[\frac{Y}{r^2}\frac{\partial}{\partial r}\left(r^2\frac{\partial R}{\partial r}\right)+\frac{R}{r^2}\left[\frac{1}{\sin\theta}\frac{\partial}{\partial\theta}\left(\sin\theta\frac{\partial Y}{\partial\theta}\right)+\frac{1}{\sin^2\theta}\frac{\partial^2 Y}{\partial\phi^2}\right]\right]$$

$$\text{(6.49)}$$

$$+ V(r)(RY) = E(RY).$$

Multiplying through by r^2/RY and rearranging gives

$$-\frac{\hbar^2}{2\mu}\left[\frac{1}{R}\frac{\partial}{\partial r}\left(r^2\frac{\partial R}{\partial r}\right)\right]+[V(r)-E]r^2$$

$$\text{(6.50)}$$

$$=\frac{\hbar^2}{2\mu}\left[\frac{1}{Y}\frac{1}{\sin\theta}\frac{\partial}{\partial\theta}\left(\sin\theta\frac{\partial Y}{\partial\theta}\right)+\frac{1}{\sin^2\theta}\frac{\partial^2 Y}{\partial\phi^2}\right].$$

Equation (6.50) expresses the desired separation: the left side is a function of r only [$f(r)$, say], while the right side is a function of θ and ϕ, [say $g(\theta,\phi)$)]. The equality can only hold if both sides are equal to some constant:

$$f(r) = g(\theta,\phi) = \text{constant}. \qquad \text{(6.51)}$$

This result can also be expressed in terms of the operator for squared total angular momentum derived in the preceding section:

$$-\frac{\hbar^2}{2\mu}\left[\frac{1}{R}\frac{\partial}{\partial r}\left(r^2\frac{\partial R}{\partial r}\right)\right]+[V(r)-E]r^2 \qquad \text{(6.52)}$$

$$=-\frac{1}{2\mu Y}[L_{op}^2 Y]=\text{constant}.$$

This expression is quite general for any central potential.

A remark concerning normalization of the eventual solution is worthwhile here. Normalization demands that the square of ψ integrated over the entire space available to the electron must evaluate to unity. We can express this as

$$\int_0^{2\pi}\int_0^{\pi}\int_0^{\infty}\psi^2(r,\theta,\phi)r^2\sin\theta\,dr\,d\theta\,d\phi = 1,$$

or

$$\left(\int_0^{\infty}R^2(r)r^2dr\right)\left(\int_0^{2\pi}\int_0^{\pi}Y^2(\theta,\phi)\sin\theta\,d\theta\,d\phi\right)=1.$$

If R and Y are separately normalized, then ψ will be as well; it is conventional to separately normalize each function in such a product wavefunction. Anticipating things somewhat, we remark that, in the following section, Y will be broken into individual functions of θ and ϕ, which means that we will have three functions to be individually normalized.

We now turn to the solution of the right side of (6.50) [or (6.52)], the angular equation for any central potential.

6.5 Angular Wavefunctions and Spherical Harmonics

We seek functions $Y(\theta, \phi)$ such that the right side of (6.50) above is a constant. It is convenient to define this constant in such a way that there are no factors of μ or $\hbar$ to carry along in the manipulations. To this end, if we define the constant appearing in the right side of (6.51) to be

$$\text{constant} = -\frac{\alpha\hbar^2}{2\mu}, \tag{6.53}$$

where α is another constant, then the angular equation can be written as

$$-\left[\frac{1}{\sin\theta}\frac{\partial}{\partial\theta}\left(\sin\theta\frac{\partial Y}{\partial\theta}\right) + \frac{1}{\sin^2\theta}\frac{\partial^2 Y}{\partial\phi^2}\right] = \alpha Y. \tag{6.54}$$

We can simplify this expression by another application of separation of variables. Assume that $Y(\theta, \phi)$ can be expressed as the product of a function of θ and a function of ϕ:

$$Y(\theta, \phi) = \Theta(\theta)\Phi(\phi). \tag{6.55}$$

With this assumption, (6.54) becomes

$$-\frac{\Phi}{\sin\theta}\frac{\partial}{\partial\theta}\left(\sin\theta\frac{\partial\Theta}{\partial\theta}\right) - \frac{\Phi}{\sin^2\theta}\frac{\partial^2\Phi}{\partial\phi^2} = \alpha\Theta\Phi, \tag{6.56}$$

where partial derivatives have converted to ordinary derivatives as they operate on functions of only one variable. Multiplying through by $\sin^2\theta/\Theta\Phi$ and rearranging gives

$$-\frac{\sin\theta}{\Theta}\frac{\partial}{\partial\theta}\left(\sin\theta\frac{\partial\Theta}{\partial\theta}\right) - \alpha\sin^2\theta = \frac{1}{\Phi}\frac{\partial^2\Phi}{\partial\phi^2}. \tag{6.57}$$

A further separation of variables has been effected: the angular equation reduces to a function of θ equal to a function of ϕ, both of which must equal some constant.

6.5.1 Solution of the Φ Equation

We arbitrarily define the constant to which both sides of (6.57) are equal to be $-m^2$. m is known as the magnetic quantum number, a designation whose meaning will be made clear later.

With this definition we have

$$\frac{d^2\Phi}{d\varphi^2} = -m^2\,\Phi, \tag{6.58}$$

with m to be determined by appropriate boundary conditions. This differential equation is identical in form to that for the infinite potential well of Sect. 3.2. The solution is

$$\Phi(\varphi) = A\,e^{\imath m\varphi} + B\,e^{-\imath m\varphi} \tag{6.59}$$

To normalize Φ we must demand

$$\int_0^{2\pi} \Phi^*\Phi\,d\varphi = 1.$$

Carrying out the integral results in a constraint involving A, B, and m:

$$2\pi(A^*A + B^*B) - \frac{A^*B}{2\imath m}(e^{-4\imath m\pi} - 1) + \frac{AB^*}{2\imath m}(e^{4\imath m\pi} - 1) = 1. \tag{6.60}$$

Now, $\Phi(\phi)$ must satisfy one very important additional condition: ϕ is a *cyclic* coordinate, that is, if the value of ϕ is changed by adding any integral multiple (positive, negative, or zero) of 2π radians, then it must describe the same direction in space as it did initially. Mathematically, this can be expressed as

$$\Phi(\phi) = \Phi(\phi + 2\pi),$$

or

$$\Phi(\varphi) = Ae^{\imath m\varphi} + Be^{-\imath m\varphi} = Ae^{\imath m(\varphi+2\pi)} + Be^{-\imath m(\varphi+2\pi)}. \tag{6.61}$$

We can rewrite the right side of this constraint as

$$A\,e^{\imath m\varphi} + B\,e^{-\imath m\varphi} = A\,e^{\imath m\varphi}(e^{2\pi\imath m}) + B\,e^{-\imath m\varphi}(e^{-2\pi\imath m}) \tag{6.62}$$

This can only be satisfied if $e^{\pm 2\pi \iota m} = 1$, that is, if

$$\cos(2\pi m) \pm \iota \sin(2\pi m) = 1,$$

which can only be satisfied if

$$m = 0, \pm 1, \pm 2, \pm 3, \dots. \tag{6.63}$$

Once again, imposition of a boundary condition has led to a quantization condition on a separation constant. On back-substituting this constraint into (6.60), we have

$$2\pi \left(A^* A + B^* B\right) = 1. \tag{6.64}$$

Having accounted for both the normalization of Φ and the cyclic nature of ϕ, we are at liberty to choose A and B as we please, provided that (6.64) is satisfied. Tradition is to take $B = 0$, giving

$$A = 1/\sqrt{2\pi}, \tag{6.65}$$

and hence

$$\Phi(\varphi) = \frac{1}{\sqrt{2\pi}} e^{\iota m \varphi}, \quad m = 0, \pm 1, \pm 2, \pm 3, \dots. \tag{6.66}$$

Another way of looking at this choice is that since m is allowed to be both positive and negative, either the A or B term in (6.64) is redundant and can be dropped; we choose to drop the B term. Note that $|\Phi|^2$ is independent of ϕ; this feature will be useful when we come to constructing and interpreting plots of hydrogen-atom probability densities in Chap. 7.

The integral restriction on m implies that something physical is being quantized, but what? To elucidate this, consider the operator for the z-component of angular momentum, (6.46):

$$L_z = -\iota \hbar \frac{\partial}{\partial \varphi}.$$

Applying L_z to Φ gives

$$L_z \Phi(\varphi) = -\iota \hbar \frac{\partial}{\partial \varphi} (A e^{\iota m \varphi}) = -\iota^2 m \hbar (A e^{\iota m \varphi}) = (m \hbar) \, \Phi(\varphi). \tag{6.67}$$

This is an eigenvalue equation. *This result indicates that for a particle moving in a central potential, the z-component of its angular momentum is quantized in integral multiples of $\hbar$.* In the case of the Coulomb potential, the angular momentum is that of the proton/electron system "orbiting" about their mutual center of mass.

The forms of (6.35)–(6.37) should convince you that none of the x, y, and z components of L is any more "special" than either of the others. Consequently, as discussed following (6.46), *all* components of L are quantized in this way, not just the z-component. The compact form of (6.67) is a result of the way that the angle ϕ is defined in spherical coordinates; it has become conventional (if misleading) to consider the z-component as the "one" that is quantized. Ultimately, L will be quantized along a direction defined by an experimental arrangement; see Chap. 8. Also, bear in mind that (6.67) is a statement regarding eigenvalues, not expectation values: a system might well find itself in a superposition of angular momentum states.

Example 6.3 Equation (6.66) clearly satisfies (6.58). However, if $m = 0$, the normalization condition, (6.60), would appear to involve infinities. Without invoking this form of normalization, show that (6.66) is valid for $m = 0$.

With $m = 0$, (6.58) becomes

$$\frac{d^2\Phi}{d\phi^2} = 0,$$

which has the solution $\Phi = A\phi + B$. Normalization then demands

$$\int_0^{2\pi} \Phi^2 d\varphi = 1 \Rightarrow \frac{(2\pi)^3}{3} A^2 + (2\pi)^2 AB + (2\pi) B^2 = 1.$$

Now, the cyclic nature of ϕ demands $\Phi(\phi) = \Phi(\phi + 2\pi)$, or

$$A(\phi + 2\pi) + B = A\phi + B.$$

This can only be satisfied if $A = 0$. The normalization condition then gives $B = 1/\sqrt{2\pi}$, leading to an identical solution for Φ as (6.66) when $m = 0$. As a further exercise, you should be able to show that (6.60) does not in fact diverge as $m \to 0$.

6.5.2 Solution of the Θ Equation

From (6.57) in combination with the result for Φ derived in the preceding subsection, we have

$$-\frac{\sin\theta}{\Theta}\frac{d}{d\theta}\left(\sin\theta\frac{d\Theta}{d\theta}\right) - \alpha\sin^2\theta = -m^2. \tag{6.68}$$

Recall that m is the magnetic quantum number, not a mass. Multiplying through by $-\Theta/\sin^2\theta$ casts this equation into a somewhat neater form:

$$\frac{1}{\sin\theta}\frac{d}{d\theta}\left(\sin\theta\frac{d\Theta}{d\theta}\right)+\left(\alpha-\frac{m^2}{\sin^2\theta}\right)\Theta=0. \tag{6.69}$$

The presence of $\sin\theta$ and $\sin^2\theta$ in this differential equation hint that the solution likely involves a polynomial series in sines and cosines of θ. Because the algebra involved in handling infinite series of trigonometric functions is awkward, it is convenient to introduce an independent variable which transforms the differential equation into a linear form. Define

$$\chi=\cos\theta. \tag{6.70}$$

With this definition we have

$$d\chi=-\sin\theta\,d\theta=-\sqrt{1-\chi^2}\,d\theta$$

or

$$\frac{d}{d\theta}=-\sqrt{1-\chi^2}\frac{d}{d\chi}. \tag{6.71}$$

Substituting (6.70) and (6.71) into (6.69) gives

$$\frac{d}{d\chi}\left[(1-\chi^2)\frac{d\Theta}{d\chi}\right]+\left[\alpha-\frac{m^2}{(1-\chi^2)}\right]\Theta=0, \tag{6.72}$$

where Θ is now regarded as a function of χ.

At this point it would seem reasonable to attempt a series expansion of Θ in terms of powers of χ. Such an attempt leads to a problem, however: the recursion relation involves more than two coefficients, making it impossible to terminate the series by the sort of artifice employed in the solution of the harmonic oscillator. However, it is possible to transform (6.72) into a form wherein a series solution yields a more tractable recursion relation. Define $\Theta(\chi)$ in terms of another function $G(\chi)$ according as

$$\Theta(\chi)=(1-\chi^2)^{|m|/2}G(\chi). \tag{6.73}$$

With this definition, (6.72) transforms into a differential equation for G:

$$(1-\chi^2)\frac{d^2G}{d\chi^2}-2\chi(|m|+1)\frac{dG}{d\chi}+[\alpha-|m|(|m|+1)]G(\chi)=0. \tag{6.74}$$

Equation (6.73) is by no means an obvious substitution, and deserves a comment. Readers familiar with advanced techniques of solving differential equations will recognize this substitution as one that renders the Θ-equation in the form of a hypergeometric equation, a standard form for a second-order differential equation.

We now attempt a series solution of the form

$$G(\chi) = \sum_{n=0}^{\infty} a_n \chi^n. \tag{6.75}$$

Substituting this trial solution into (6.74) gives

$$\sum_{n=0}^{\infty} n(n-1)a_n \chi^{n-2} + \sum_{n=0}^{\infty} [\beta - 2(|m|+1)n - n(n-1)] a_n \chi^n = 0, \tag{6.76}$$

where

$$\beta = [\alpha - |m|(|m|+1)]. \tag{6.77}$$

The first term of (6.76) vanishes for $n = 0$ and $n = 1$. For the remaining values of n, we can write

$$\sum_{n=0}^{\infty} n(n-1)a_n \chi^{n-2} = \sum_{n=2}^{\infty} n(n-1)a_n \chi^{n-2} = \sum_{j=0}^{\infty} (j+1)(j+2)a_{j+2}\chi^j,$$

where $j = n - 2$. Since j is a dummy index, we are at liberty to re-label it as n, just as we did in the case of the harmonic oscillator. The result of this is to reduce (6.76) to the form

$$\sum_{n=0}^{\infty} \{[\beta - 2(|m|+1)n - n(n-1)] a_n + (n+1)(n+2)a_{n+2}\} \chi^n = 0. \tag{6.78}$$

Since $\chi \,(= \cos\theta)$ is not in general equal to zero, we are forced to establish the recursion relation

$$a_{n+2} = -\frac{[\beta - 2(|m|+1)n - n(n-1)]}{(n+1)(n+2)} a_n, \tag{6.79}$$

or, on substituting for β,

$$a_{n+2} = \frac{[(n+|m|)(n+|m|+1) - \alpha]}{(n+1)(n+2)} a_n. \tag{6.80}$$

Again we have a recursion relation in which there are two arbitrary constants, a_0 and a_1, corresponding to even and odd series, respectively.

Now, refer back to (6.73). Since χ was defined as $\chi = \cos\theta$ and since $0 \le \theta \le \pi$, then $-1 \le \chi \le 1$, that is, χ itself can never diverge. Hence, $1 - \chi^2$ will never diverge. But what about the convergence of $G(\chi)$? We will not investigate this rigorously, but we will give an informal argument to show that $G(\chi)$ does diverge for certain values of χ.

First we note that, for large n, the recursion relation gives

$$a_{n+2} \sim a_n, \tag{6.81}$$

that is, in either the even or the odd series, all of the high-n coefficients are of the same sign and approximately the same value. Hence, $G(\chi)$ would appear to converge since χ^n diminishes rapidly for large n. A problem occurs, however, when $\chi = \pm 1$ ($\theta = 0$ or π), for which the high-order terms in the series for $G(\chi)$ behave as

$$\pm \text{constant}(\cdots + 1 + 1 + 1 + 1 \ldots),$$

which is clearly divergent. [Strictly, as $\chi \to \pm 1$, $\Theta(\chi) \to (0)(\infty)$, which is indeterminate. But it is possible to show that $\theta(\chi)$ diverges as $\chi \to \pm 1$: see Problem 6.15.] To avoid this catastrophe, it is necessary to terminate the solution for $G(\chi)$ at some maximum value of n; that is, to demand that

$$a_n = 0, n > n_{\max}. \tag{6.82}$$

Since the recursion relation involves every second coefficient, either the even or the odd series will also have to be suppressed by assuming in addition that either a_0 or a_1 vanishes.

For (6.82) to be satisfied, the recursion relation demands

$$\alpha = [(n_{\max} + |m|)(n_{\max} + |m| + 1)]. \tag{6.83}$$

The choice of $n_{\max}$ is as yet arbitrary, except that it must be an integer ≥ 0: as long as the series for $G(\chi)$ contains a finite (even if very large) number of terms, it can never diverge. Since $|m| = 0, 1, 2, \ldots$, then $(n_{\max} + |m|) = 0, 1, 2, \ldots$. Defining $\ell = (n_{\max} + |m|)$, we can write

$$\alpha = \ell(\ell + 1), \quad \ell = 0, 1, 2, \ldots. \tag{6.84}$$

ℓ is known as the orbital angular momentum quantum number. To understand its physical significance, recall (6.52):

$$-\frac{1}{2\mu Y}[L_{op}^2 Y] = \text{constant}.$$

Recalling the definition of the constant from (6.53) casts this into the form

$$L_{op}^2 Y(\theta, \phi) = \hbar^2 \alpha Y(\theta, \phi) = [\hbar^2 \ell(\ell + 1)] Y(\theta, \phi). \tag{6.85}$$

Once again we have an eigenvalue equation. The interpretation of this result is that the square of the magnitude of the angular momentum of the system is quantized in units of $\hbar^2$ or,

$$L = \sqrt{\ell(\ell + 1)}\hbar \quad \ell = 0, 1, 2, 3, \ldots . \tag{6.86}$$

In the Coulomb case, the total angular momentum is essentially the electron's orbital angular momentum. We can combine this result with our previous conclusion that $L_z = m\hbar$. Since we must obviously have $L_z \leq |L|$, the possible values that m can take are restricted once ℓ is chosen: $m \leq \sqrt{\ell(\ell + 1)}$. Since m and ℓ are integers, this corresponds to $|m| \leq \ell$, hence

$$L_z = m\hbar, \quad (m \leq \ell). \tag{6.87}$$

The essential physical conclusion here is that the lengths of both L and L_z are quantized according as (6.86) and (6.87).

As an example, consider the case of $\ell = 2$, that is, $L = \sqrt{6}\hbar$, for which $L_z = m\hbar$, $-2 \leq m \leq 2$. In the convention where the z-component of L is taken to be quantized as $m\hbar$, the L vector can lie on any one of five cones concentric with the z-axis consistent with its z-component being an integral multiple of $\hbar$ lying between -2 and $+2$, inclusive. This is illustrated in Figs. 6.5 and 6.6.

Quantization of angular momentum components is another purely quantum-mechanical effect; there is no analog in classical mechanics. Experimental proof of

Fig. 6.5 Possible orientations for L for $\ell = 2$, $|L| = \sqrt{6}\hbar$

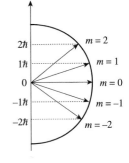

Fig. 6.6 For $\ell = 2, m = 2$, L can lie anywhere on the cone

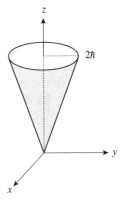

this effect is powerful evidence in support of Schrödinger's wave mechanics. Ironically, the experimental verification, due to Stern and Gerlach, was in hand before the arrival of Schrödinger's equation. So as not to disturb the flow of the present chapter, a discussion of this important experiment and its interpretation appears in Chap. 8. An excellent article on the history of the Stern-Gerlach experiment can be found in [4].

6.5.3 Spherical Harmonics

What of the detailed solutions for $G(\chi)$ in (6.73)? The series solution gave

$$\Theta_{\ell,m}(\theta) = (1 - \chi^2)^{|m|/2} G(\chi) = (\sin\theta)^{|m|} G(\cos\theta).$$

with

$$G(\cos\theta) = \sum_{n=0}^{\ell-|m|} a_n (\cos\theta)^n,$$

with coefficients a_n dictated by the recursion relation (6.80). The solution for $\Theta(\theta)$ now contains two indices, ℓ and m, which together designate a particular (L, L_z) state; the upper limit of the sum for $G(\chi)$ in (6.75) is consequently written as $n_{max} = \ell - |m|$ in view of the discussion following (6.83).

With a view to the fact that we will eventually wish to compute probabilities, it is useful to choose a_0 or a_1 (as applicable) for a given (ℓ, m) to be such that $\Theta_{\ell,m}(\theta)$ is normalized, that is, such that

$$\int_0^\pi \left|\Theta_{\ell,m}(\theta)\right|^2 \sin\theta d\theta = 1.$$

The details of this normalization are rather involved (see Appendix VI of Pauling and Wilson); we omit them in favor of quoting the final results:

$$\Theta_{\ell,m}(\theta) = \sqrt{\frac{2\ell+1}{2}\frac{(\ell-m)!}{(\ell+m)!}} P_{\ell,m}(\cos\theta) \quad (|m| \leq \ell). \tag{6.88}$$

$P_{\ell,m}(\cos\theta)$ is known as an Associated Legendre function, which is expressed in terms of a general argument x by

$$P_{\ell,m}(x) = \frac{1}{2^\ell \ell!}(1-x^2)^{m/2}\frac{d^{\ell+m}}{dx^{\ell+m}}(x^2-1)^\ell. \tag{6.89}$$

In our case, $x \equiv \cos\theta$. Both (6.88) and (6.89) are valid for m positive, negative, or zero; further details on the properties of these functions are given in Chap. 11 of Weber and Arfken.

For computational purposes, this form for the Associated Legendre functions is not convenient as it involves computing a derivative of $(x^2 - 1)^\ell$. More practical would be a form wherein $\Theta(\theta)$ is expressed directly as a series in terms of, say, $\sin\theta$ or $\cos\theta$. This can be achieved via a binomial-expansion method. The details of this manipulation are given in Appendix A3; the result is

$$\Theta(\theta) = A_{\ell,m}(\sin^m\theta) \sum_{k=0}^{[\frac{\ell-m}{2}]} \left[(-1)^k \binom{\ell}{k} \binom{2\ell - 2k}{\ell + m} (\cos^{\ell-m-2k}\theta) \right], \quad (6.90)$$

where the prefactor $A_{\ell,m}$ is given by

$$A_{\ell,m} = \frac{1}{2^\ell \ell!} \sqrt{\frac{2\ell + 1}{2}(\ell + m)!(\ell - m)!}. \quad (6.91)$$

A few remarks on the nature of $\Theta(\theta)$ that will be useful when we come to interpret plots of hydrogen atom wavefunctions can be made at this point. By studying (6.90), you should be able to convince yourself that the sum appearing therein is a polynomial in $\cos\theta$ whose highest order (when $k = 0$) is always $\ell - m$ and whose lowest order depends on the evenness or oddness of $\ell - m$. Neglecting numerical coefficients, the series behave as

$$(\ell - m)_{even} \Rightarrow \Theta_{\ell,m}(\theta) \sim \sin^m\theta[\cos^{\ell-m}\theta + \cos^{\ell-m-2}\theta + \cdots + 1] \quad (6.92)$$

$$(\ell - m)_{odd} \Rightarrow \Theta_{\ell,m}(\theta) \sim \sin^m\theta[\cos^{\ell-m}\theta + \cos^{\ell-m-2}\theta + \cdots + \cos\theta]. \quad (6.93)$$

Consider the factor of $\sin^m\theta$ that appears in these expressions. If $m \neq 0$, this term will force $\Theta_{\ell,m} = 0$ along $\theta = 0$, that is, there will be what we can call an "angular node" along the z-axis. Conversely, if $m = 0$, there will never be an angular node along the z-axis.

Now consider the terms in the square brackets in (6.92) and (6.93). In each case the polynomials in $\cos\theta$ will have $(\ell - m)$ zeros, that is, there will be $(\ell - m)$ angular nodes. In the case of $(\ell - m)$ even, $\Theta_{\ell,m}$ will be non-zero along $\theta = \pi/2$. If $(\ell - m)$ is odd, the square bracket will lead to an angular node along $\theta = \pi/2$. We can summarize these results as follows:

(i) If in a plot of $\Theta(\theta)$ there is an angular node along z axis, then $m \neq 0$, and the total number of angular nodes will be $(\ell - m + 1)$, including the one along the z-axis. If there is not an angular node along z axis then we must have $m = 0$, and the total number of angular nodes will be $(\ell - m)$, or simply ℓ.

(ii) If there is an angular node along $\theta = \pi/2$, then $(\ell - m)$ must be odd; if there is not, then $(\ell - m)$ must be even. This rule is independent of and in addition to rule (i).

Finally, the complete solution to the angular part of Schrödinger's equation for a central potential, $Y_{\ell,m}(\theta, \phi) = \Theta(\theta)\Phi(\phi)$, is given by the product of (6.66) and (6.90). Traditionally, a further multiplicative factor of $(-1)^m$ is included in this definition: $Y_{\ell,m}(\theta, \phi) = (-1)^m\Theta(\theta)\Phi(\phi)$. This factor, known as the Condon-Shortley phase, is a convention that serves to define spherical harmonics for negative values of m (see Chap. 11 of Weber and Arfken). Hence we have

$$Y_{\ell,m}(\theta, \varphi)$$

$$= \frac{(-1)^m}{\sqrt{2\pi}}\left\{A_{\ell,m}(\sin^m\theta)\sum_{k=0}^{\left[\frac{\ell-m}{2}\right]}\left[(-1)^k\binom{\ell}{k}\binom{2\ell-2k}{\ell+m}(\cos^{\ell-m-2k}\theta)\right]\right\}e^{\iota m\varphi}$$

$$(6.94)$$

where the prefactor $A_{\ell,m}$ is as in (6.91).

These functions are known as *spherical harmonics*. Since both Θ and Φ were independently normalized, the $Y_{\ell,m}$ are as well. Spherical harmonics up to $\ell = 3$ are listed in Table 6.1.

Figure 6.7 shows pseudo-three-dimensional plots of $|Y_{\ell,m}|^2$ for various (ℓ, m). Study these figures carefully and convince yourself of the veracity of the foregoing remarks concerning angular nodes; recall that θ is measured from the vertically-upward z-axis (north pole). Since $|Y|^2$ is independent of ϕ, all of these figures are rotationally symmetric around the z-axis.

In Chap. 7 we will see that these functions manifest themselves in the probability distribution for finding the electron in an infinitesimal volume of space dV at location (r, θ, ϕ) in a hydrogen atom:

Table 6.1 Spherical harmonics

$Y_{0,0}(\theta, \varphi) = \frac{1}{\sqrt{4\pi}}$
$Y_{1,0}(\theta, \varphi) = \sqrt{\frac{3}{4\pi}}\cos\theta$
$Y_{1,\pm1}(\theta, \varphi) = \mp\sqrt{\frac{3}{8\pi}}\sin\theta e^{\pm\iota\varphi}$
$Y_{2,0}(\theta, \varphi) = \sqrt{\frac{5}{4\pi}}\left(\frac{3}{2}\cos^2\theta - \frac{1}{2}\right)$
$Y_{2,\pm1}(\theta, \varphi) = \mp\sqrt{\frac{15}{8\pi}}\sin\theta\cos\theta e^{\pm\iota\varphi}$
$Y_{2,\pm2}(\theta, \varphi) = \sqrt{\frac{15}{32\pi}}\sin^2\theta e^{\pm2\iota\varphi}$
$Y_{3,0}(\theta, \varphi) = \sqrt{\frac{7}{16\pi}}(5\cos^3\theta - 3\cos\theta)e^{\pm2\iota\varphi}$
$Y_{3,\pm1}(\theta, \varphi) = \mp\sqrt{\frac{21}{64\pi}}\sin\theta(5\cos^2\theta - 1)e^{\pm\iota\varphi}$
$Y_{3,\pm2}(\theta, \varphi) = \sqrt{\frac{105}{32\pi}}\sin^2\theta\cos\theta e^{\pm2\iota\varphi}$
$Y_{3,\pm3}(\theta, \varphi) = \mp\sqrt{\frac{35}{64\pi}}\sin^3\theta e^{\pm3\iota\varphi}$

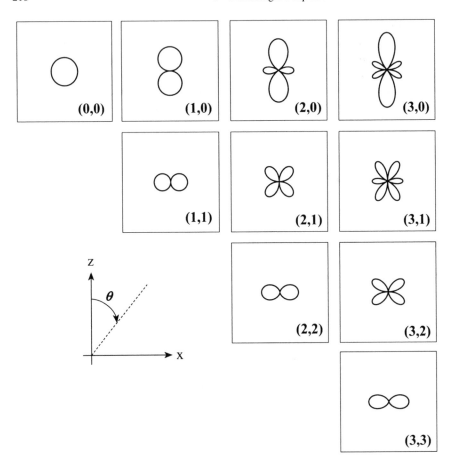

Fig. 6.7 Plots of the absolute values of spherical harmonics for various (ℓ, m) up to $\ell = 3$ in the xz-plane (x = horizontal; z = vertical). All panels cover the range ± 1 in both directions

$$P(r, \theta, \varphi) \, dV = \psi^* \psi \, dV = |R(r)|^2 \left| Y_{\ell, m}(\theta, \phi) \right|^2 r^2 \sin\theta \, dr \, d\theta \, d\phi. \qquad (6.95)$$

The forms of the graphs in Fig. 6.7 do not wholly dictate the probability distribution of the electron, as we have yet to determine the solution of the radial function $R(r)$ for the Coulomb potential. However, it is clear that the probability distributions will not be spherically symmetric; there will be regions of space wherein one is more or less likely to find the electron, depending on its (ℓ, m) state. Fundamentally, it is the shape of these $|Y|^2$ functions that dictate bonding orientations in molecules.

Finally, while it is not obvious from inspection of (6.94), it can be shown that $Y_{\ell, -m} = (-1)^m Y_{\ell, m}$; a proof appears in Appendix A4.

Example 6.4 Show that for $(\ell, m) = (3, 0)$, expression (6.94) yields $Y_{3,0}$ as given in Table 6.1.

First compute the prefactor:

$$A_{3,0} = \frac{1}{2^3 3!}\sqrt{\frac{2(3)+1}{2}}(3+0)!(3-0)! = \frac{1}{48}\sqrt{\frac{7}{2}(3!)^2} = \frac{1}{8}\sqrt{\frac{7}{2}}.$$

With $(\ell, m) = (3, 0)$, the upper index of summation in (6.94) will be 1, so we have, on incorporating the factor of $\sqrt{2\pi}$ into the prefactor,

$$Y_{3,0} = \frac{1}{8}\sqrt{\frac{7}{4\pi}}\sum_{k=0}^{1}\left[(-1)^k\binom{3}{k}\binom{6-2k}{3}\right](\cos^{3-2k}\theta)$$

$$= \frac{1}{8}\sqrt{\frac{7}{4\pi}}\left[\binom{3}{0}\binom{6}{3}\cos^3\theta - \binom{3}{1}\binom{4}{3}\cos\theta\right].$$

Recalling that

$$\binom{a}{b} = \frac{a!}{b!(a-b)!},$$

we have

$$Y_{3,0} = \frac{1}{8}\sqrt{\frac{7}{4\pi}}\left(20\cos^3\theta - 12\cos\theta\right).$$

Extract a factor of 4 from within the bracket and combine it with the factors outside to give

$$Y_{3,0} = \sqrt{\frac{7}{16\pi}}\left(5\cos^3\theta - 3\cos\theta\right),$$

in agreement with Table 6.1.

Summary

The central concepts introduced in this chapter are separation of variables in Cartesian and spherical coordinates, degeneracy, angular momentum operators, and the solution of the angular part of Schrödinger's equation for central potentials in terms of spherical harmonics. Readers should be aware that definitions and notation for some of the special mathematical functions involved here can vary; the definitions of the Associated Legendre functions and spherical harmonics used here are identical to those in Weber and Arfken.

Separation of variables is a standard technique used in the solution of partial differential equations, and is treated in any good mathematical physics text. Solutions of Schrödinger's equation in more than one dimension prove to be characterized by the presence of one quantum number for each dimension. Depending on the boundary conditions, different choices for the quantum numbers can lead to the same energy, a condition known as *degeneracy*.

The energy of a particle of mass m trapped in an infinite potential box of dimensions (a, b, c) is given in terms of the (n_x, n_y, n_z) quantum numbers by

$$E = \frac{h^2}{8m}\left(\frac{n_x^2}{a^2} + \frac{n_y^2}{b^2} + \frac{n_z^2}{c^2}\right), \quad (n_x, n_y, n_z) = 1, 2, 3, \ldots.$$

The number of individual energy states lying below energy E is

$$N(\leq E) = \frac{8^{3/2}\pi}{6}\frac{V\,m^{3/2}}{h^3}E^{3/2},$$

where $V = abc$ is the volume of the box.

Angular momentum L is a key concept in quantum mechanics, particularly in multi-body systems where a careful accounting of the vector nature of L is necessary. Operators for the x, y, and z components of angular momentum appear in Cartesian and spherical coordinates as

$$L_x \equiv -\iota\hbar\left(y\frac{\partial}{\partial z} - z\frac{\partial}{\partial y}\right) \equiv -\iota\hbar\left(-\sin\phi\frac{\partial}{\partial\theta} - \cot\theta\cos\phi\frac{\partial}{\partial\phi}\right),$$

$$L_y \equiv -\iota\hbar\left(z\frac{\partial}{\partial x} - x\frac{\partial}{\partial z}\right) \equiv -\iota\hbar\left(\cos\phi\frac{\partial}{\partial\theta} - \cot\theta\sin\phi\frac{\partial}{\partial\phi}\right),$$

and

$$L_z \equiv -\iota\hbar\left(x\frac{\partial}{\partial y} - y\frac{\partial}{\partial x}\right) \equiv -\iota\hbar\frac{\partial}{\partial\phi}.$$

The operator for the magnitude of the square of total angular momentum is

$$L^2 \equiv -\hbar^2\left[\frac{1}{\sin\theta}\frac{\partial}{\partial\theta}\left(\sin\theta\frac{\partial}{\partial\theta}\right) + \frac{1}{\sin^2\theta}\frac{\partial^2}{\partial\phi^2}\right].$$

With this result, the Hamiltonian (energy) operator for a system of mass μ can be written in spherical coordinates as

$$H \equiv -\frac{\hbar^2}{2\mu r^2}\frac{\partial}{\partial r}\left(r^2\frac{\partial}{\partial r}\right) + \frac{L_{op}^2}{2\mu r^2} + V(r, \theta, \phi).$$

Schröinger's equation in spherical coordinates for a particle of mass μ moving in a central potential $V(r)$ appears as

$$\frac{\hbar^2}{2\mu}\left\{\frac{1}{r^2}\frac{\partial}{\partial r}\left(r^2\frac{\partial}{\partial r}\right)+\frac{1}{r^2\sin\theta}\frac{\partial}{\partial\theta}\left(\sin\theta\frac{\partial}{\partial\theta}\right)+\frac{1}{r^2\sin^2\theta}\left(\frac{\partial^2}{\partial\phi^2}\right)\right\}\psi(r,\theta,\phi)$$

$$+V(r)\psi(r,\theta,\phi)=E\psi(r,\theta,\phi).$$

This equation is separable into a radial part $R(r)$ and an angular part $Y(\theta,\phi)$ on assuming

$$\psi(r,\theta,\phi)=R(r)Y(\theta,\phi).$$

The angular wavefunctions Y are given by

$$Y_{\ell,m}(\theta,\varphi)$$

$$=\frac{(-1)^m}{\sqrt{2\pi}}\left\{A_{\ell,m}(\sin^m\theta)\sum_{k=0}^{\left[\frac{\ell-m}{2}\right]}\left[(-1)^k\binom{\ell}{k}\binom{2\ell-2k}{\ell+m}(\cos^{\ell-m-2k}\theta)\right]\right\}e^{im\varphi}$$

with

$$A_{\ell,m}=\frac{1}{2^\ell\ell!}\sqrt{\frac{2\ell+1}{2}(\ell+m)!(\ell-m)!}.$$

The $Y_{\ell,m}$ are known as spherical harmonics; their shapes relate to bonding orientations of molecules. In these expressions, ℓ is the orbital angular momentum quantum number and m is the magnetic quantum number. Respectively, these two integers quantize the total and (conventionally) z – components of angular momentum of the system according as

$$L=\sqrt{\ell(\ell+1)}\hbar\quad(\ell=0,1,2,\ldots),$$

and

$$L_z=m\hbar\quad(|m|\leq\ell).$$

Problems

6.1(E) Consider an electron trapped in an infinite potential box of dimensions (1.0, 1.5, 1.9) Å. Tabulate all possible energy levels (in eV) for $(n_x, n_y, n_z) = 1$ to 3.

6.2(E) Consider an electron trapped in a box of volume 1 cubic centimeter. How many quantum states are available to the electron below $E = 10$ eV? How many states would lie between 10 eV and 10.01 eV?

6.3(E) For which of the following three-dimensional potentials would Schrödinger's equation be separable?

(a) $V(x, y, z) = x^2 y + \sin(z)$
(b) $V(x, y, z) = x^2 + y + Tan^{-1}(\sqrt{z})$
(c) $V(x, y, z) = e^x\, y^{7/2}\, z^2$
(d) $V(x, y, z) = y\sin(x) + z\cos(y) + y\tan(z)$
(e) $V(x, y, z) = y^{-4} + Sin^{-1}(e^{\sqrt{x}}) + \tan(z)$
(f) $V(x, y, z) = e^{xy\sqrt{z}}$
(g) $V(x, y, z) = e^{x+y-\sqrt{z}}$

6.4(E) In his development of thermal radiation theory, Planck assumed that photons in a blackbody cavity were of a continuum of frequencies/wavelengths as opposed to the discrete ones that the cavity could actually support on account of its finite dimensions. Planck's expression for the number of photon states between frequencies v and $v + dv$ that a cavity of volume V could support is

$$N(v)dv = \frac{8\pi V}{c^3} v^2 dv.$$

Transform this expression into an equivalent wavelength form. Now consider a tiny volume, say 1 mm³. How many photon states are possible between $\lambda = 5000$ Å and $\lambda = 5001$ Å, typical visible-light wavelengths? Was Planck justified in his assumption of an essentially continuous distribution of frequencies/wavelengths?

6.5(I) Verify the expressions given in the text for the derivatives of the spherical-coordinate unit vectors, (6.30).

6.6(I) Using the method demonstrated in (6.41), evaluate the other three terms in (6.40), and hence verify the expressions (6.42) and (6.43) for L_{op}^2.

6.7(I) Following the method that was used to derive the expression for $(L_z)_{op}$ in Sect. 6.3, show that the operators for the x and y components of angular momentum in spherical coordinates are given by

$$L_x \equiv -\iota\hbar\left(-\sin\phi\,\frac{\partial}{\partial\theta} - \cot\theta\cos\phi\,\frac{\partial}{\partial\phi}\right)$$

and

$$L_y \equiv -\iota\hbar \left(\cos\phi \frac{\partial}{\partial\theta} - \cot\theta \sin\phi \frac{\partial}{\partial\phi} \right).$$

6.8(E) In Chap. 8, the *angular momentum raising and lowering operators* $L_\pm = L_x \pm \iota L_y$ will be introduced. From your results in Problem 6.7, show that in spherical coordinates these appear as

$$L_\pm \equiv \hbar e^{\pm\iota\phi} \left\{ \pm\frac{\partial}{\partial\theta} + \iota\cot\theta \frac{\partial}{\partial\phi} \right\}.$$

6.9(A) Following the procedures of Sects. 6.2 and 6.3, derive expressions for the L^2_{op} operator in (a) Cartesian and (b) cylindrical coordinates.

6.10(I) In Example 6.2 it was shown that $[L_x, L_y] = (\iota\hbar)L_z$. Prove likewise that

$$\left[L_y, L_z \right] \equiv (\iota\hbar)L_x$$

and

$$\left[L_z, L_x \right] \equiv (\iota\hbar)L_y.$$

6.11(I) Prove that $[L_x, p_x] \equiv 0$, $[L_x, p_y] \equiv (\iota\hbar)p_z$, and $[L_x, p_z] \equiv -(\iota\hbar)p_y$. Permuting the coordinates leads to the identities $[L_y, p_x] \equiv -(\iota\hbar)p_z$, $[L_y, p_y] \equiv 0$, $[L_y, p_z] \equiv (\iota\hbar)p_x$, $[L_z, p_x] \equiv (\iota\hbar)p_y$, $[L_z, p_y] \equiv -(\iota\hbar)p_x$, and $[L_z, p_z] \equiv 0$.

6.12(I) Prove that $[L_x, x] \equiv 0$, $[L_x, y] \equiv (\iota\hbar)z$, and $[L_x, z] \equiv -(\iota\hbar)y$. Permuting the coordinates leads to the identities $[L_y, x] \equiv -(\iota\hbar)z$, $[L_y, y] \equiv 0$, $[L_y, z] \equiv (\iota\hbar)x$, $[L_z, x] \equiv (\iota\hbar)y$, $[L_z, y] \equiv -(\iota\hbar)x$, and $[L_z, z] \equiv 0$.

6.13(E) Show that (6.52) follows from (6.50) and the discussion in Sect. 6.3.

6.14(E) Prove that (6.60) does not diverge for $m \to 0$.

6.15(I) In Sect. 6.5.2 it was remarked that the angular function $\Theta(\chi) = (1 - \chi^2)^{|m|/2}G(\chi)$ is indeterminate for $\chi \to \pm 1$. Use L'Hospitals rule to prove this. Then try a different factorization for $\Theta(\chi)$, namely $\Theta(\chi) = (1 - \chi^2)^{-|m|/2}U(\chi)$. Show that the high-order terms in the resulting series solution for $U(\chi)$ behave in the same manner as do those for $G(\chi)$, and hence that $U(\chi)$ diverges for $\chi \to \pm 1$. From the uniqueness theorem for solutions of differential equations, what can you conclude about the way $G(\chi)$ was terminated in Sect. 6.5.2?

6.16(E) Verify the forms of the expressions for $\Theta_{\ell,m}(\theta)$ given in (6.92) and (6.93).

6.17(I) Working directly from (6.90) and (6.91), verify the expression given for $Y_{3,1}$ in Table 6.1.

6.18(I) Using spherical coordinates, show that $[L^2, L_z]\psi = 0$ if ψ is of the form $\psi = \Theta(\theta)\Phi(\phi)$, that is, if ψ is a separable function of θ and ϕ. If you are ambitious, show as well that $[L^2, L_x]\psi = 0$ and $[L^2, L_y]\psi = 0$, but be warned that the algebra is messy. An easier method is explored in Sect. 8.1.

6.19(I) Show that if the potential function is a function only of the radial coordinate r [that is, $V = V(r)$ only] and that the wavefunction can be separated as $\psi(r, \theta, \phi) = R(r)g(\theta, \phi)$, then it follows that that $[H, L^2] = 0$.

6.20(A) Consider a mass m moving in a spherical-coordinate potential of the form

$$V(r, \theta, \varphi) = V_r(r) + \frac{1}{r^2}\left[V_\theta(\theta) + \frac{V_\varphi(\varphi)}{(1 - \chi^2)}\right],$$

where $\chi = \cos\theta$. Show that on assuming a solution of the form

$$\psi(r, \theta, \varphi) = \frac{1}{r}R(r)\,\Theta(\theta)\,\Phi(\varphi),$$

Schrödinger's equation separates as

$$\begin{cases} \varepsilon\frac{d^2R}{dr^2} + \left(E - V_r - \frac{E_\theta}{r^2}\right)R = 0 \\[2mm] \varepsilon\left[(1 - \chi^2)\frac{d^2\Theta}{d\chi^2} - 2\chi\frac{d\Theta}{d\chi}\right] + \left[E_\theta - V_\theta - \frac{E_\varphi}{(1-\chi^2)}\right]\Theta = 0 \\[2mm] \varepsilon\frac{d^2\Phi}{d\varphi^2} + \left(E_\varphi - V_\varphi\right)\Phi = 0, \end{cases}$$

where $\epsilon = \hbar^2/2m$ and where E_θ and E_φ are separation constants defined as follows at appropriate points in the derivation:

$$f(r) = g(\theta, \varphi) = -E_\theta$$

and

$$h(\theta) = k(\varphi) = +E_\varphi.$$

This potential arises in the study of the motion of an electron in the field of a molecule with an electric dipole moment; see [5].

6.21(I) In more than one dimension, the "uncertainty packet" $\Delta r\,\Delta p$ is given by computing $\Delta r\,\Delta p = \sqrt{(\langle r^2\rangle - \langle r\rangle^2)(\langle p^2\rangle - \langle p\rangle^2)}$, where the square of a vector expectation value is given by a dot product as in $\langle r\rangle^2 = \langle r\rangle^2 = \langle r\rangle \cdot \langle r\rangle$. In three dimensions it can be shown that the uncertainty principle appears as $\Delta r\,\Delta p \geq (\sqrt{3}/2)\hbar$; see [6]. For a centrally symmetric potential in three dimensions we can put $\psi(r, \theta, \varphi) = R(r)Y_{\ell m}(\theta, \varphi)$ where the $Y_{\ell m}$ are the spherical harmonics of Sect. 6.5.3. Show that in such a case, $\langle r\rangle = 0$ always. [HINT: Note that the $Y_{\ell m}$ are explicitly complex. Use spherical coordinates; write $\mathbf{r} = r\hat{\mathbf{r}}$ where $\hat{\mathbf{r}}$ is the radial unit vector in spherical coordinates, (6.28)]. Since $\langle p \rangle = 0$ for any time-dependent wavefunction (Sect. 4.2), the uncertainty product for such wavefunctions reduces to $\Delta r \; \Delta p = \sqrt{\langle r^2 \rangle \langle p^2 \rangle}$.

6.22(I) In ordinary vector algebra, the cross-product of a vector with itself is always zero: $A \times A = 0$. When the vectors are operators, however, things can be different. Using spherical coordinates, show that the angular momentum operator behaves as $L \times L = (\iota \hbar) L$.

6.23(I) Show that $L \times r = (2\iota \hbar) r$.

References

1. L.P. Eisenhart, Phys. Rev. **45**, 427 (1934)
2. D.B. Beard, G.B. Beard, *Quantum Mechanics with Applications* (Allyn and Bacon, Boston, 1970), Chap. 1
3. P.T. Leung, Am. J. Phys. **54**(12), 1148 (1986)
4. B. Friedrich, D. Hershbach, Phys. Today **56**(12), 53 (2003)
5. A.D. Alhaidari, H. Bahlouli, Phys. Rev. Lett. **100**(11), 110401 (2008)
6. R.L. Liboff, *Introductory Quantum Mechanics*, 2nd edn. (Addison-Wesley, Reading, Massachusetts, 1992), p.147

Chapter 7
Central Potentials

Summary Systems wherein particles interact with each other via forces that depend only on their separation and not on their orientation with respect to each other are said be be "central force" systems, with corresponding potentials $V(r)$. This chapter begins by examining three-dimensional spherical analogs of the infinite and finite rectangular wells, and then moves on to the crown jewel of early quantum mechanics, the solution of Schrödinger's equation for the Coulomb potential of the hydrogen atom. The solution of this problem will bring us to an understanding of how the shapes of electron orbitals arise. Discussion is also devoted to how the wave-mechanical solution for the hydrogen atom compares to that of Niels Bohr's planetary model.

7.1 Introduction

In the preceding chapter it was shown how solutions of Schrödinger's equation in spherical coordinates (r, θ, ϕ) for a mass μ moving in a central potential $V(r)$,

$$-\frac{\hbar^2}{2\mu}\left\{\frac{1}{r^2}\frac{\partial}{\partial r}\left(r^2\frac{\partial}{\partial r}\right) + \frac{1}{r^2\sin\theta}\frac{\partial}{\partial\theta}\left(\sin\theta\frac{\partial}{\partial\theta}\right) + \frac{1}{r^2\sin^2\theta}\left(\frac{\partial^2}{\partial\phi^2}\right)\right\}\psi(r,\theta,\phi)$$
$$+ V(r)\psi(r,\theta,\phi) = E\psi(r,\theta,\phi), \tag{7.1}$$

can be separated into radial and angular parts as $\psi(r,\theta,\phi) = R(r)Y(\theta,\phi)$. In terms of the operator for the squared angular momentum of the system, L^2_{op} (see 6.52), this separation takes the form

$$-\frac{\hbar^2}{2\mu}\left\{\frac{1}{R}\frac{\partial}{\partial r}\left(r^2\frac{\partial R}{\partial r}\right)\right\} + [V(r) - E]r^2 = -\frac{1}{2\mu Y}\left(L^2_{op}Y\right). \tag{7.2}$$

© The Author(s), under exclusive license to Springer Nature Switzerland AG 2022

B. C. Reed, *Quantum Mechanics*, https://doi.org/10.1007/978-3-031-14020-4_7

The solution of the angular function $Y(\theta, \phi)$ was developed in Sect. 6.5 as the spherical harmonic functions:

$$Y_{\ell,m}(\theta, \varphi) = \frac{(-1)^m}{\sqrt{2\pi}}$$

$$\left\{ A_{\ell,m}(\sin^m \theta) \sum_{k=0}^{\left[\frac{\ell-m}{2}\right]} \left[(-1)^k \binom{\ell}{k} \binom{2\ell - 2k}{\ell + m} (\cos^{\ell-m-2k} \theta) \right] \right\} e^{\iota m\varphi} \quad (7.3)$$

with

$$A_{\ell,m} = \frac{1}{2^\ell \ell!} \sqrt{\frac{2\ell + 1}{2} (\ell + m)!(\ell - m)!}, \quad (7.4)$$

where ℓ and m are respectively the orbital angular momentum and magnetic quantum numbers, with $\ell = 0, 1, 2, \ldots$ and $|m| \leq \ell$.

The eigenvalues of L_{op}^2 are, from (6.85),

$$L_{op}^2 Y_{\ell,m}(\theta, \phi) = [\hbar^2 \ell(\ell + 1)] Y_{\ell,m}(\theta, \phi). \quad (7.5)$$

In developing solutions for the radial part of Schrödinger's equation for central potentials, it is conventional to put the algebra in terms of ℓ by substituting this result into (7.2):

$$-\frac{\hbar^2}{2\mu} \left\{ \frac{1}{R} \frac{\partial}{\partial r} \left(r^2 \frac{\partial R}{\partial r} \right) \right\} + [V(r) - E] r^2 = -\frac{1}{2\mu Y} [\hbar^2 \ell(\ell + 1)] Y, \quad (7.6)$$

and then cancel the Y-functions and rearrange to form

$$-\frac{\hbar^2}{2\mu} \left\{ \frac{1}{R} \frac{\partial}{\partial r} \left(r^2 \frac{\partial R}{\partial r} \right) - \ell(\ell + 1) \right\} + V(r) r^2 = E r^2. \quad (7.7)$$

This expression is known as the *radial equation*. The magnetic quantum number m is "hidden" in the spherical harmonic $Y_{\ell,m}(\theta, \phi)$ appropriate to the angular momentum state ℓ for which this equation is to be solved. Clearly, if we can solve for $R(r)$ for some potential $V(r)$, there will be a whole family of solutions available given by choosing different combinations of the ℓ and m quantum numbers.

Two other forms of the radial equation are commonly seen. The first casts the derivative into a more compact form by noting that

$$\frac{1}{R} \frac{\partial}{\partial r} \left(r^2 \frac{\partial R}{\partial r} \right) = \frac{r}{R} \frac{\partial^2}{\partial r^2} (rR), \quad (7.8)$$

hence rendering (7.7) as

$$-\frac{\hbar^2}{2\mu}\left\{\frac{r}{R}\frac{\partial^2}{\partial r^2}(rR) - \ell(\ell+1)\right\} + V(r)r^2 = Er^2. \tag{7.9}$$

The other alternate form introduces the auxiliary function

$$U(r) = rR(r), \tag{7.10}$$

in which case we have

$$\frac{d^2U}{dr^2} + \frac{2\mu}{\hbar^2}\left\{E - V(r) - \frac{\ell(\ell+1)\hbar^2}{2\mu r^2}\right\}U(r) = 0. \tag{7.11}$$

In anticipation that the solution of (7.7), (7.9) or (7.11) will lead to a quantum number n which dictates energy levels, the overall wavefunction ψ will be of the form

$$\psi_{n\ell m}(r, \theta, \phi) = R_n(r)Y_{\ell,m}(\theta, \phi) = \frac{U_n(r)}{r}Y_{\ell,m}(\theta, \phi). \tag{7.12}$$

The purpose of this chapter is to take up solutions to these various versions of the radial equation. We will begin by looking at two relatively simple cases, solutions for the infinite and finite spherical potential wells (Sects. 7.2 and 7.3) for the specific case of $\ell = 0$. These situations represent spherically-symmetric analogs of the three-dimensional rectangular box treated in Chap. 6. Solutions along these lines are of interest in nuclear physics, where nuclei can be modeled as spherical potential wells on the order of tens of MeV deep and $\sim 10^{-14}$ meters in radius. In Sects. 7.4 and 7.5 we develop and examine a detailed solution for the hydrogenic Coulomb potential. It is difficult to overstate the importance of this solution in the history and continuing development of quantum physics; it truly is one of the great successes of quantum mechanics. Bohr's treatment of the hydrogen spectrum marked the real beginnings of quantum theory, and, in his seminal paper on quantum mechanics, Schrödinger developed the solution of the wave equation for the Coulomb potential [1]. Being the only realistic atomic potential for which Schrödinger's equation is exactly soluble analytically, solutions for the Coulomb potential form the foundation on which our understanding of the physics of more complex atomic and molecular systems rests. As solutions of Schrödinger's equation are interpreted probabilistically, we can expect the resulting detailed workings it predicts for hydrogen to be very different from those of the Bohr model. But the Bohr and Schrödinger pictures must agree on at least one point: the permissible energy levels of the electron/proton system as revealed by spectroscopic analysis.

7.2 The Infinite Spherical Well

The infinite spherical well is an analog of the three-dimensional box potential discussed in Sect. 6.1. The particle (mass μ) is trapped in a spherical region of space of radius a within an impenetrable barrier defined by

$$V(r) = \begin{cases} \infty & (r \geq a) \\ 0 & (r < a). \end{cases} \tag{7.13}$$

This is illustrated in Fig. 7.1.

The simplest case is a particle with zero angular momentum, $\ell = 0$. Solving the radial equation for $\ell \neq 0$ involves functions known as spherical Bessel and Neumann functions, exploration of which lies beyond the scope of this book.

In this case, it proves most straightforward to work with version (7.11) of the radial equation. A subscript 0 is appended to U as a reminder that the solution is valid only for $\ell = 0$. Inside the well we have

$$\frac{d^2 U_0}{dr^2} + k^2 U_0 = 0, \tag{7.14}$$

where

$$k^2 = \frac{2\mu E}{\hbar^2}. \tag{7.15}$$

Equations (7.14) and (7.15) are formally identical to those describing a one-dimensional infinite potential well; this is why we chose version (7.11) of the radial equation. We can immediately write the solution as

$$U_0(r) = A \sin(kr) + B \cos(kr). \tag{7.16}$$

What boundary conditions apply in this case? It is important to bear in mind that the solution to the radial equation is not given by U directly but by $R = U/r$. To

Fig. 7.1 Infinite spherical well

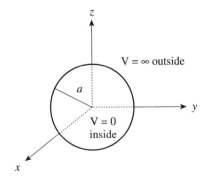

prevent divergence as $r \to 0$, we must have $U(r = 0) = 0$, which forces $B = 0$ in (7.16). Furthermore, since the barrier is impenetrable, we must have $U(r = a) = 0$, or $ka = n\pi$. Hence

$$E_n(\ell = 0) = \frac{n^2\pi^2\hbar^2}{2\mu a^2}. \tag{7.17}$$

The energy levels for a particle with $\ell = 0$ in an infinite spherical well are identical to those of a one-dimensional infinite rectangular well with the width of the rectangular well replaced by the radius of the spherical well.

Full disclosure: The argument given here regarding the vanishing of $U(r)$ at $r = 0$ is too simplistic, although adequate for our purposes. Deeper details are covered in Shankar [2].

The total wavefunction in this case is

$$\psi_{n00}(r, \theta, \phi) = R_n(r)Y_{0,0}(\theta, \phi) = \frac{A}{\sqrt{4\pi r}} \sin\left(\frac{n\pi r}{a}\right).$$

Normalizing demands

$$\frac{A^2}{4\pi} \int_0^a \int_0^\pi \int_0^{2\pi} \left[\frac{1}{r^2} \sin^2\left(\frac{n\pi r}{a}\right)\right] r^2 \sin\theta \, d\phi \, d\theta \, dr = 1,$$

which gives $A = \sqrt{2/a}$. Hence we have

$$\psi_{n00} = \frac{1}{\sqrt{2\pi a} \, r} \sin\left(\frac{n\pi r}{a}\right). \tag{7.18}$$

The expectation value of r for a particle in such a potential is given by

$$\langle r \rangle = \int_0^a \int_0^\pi \int_0^{2\pi} [\psi_{n00} r \psi_{n00}] r^2 \sin\theta \, d\phi \, d\theta \, dr$$

$$= \frac{1}{2\pi a} \left\{\int_0^a r \sin^2(n\pi r/a) \, dr\right\} \left\{\int_0^\pi \sin\theta \, d\theta\right\} \left\{\int_0^{2\pi} d\varphi\right\} \tag{7.19}$$

$$= \frac{2}{a} \int_0^a r \sin^2(n\pi r/a) \, dr$$

$$= \frac{2}{a} \left[\frac{r^2}{4} - \frac{r \sin(n\pi r/a)}{4(n\pi/a)} - \frac{\cos(n\pi r/a)}{8(n\pi/a)^2}\right]_0^a = \frac{a}{2}.$$

$\langle r \rangle$ is independent of n: increasing the energy of the particle alters its probability distribution, but not its average position. If $\ell \neq 0$, $\langle r \rangle$ would be a function of both n and ℓ.

7.3 The Finite Spherical Well

The spherical analog of the one-dimensional finite well treated in Chap. 3 is defined by

$$V(r) = \begin{cases} V_o & (r \geq a) \\ 0 & (r < a). \end{cases} \qquad (7.20)$$

Such a potential might describe a particle trapped inside a nucleus along the lines of the alpha-decay problem treated in Chap. 3. We seek bound-state ($E < V_o$) solutions of the radial equation for this potential; the solution for the energy levels of this potential closely parallels that for the one-dimensional finite well studied in Sects. 3.3–3.5

Once again, we consider the $\ell = 0$ case. Inside the well ($r < a$), the radial equation is the same as that for the infinite spherical well (7.16). Applying the boundary condition that $U(r = 0) = 0$ leads to the solution

$$R_0^{in}(r) = \frac{A \sin(k_1 r)}{r}, \qquad k_1^2 = 2\mu E/\hbar^2 \quad (r < a), \qquad (7.21)$$

where μ is again the mass of the particle involved. Outside the well ($r > a$), the radial equation becomes

$$\frac{d^2 U_0^{out}(r)}{dr^2} = k_2^2 U_0^{out}(r), \qquad (7.22)$$

where

$$k_2^2 = \frac{2m}{\hbar^2} (V_o - E). \qquad (7.23)$$

The general form of the solution is

$$U_0^{out}(r) = C e^{k_2 r} + D e^{-k_2 r},$$

that is

$$R_0^{out}(r) = \frac{C e^{k_2 r}}{r} + \frac{D e^{-k_2 r}}{r}. \qquad (7.24)$$

Requiring $U(r) \rightarrow 0$ as $r \rightarrow \infty$ demands $C = 0$, reducing the solution to

$$R_0^{out}(r) = \frac{De^{-k_2 r}}{r}. \tag{7.25}$$

Forcing the continuity of R and dR/dr at $r = a$ leads to two conditions:

$$A \sin(k_1 a) = De^{-k_2 a}, \tag{7.26}$$

and

$$A[ak_1 \cos(k_1 a) - \sin(k_1 a)] = -De^{-k_2 a}[ak_2 + 1]. \tag{7.27}$$

Dividing (7.27) by (7.26) eliminates A and D and yields a transcendental equation for the energy eigenvalues:

$$\xi \cot \xi = -\eta, \tag{7.28}$$

where we have defined

$$\xi = k_1 a \tag{7.29}$$

and

$$\eta = k_2 a. \tag{7.30}$$

From the definitions of k_1 and k_1, we can form a strength parameter for the spherical potential much like that for the one-dimensional finite potential of Chap. 3:

$$\xi^2 + \eta^2 = a^2(k_1^2 + k_2^2) = \frac{2\mu V_o a^2}{\hbar^2} = \text{constant} = R^2. \tag{7.31}$$

Equation (7.31) is that of a circle of radius R in the (ξ, η) plane: $\xi^2 + \eta^2 = R^2$; this is why the strength parameter is designated as R^2. The permissible bound-state energies are given by finding the values of ξ and η that satisfy Eqs. (7.28) and (7.31) simultaneously, that is, by finding the points of intersection of these equations in the (ξ, η) plane. From the definitions of k_1 and k_2, ξ and η must be positive, so our attention can be restricted to the first quadrant.

The situation is shown in Fig. 7.2.

The quarter-circular-like arcs correspond to wells of different depths, specifically, $R^2 = 1, 2, 4$, and 6 in (7.31). The points of intersection between the circles and the almost-vertical cotangent curves ($\eta = -\xi \cot \xi$) correspond to the quantized energy levels. The $R = 1$ well possesses no bound states, while the others bind, one, one, and two states, respectively. Since the zeros of the cotangent function occur at $(2n - 1)\pi/2, n = 1, 2, 3, \ldots$, it follows that for a spherical well to possesses n bound states it must have $R > (2n - 1)\pi/2$, or

Fig. 7.2 Illustration of the
solution for energy
eigenvalues for a finite
spherical well

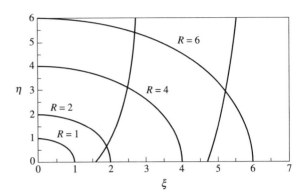

$$V_o a^2 \geq \frac{(2n-1)^2 h^2}{32\mu} \quad (\ell = 0). \tag{7.32}$$

It is instructive to compare this example with the one-dimensional finite potential well. In the present case there are no even and odd solutions to keep track of; that complication arises in one dimension because the boundary conditions have to be applied at both sides of the well $(x = \pm L)$, whereas r is not defined for $r < 0$. While a one-dimensional well will always have at least one bound state no matter how shallow it is, this is not the case for a spherical well.

Example 7.1 How many energy states are available to an alpha-particle trapped in a finite spherical well of depth 50 MeV and radius 10^{-14} meters? Assume $\ell = 0$.

From (7.32),

$$(2n-1)^2 < \frac{32\mu a^2 V_o}{h^2}.$$

On substituting the appropriate numerical values we find

$$(2n-1)^2 < \frac{32(6.646 \times 10^{-27} \text{ kg})(10^{-15} \text{ m})^2(8.010 \times 10^{-12} \text{ J})}{(6.626 \times 10^{-34} \text{ J-s})^2},$$

or

$$(2n-1)^2 < 388.$$

From this we find $n < 10.4$: the alpha-particle has available only ten bound states for $\ell = 0$.

We can determine the energy of the lowest-energy bound state for this system as follows. From (7.31),

$$\xi^2 + \eta^2 = \frac{2\mu a^2 V_o}{\hbar^2} = 956.6.$$

Using this result to eliminate η in (7.28) provides a constraint on ξ:

$$\xi \cot \xi + \sqrt{956.6 - \xi^2} = 0.$$

The lowest-valued root of this equation is $\xi = 3.04305$, which gives, via (7.29) and the definition of k_1 in (7.21),

$$E = \frac{\xi^2 \hbar^2}{2\mu a^2} = 0.48 \text{ MeV}.$$

7.4 The Coulomb Potential

As emphasized in the introduction to this chapter, the hydrogenic or Coulomb potential is perhaps the single most important potential for which an exact solution of Schrödinger's equation can be obtained. Our model for the hydrogen atom consists of a single electron of charge $-e$ and a single proton of charge $+e$ separated by distance r. From electromagnetic theory we know that the potential energy of such an arrangement is

$$V(r) = -\frac{e^2}{4\pi \varepsilon_o r}. \tag{7.33}$$

The importance of this potential demands that a full, rigorous solution be presented. In the analysis of the Bohr atom in Chap. 1, it was assumed that the proton's position could be considered as fixed in view of it's much greater mass than that of the "orbiting" electron. This is not strictly true, of course, and leads to an analysis that cannot be generally applied to two-body systems. Hence, we open this section with a proof that Schrödinger's equation for any two-body system can be separated into a part that describes the motion of the center of mass (CM) of the system as a whole plus a part which describes the "internal" energy levels of the system. The second part reduces the two-body problem to an equivalent one-body problem in a

Fig. 7.3 Reduction of
two-body system to
equivalent one-body system

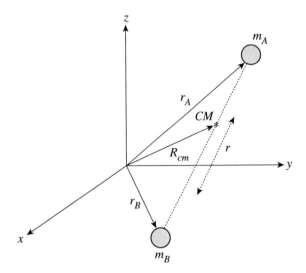

formulation that can be applied to any two body central-potential problem, and thus becomes the primary focus for the rest of the chapter.

Consult Fig. 7.3, which shows two masses, m_A and m_B. [We revert here to using the symbol "m" to designate mass as no confusion with the magnetic quantum number is likely to arise. μ is reserved for the reduced mass of this two-body system (7.36) below.] With respect to the coordinate origin shown, they lie at vector positions r_A and r_B. R_{CM} is the position of the center of mass of the two-mass system, which we know must lie along the line joining the two masses. From classical mechanics, this is given by

$$R_{CM} = \frac{m_A r_A + m_B r_B}{m_A + m_B}. \tag{7.34}$$

It will prove helpful to have a symbol to represent the sum of the masses,

$$M = m_A + m_B, \tag{7.35}$$

as well as one to represent what is known as the reduced mass of the system:

$$\mu = \frac{m_A m_B}{m_A + m_B}. \tag{7.36}$$

A crucial concept here is that of the relative coordinate of the two masses, defined as

$$r_{rel} = r_A - r_B. \tag{7.37}$$

By imagining a vector that goes from m_B to m_A, you should be able to convince yourself that the separation between the two masses is given by the magnitude of the vector $r_A - r_B$, or simply r_{rel}. If the potential energy of the system is a function only the separation of the two particles, we can write it as $V(r_{rel})$.

Now, Schrödinger's equation for this system is

$$-\frac{\hbar^2}{2m_A}\nabla_A^2\,\psi(r_A, r_B) - \frac{\hbar^2}{2m_B}\nabla_B^2\psi(r_A, r_B) + V(r_{rel})\psi(r_A, r_B) = E\psi(r_A, r_B).$$

(7.38)

In writing this, note that ∇_A^2 takes derivatives only with respect to the coordinates of mass A; similarly for ∇_B^2 and mass B. $\psi(r_A, r_B)$ is a "joint probability" wavefunction which gives the probability of mass A being in a small volume of space at r_A while mass B is simultaneously in a small volume of space at r_B.

Given the success of the separation-of-variables approach of Chap. 6, one might be tempted to try a separation of the form $\psi(r_A, r_B) = \psi_A(r_A)\psi_B(r_B)$. However, you would discover that this would not work because the potential energy is a function of the form $V(r_{rel}) = V(|r_A - r_B|)$ and is not separable into "A" and "B" coordinates. If the derivatives were with respect to r_{rel}, however, we might have some hope of separability. Further, since there are two masses, $r_{rel} = r_A - r_B$ gives information only on their relative positions; we need some other piece of information to pin down their "absolute" positions with respect to the origin. This second coordinate is taken to be R_{CM} as defined in (7.34). The point here is that we are going to transform (7.38) from (r_A, r_B) coordinates to (r_{rel}, R_{CM}) coordinates. In doing so we will find that Schrödinger's equation splits naturally into separate r_{rel} and R_{CM} sub-equations.

The Laplacian operators in (7.38) need to be transformed to the new coordinates. We will illustrate this with the x-component of ∇_A^2; the other components and those of ∇_B^2 will then follow by inspection.

From multivariable calculus, this sort of transformation goes as

$$\frac{\partial}{\partial x_A} \equiv \left(\frac{\partial x_{CM}}{\partial x_A}\right)\frac{\partial}{\partial x_{CM}} + \left(\frac{\partial x_{rel}}{\partial x_A}\right)\frac{\partial}{\partial x_{rel}} \equiv \left(\frac{m_A}{M}\right)\frac{\partial}{\partial x_{CM}} + \frac{\partial}{\partial x_{rel}}.$$

(7.39)

$\partial/\partial x_B$ follows similarly but with a change of sign due to the second term in the right side of (7.37):

$$\frac{\partial}{\partial x_B} \equiv \left(\frac{m_B}{M}\right)\frac{\partial}{\partial x_{CM}} - \frac{\partial}{\partial x_{rel}}.$$

(7.40)

On extending these to (y, z) components and then to second derivatives, the negative sign in (7.40) disappears due to a second application of ∇_B. The results are

$$\nabla_A^2 \equiv \left(\frac{m_A}{M}\right)\nabla_{CM}^2 + \nabla_{rel}^2$$

(7.41)

and

$$\nabla_B^2 \equiv \left(\frac{m_B}{M}\right) \nabla_{CM}^2 + \nabla_{rel}^2. \tag{7.42}$$

Now assume separability of the form $\psi(r_A, r_B) = \psi_{CM}(\boldsymbol{R}_{CM})\psi_{rel}(\boldsymbol{r}_{rel})$ and substitute (7.41) and (7.42) into (7.38). Remembering that derivatives with respect to CM coordinates do not affect rel coordinates and vice-versa, we find

$$
\begin{aligned}
&-\frac{\hbar^2}{2}\frac{m_A}{M^2}\psi_{rel}\nabla_{CM}^2\,\psi_{CM} - \frac{\hbar^2}{2}\frac{m_B}{M^2}\psi_{rel}\nabla_{CM}^2\,\psi_{CM} - \frac{\hbar^2}{2m_A}\psi_{CM}\nabla_{rel}^2\,\psi_{rel} \\
&-\frac{\hbar^2}{2m_B}\psi_{CM}\nabla_{rel}^2\,\psi_{rel} + V(r_{rel})\,\psi_{CM}\psi_{rel} = E\,\psi_{CM}\psi_{rel}.
\end{aligned}
\tag{7.43}
$$

On dividing through by $\psi_{CM}\,\psi_{rel}$, this reduces to

$$\left[-\frac{\hbar^2}{2M}\frac{1}{\psi_{CM}}\nabla_{CM}^2\,\psi_{CM}\right] + \left[-\frac{\hbar^2}{2\mu}\frac{1}{\psi_{rel}}\nabla_{rel}^2\,\psi_{rel} + V(r_{rel})\right] = E. \tag{7.44}$$

In this expression, the term in the first set of square bracket deals with only CM coordinates, whereas that in the second set involves only relative coordinates: Schrödinger's equation has been separated! Note that the reduced mass appears in the relative-coordinates part of (7.44). Writing $E = E_{CM} + E_{rel}$, we then have two independent equations:

$$-\frac{\hbar^2}{2M}\nabla_{CM}^2\,\psi_{CM} = E_{CM}\,\psi_{CM} \tag{7.45}$$

and

$$-\frac{\hbar^2}{2\mu}\nabla_{rel}^2\,\psi_{rel} + V(r_{rel})\,\psi_{rel} = E_{rel}\,\psi_{rel}. \tag{7.46}$$

Study (7.45) and (7.46) carefully. Equation (7.45) is Schrödinger's equation for a "free" particle, that is, one where $V = 0$. This equation describes the "motion" of the CM of our two-particle system and tells us that the kinetic energy of the CM is a continuous variable. If the coordinate origin in Fig. 7.3 had been put at the CM, only (7.46) would have emerged from the derivation.

Equation (7.46) is of the form of Schrödinger's equation for a single particle but with the mass of the particle being the reduced mass of the two-particle system. By separating into "CM" and "relative" coordinates, we have reduced the two-mass system to an equivalent single-particle system.

This approach is powerful: it applies to any two-body central-force potential. Because we have recovered Schrödinger's equation, the angular/radial separation carried out in Chap. 6 and all that follows from it—in particular the various forms of the radial equation given in Sect. 7.1—retain their validity: we need only replace the single-particle mass with reduced mass of (7.36). In practice, it is customary to drop the "rel" subscripts from (7.46).

Section **??** deals with some further separation-of-variables issues in multiparticle systems, particularly with regard to the role of distinguishable and indistinguishable particles. At present, our task is to solve the radial equation of the hydrogenic electron/proton system. The reduced mass of this system is

$$\mu = \frac{m_e\, m_p}{m_e + m_p} = \frac{m_e}{1 + m_e/m_p}, \tag{7.47}$$

which will be only very slightly less than the electron mass. For this reason, this expression is also sometimes called the "reduced electron mass", designated μ_e.

As an example of the power of this two-body-reduction approach, consider a hydrogen atom where the proton is replaced by a positive electron, that is, a positron. Such a species is known as positronium. The same radial wavefunctions and energies that we derive below for the hydrogenic case will also be valid for positronium upon substituting the appropriate reduced mass, which would be $m_e/2$. We are, in effect, about to solve many problems at once.

The form of the radial equation we will use here is that of (7.11), rearranged slightly:

$$-\frac{\hbar^2}{2\,\mu_e}\left\{\frac{d^2}{dr^2} - \frac{\ell(\ell+1)}{r^2}\right\} U(r) + V(r)U(r) = E U(r). \tag{7.48}$$

We seek solutions for the particular case of the Coulomb potential (7.33) in the present section:

$$\left\{-\frac{\hbar^2}{2\,\mu_e}\frac{d^2}{dr^2} + \frac{\hbar^2\ell(\ell+1)}{2\mu_e r^2} - \frac{e^2}{4\pi\varepsilon_o r}\right\} U(r) = E U(r). \tag{7.49}$$

The various physical constants appearing in (7.49) threaten to make subsequent algebra awkward. In Chap. 5 we saw that the algebra attending solutions of Schrödinger's equation can be simplified by casting it in a dimensionless form by finding combinations of the constants involved in a problem that have dimensions of length and energy. In the present case, such combinations are already in hand: the Bohr radius and the Rydberg energy for hydrogen from Chap. 1, equations (1.24) and (1.30), modified to have the reduced electron mass in place of the normal electron mass. Labeling these as a_o and E_r, respectively, we define dimensionless coordinates ρ and ϵ according as

$$r = \left(\frac{4\pi\varepsilon_o\hbar^2}{\mu_e e^2}\right)\rho = a_o\rho, \tag{7.50}$$

and

$$E = \left(\frac{\mu_e e^4}{32\pi^2 \varepsilon_o^2 \hbar^2} \right) \epsilon = E_r \epsilon. \tag{7.51}$$

Such dimensionless constants are also known as reduced variables, as in reduced of units, hence the notation E_r for the ground-state energy. Be careful not to confuse the electron charge e with the permittivity constant ε_o and the variable ϵ. With these definitions we have

$$\frac{d}{dr} = \frac{1}{a_o} \frac{d}{d\rho} \quad \text{and} \quad \frac{d^2}{dr^2} = \frac{1}{a_o^2} \frac{d^2}{d\rho^2}. \tag{7.52}$$

On substituting for r and E in terms of ρ and ϵ in (7.49), we find that no physical constants appear in the transformed equation:

$$-\frac{d^2 U}{d\rho^2} + \frac{\ell(\ell+1)}{\rho^2} U - \frac{2}{\rho} U = \epsilon U. \tag{7.53}$$

This is a second-order linear differential equation with non-constant coefficients; evidently we will have to seek a series solution for U. However, as in the case of the harmonic oscillator, we can simplify the ultimate form of the final result by establishing asymptotic solutions for U. Here we have two cases: $U(\rho)$ as $\rho \to \infty$ and as $\rho \to 0$. We examine these two extremes separately.

Case of $\rho \to 0$.

As $\rho \to 0$, the second term in (7.53) dominates by virtue of the factor of ρ^{-2}, leaving

$$\frac{d^2 U}{d\rho^2} \sim \frac{\ell(\ell+1)}{\rho^2} U \quad (\rho \to 0). \tag{7.54}$$

Now, as $\rho \to 0$, we must have $U \to 0$ to keep the wavefunction finite. We might therefore expect some power-law solution of the form $U(\rho) = A\rho^k$. This would demand

$$k(k-1) A\rho^{k-2} = \ell(\ell+1)A\rho^{k-2},$$

or $k(k-1) = \ell(\ell+1)$, which gives $k = -\ell$ or $k = \ell+1$. Since ℓ is restricted to positive integers or zero, the $k = -\ell$ solution would lead to negative values for k. This is not physically acceptable as it would cause U to diverge at the origin and hence be non-normalizable. Consequently, we take

$$U(\rho) \propto \rho^{\ell+1}, \quad (\rho \to 0). \tag{7.55}$$

[Aside: The behavior of $U(\rho)$ as $\rho \to 0$ is more subtle than the above argument suggests. The function that is ultimately to be normalized is $R = U/r$, that is,

$\int R^2 r^2 dr = \int U^2 dr$. It is possible for a function to have an infinite value at some point but still have a finite integral over a domain which includes that point, for example, $U(\rho) = 1/(\rho - a)$ with $a > 0$. But a simple inverse power of ρ such as $\rho^{-\ell}$ is not such a function and so must be rejected. See [2] for more details. Further, the above argument appears to fail for $\ell = 0$ since the ρ^{-2} term in (7.53) would vanish, leaving $d^2 U/d\rho^2 \sim -(2/\rho)U$ as the dominant behavior. No consistent solution of the form $U(\rho) = A\rho^k$ can be found for this case; it is only because both factors of ρ in the equation between (7.54) and (7.55) are raised to the same power and hence cancel that the above derivation works. If $\ell = 0$, however, a series solution of the form $U \sim \Sigma \rho^n$ is workable. The result is a series that can be folded into the general series solution of (7.58) and (7.63) below, so no separate derivation need be made for this case.]

Case of $\rho \to \infty$.

In this case the second and third terms in (7.53) vanish to leave

$$\frac{d^2 U}{d\rho^2} \sim -\varepsilon U \quad (\rho \to \infty). \tag{7.56}$$

This simple differential equation has two possible solutions: $U(\rho) \propto e^{\pm\sqrt{-\epsilon}\rho}$. Since we are seeking bound-state energies, $\epsilon < 0$, that is, $\sqrt{-\epsilon}$ is positive real. The solution corresponding to the positive sign in the exponential is unacceptable as it would cause U to diverge as $\rho \to \infty$, so we take

$$U(\rho) \propto e^{-\sqrt{-\epsilon}\rho} \quad (\rho \to \infty). \tag{7.57}$$

General Case

Combining (7.55) and (7.57), we infer that, in general,

$$U(\rho) = F(\rho)\rho^{\ell+1}e^{-\sqrt{-\epsilon}\rho}, \tag{7.58}$$

where $F(\rho)$ is some polynomial in ρ.

Substituting (7.58) into (7.53) yields a differential equation for $F(\rho)$:

$$\frac{d^2 F}{d\rho^2} + \left\{\frac{2D}{\rho} + 2\gamma\right\}\frac{dF}{d\rho} + \left\{\frac{2\gamma D}{\rho} + \frac{2}{\rho}\right\}F(\rho) = 0, \tag{7.59}$$

where

$$\gamma = -\sqrt{-\epsilon} \tag{7.60}$$

and

$$D = \ell + 1. \tag{7.61}$$

Multiplying through (7.59) by ρ renders it as

$$\rho\frac{d^2 F}{d\rho^2} + 2(D + \gamma\rho)\frac{dF}{d\rho} + 2(\gamma D + 1)F = 0. \tag{7.62}$$

We now attempt a series solution for $F(\rho)$ of the form

$$F(\rho) = \sum_{k=0}^{\infty} b_k \rho^k. \tag{7.63}$$

Substituting this trial sum into (7.62) yields, after the usual manipulations to bring all of the terms in ρ to the same power,

$$[2Db_1 + 2b_o(\gamma D + 1)]$$
$$\tag{7.64}$$
$$+ \sum_{k=1}^{\infty} [b_{k+1}(k + 1)(k + 2D) + 2b_k(\gamma k + \gamma D + 1)]\rho^k = 0.$$

Since ρ is not in general equal to zero, the quantities in square brackets must both vanish. This gives

$$b_1 = -\frac{(\gamma D + 1)}{D}b_o \tag{7.65}$$

and

$$b_{k+1} = -\frac{2[\gamma(k + \ell + 1) + 1]}{(k + 1)[k + 2(\ell + 1)]}b_k. \tag{7.66}$$

In writing (7.66) we have used (7.61) to eliminate D. Note that (7.65) is just (7.66) for $k = 0$. Equation (7.66) is our recursion relation.

Now, the essential question is "Does the series converge for all values of ρ?" We explore this by examining the behavior of (7.66) for large values of k:

$$b_{k+1} \approx -\left(\frac{2\gamma}{k}\right)b_k. \tag{7.67}$$

The ratio of two successive terms in the series for $F(\rho)$ then behaves as

$$\left[\frac{\text{Term}(n + 1)}{\text{Term}(n)}\right]_{\substack{series \\ solution}} = \frac{b_{k+1}}{b_k}\rho \sim -\left(\frac{2\gamma}{k}\right)\rho. \tag{7.68}$$

Now, from (5.32), the ratio of successive terms of the general exponential series for $(\exp)^{\eta\rho^p}$ is, for large values of the expansion index n,

$$\left[\frac{\text{Term}(n+1)}{\text{Term}(n)}\right]_{exponential} \sim \frac{\eta \, \rho^p}{n}. \tag{7.69}$$

Comparing (7.68) and (7.69) shows that we have $p = 1$ and $\eta = -2\gamma$. This means that, asypmtotically,

$$F(\rho) \sim e^{-2\gamma\rho}. \tag{7.70}$$

Since $\gamma = -\sqrt{-\epsilon}$, we must have $\gamma < 0$, which means that $F(\rho)$ diverges. However, the solution to the radial equation, $U(\rho)$, is the product of $F(\rho)$ and two other functions (see 7.58), so the question of the overall convergence or divergence of $U(\rho)$ remains. Looking at the asymptotic behavior of all three terms together gives

$$U(\rho) = F(\rho)\rho^{\ell+1}e^{-\sqrt{-\epsilon}\rho} \sim e^{-2\gamma\rho}\rho^{\ell+1} \, e^{\gamma\rho} \sim e^{-\gamma\rho}\rho^{\ell+1} \sim \rho^{\ell+1}e^{+\sqrt{-\epsilon}\rho},$$

which is clearly divergent. The only way to prevent this catastrophe is to assume that the series for $F(\rho)$ terminates at some finite n. Assuming $b_k = 0$ for values of k greater than some limiting value N, the recursion relation demands

$$\gamma N + \gamma(\ell+1) + 1 = 0,$$

or

$$\gamma(N + \ell + 1) = -1. \tag{7.71}$$

Inasmuch as $\ell = 0, 1, 2, \ldots$, $(N + \ell + 1) = 1, 2, 3, \ldots$. If we define

$$n = N + \ell + 1, \tag{7.72}$$

then the termination condition imposes a restriction on γ:

$$\gamma = -\sqrt{-\epsilon} = -\frac{1}{n},$$

or

$$\epsilon = -\frac{1}{n^2}. \tag{7.73}$$

Since the minimum value that N can take is 0 (see 7.72), we can write

$$(\ell + 1) \leq n, \quad (n \geq 1). \tag{7.74}$$

Recalling the definition of ϵ in terms of the energy E (7.51), we find

$$E_n = E_r \epsilon = -\frac{E_r}{n^2} = -\frac{\mu_e e^4}{32\pi^2 \epsilon_o^2 \, \hbar^2 \, n^2} \quad (n \geq 1). \tag{7.75}$$

This expression gives the hydrogen-atom bound-state energies predicted by Schrödinger's equation: *They are identical to those predicted by the Bohr model when the reduced mass of the electron is used in the latter.*

The solution for the radial part of the hydrogen-atom wavefunction can now be expressed as

$$R_{n\ell}(r) = \frac{1}{r}U(r) = \frac{1}{r}F(\rho)\rho^{\ell+1}e^{-\sqrt{-\varepsilon}\rho}$$

$$= \frac{1}{r} \, F(r/a_o)(r/a_o)^{\ell+1} \exp(-r/na_o) \tag{7.76}$$

$$= \frac{1}{r} \, F(r/a_o)(r/a_o)^{\ell+1} \exp(-r/na_o).$$

There is one arbitrary constant, b_0. Selecting this constant such that $R_{n\ell}$ is normalized for a given (n, ℓ),

$$\int\limits_0^\infty R_{n\ell}^2(r)r^2 \mathrm{d}r = 1,$$

leads to the following results [3]

$$R_{n\ell}(r) = A_{n\ell}e^{-r/na_o}(2r/na_o)^\ell L_{n-\ell-1}^{2\ell+1}(2r/na_o), \tag{7.77}$$

where $A_{n\ell}$ is given by

$$A_{n\ell} = \sqrt{\frac{4(n-\ell-1)!}{a_o^3 n^4 (n+\ell)!}}. \tag{7.78}$$

$L_{n-\ell-1}^{2\ell+1}$ denotes an *Associated Laguerre polynomial* in argument x:

$$L_{n-\ell-1}^{2\ell+1}(x) = \sum_{k=0}^{n-\ell-1} (-1)^k \binom{n+\ell}{n-\ell-1-k} \frac{x^k}{k!}$$

$$= \sum_{k=0}^{n-\ell-1} \frac{(-1)^k(n+\ell)!}{(n-\ell-1-k)!(2\ell+1+k)!\,k!}x^k. \tag{7.79}$$

In our case, $x = (2r/na_o)$. Readers are warned that definitions of the Laguerre polynomials differ from text to text; that quoted here is consistent with the usages

Table 7.1 Hydrogen radial wavefunctions

$R_{10}(r) = \frac{2}{a_o^{3/2}} e^{-r/a_o}$

$R_{20}(r) = \frac{1}{(2a_o)^{3/2}} (2 - r/a_o) e^{-r/2a_o}$

$R_{21}(r) = \frac{1}{\sqrt{3}(2a_o)^{3/2}} (r/a_o) e^{-r/2a_o}$

$R_{30}(r) = \frac{2}{(3a_o)^{3/2}} \left[1 - \frac{2r}{3a_o} + \frac{2r^2}{27a_o^2}\right] e^{-r/3a_o}$

$R_{31}(r) = \frac{4\sqrt{2}}{9(3a_o)^{3/2}} (r/a_o) \left[1 - \frac{r}{6a_o}\right] e^{-r/3a_o}$

$R_{32}(r) = \frac{2\sqrt{2}}{27\sqrt{5}(3a_o)^{3/2}} (r/a_o)^2 e^{-r/3a_o}$

in Weber and Arfken (Sect. 13.2) and in Pauling and Wilson. The first few radial wavefunctions are given in Table 7.1. In these expressions, the radial coordinate r always appears divided the Bohr radius, a natural unit of length for hydrogenic systems.

To summarize to this point, three quantum numbers, (n, ℓ, m), are needed for a complete description of the hydrogenic wavefunctions. Quantum number n (the principal quantum number) dictates the energy level of the system. A given energy level can be realized by n states of total orbital angular momentum given by $L = \sqrt{\ell(\ell + 1)}\hbar$, $0 \leq \ell \leq (n - 1)$, as deduced in Chap. 6, and each of these realizations can be obtained for a number of L_z states enumerated by $-\ell \leq m \leq \ell$, that is, there are $(2\ell + 1)$ possible values of m for each value of ℓ. To get the total number of substates for a given energy level n, we sum the values of $(2\ell + 1)$ over all possible values of ℓ:

$$\text{degeneracy of level } E_n = \sum_{\ell=0}^{n-1} (2\ell + 1) = 2\sum_{\ell=0}^{n-1} \ell + \sum_{\ell=0}^{n-1} (1)$$

(7.80)

$$= (n)(n - 1) + n = n^2.$$

Upon accounting for electron spin (Chap. 8), this degeneracy actually proves to be $2n^2$.

Example 7.2 In writing the solution of (7.62),

$$\rho \frac{d^2 F}{d\rho^2} + 2(D + \gamma\rho)\frac{dF}{d\rho} + 2(\gamma D + 1)F = 0,$$

as a series of the form

$$F(\rho) = \sum_{k=0}^{\infty} b_k \rho^k,$$

we neglected to account for the possibility of any indicial term in the power of ρ, that is, for a solution of the form

$$F(\rho) = \sum_{k=0}^{\infty} b_k \rho^{k+c}.$$

Rework the solution with such an indicial term.

Substituting this new expression for F into the differential equation gives

$$\sum_{k=0}(k+c)(k+c-1+2D)b_k\rho^{k+c-1} + 2\sum_{k=0}[\gamma(k+c+D+1)]b_k\rho^{k+c} = 0.$$

Expanding out the first term by one step gives

$$c(c-1+2D)b_0\rho^{c-1} + \sum_{k=1}(k+c)(k+c-1+2D)\,b_k\rho^{k+c-1}$$
$$+ 2\sum_{k=0}[\gamma(k+c+D) + 1]\,b_k\rho^{k+c} = 0.$$

In the first sum, define a new index $j = k - 1$ in the usual way. This gives

$$c(c-1+2D)b_0\rho^{c-1} + \sum_{j=0}(j+c+1)(j+c+2D)\,b_{j+1}\rho^{j+c}$$
$$+ 2\sum_{k=0}[\gamma(k+c+D) + 1]\,b_k\rho^{k+c} = 0.$$

The two summations can now be combined as

$$c(c-1+2D)b_0\rho^{c-1}$$
$$+ \sum_{k=0}\{(k+c+1)(k+c+2D)b_{k+1} + 2[\gamma(k+c+D)+1]b_k\}\rho^{k+c} = 0.$$

With $D = \ell + 1$, we have the indicial equation

$$c(c+2\ell+1)b_0 = 0$$

and the recursion relation

$$b_{k+1} = -\frac{2[\gamma(k+c+\ell+1)+1]}{(k+c+1)(k+c+2\ell+2)}b_k.$$

This last expression is identical to (7.66) with $c = 0$. Presuming that $b_0 \neq 0$ (otherwise we would have a null solution), the second-last expression indicates that $c = 0$ or $c = -(2\ell+1)$. $c = 0$ is the standard solution already derived; let us examine the consequences of taking $c = -(2\ell+1)$. In this case, the recursion relation will appear as

$$b_{k+1} = -\frac{2[\gamma(k-\ell)+1]}{(k-2\ell)(k+1)}b_k.$$

While this relation has the same asymptotic form as that for $c = 0$, $b_{k+1} \sim -(2\gamma/k)b_k$, it has a serious problem: A factor of $k - 2\ell$ appears in the denominator. If $k = 2\ell$, then $b_{k+1} = \infty$, and the resulting series would be impossible to normalize. Consequently, we are forced to discard $c \neq 0$ as unphysical.

7.5 Hydrogen Atom Probability Distributions

As can be seen from Tables 6.1 and 7.1, the complete hydrogen-atom wavefunctions $\psi_{n\ell m}(r, \theta, \phi) = R_{n\ell}(r)Y_{\ell m}(\theta, \phi)$ are quite involved. The first few of these are recorded in Table 7.2. In this table, we have suppressed the fact that some of the spherical harmonics are negative.

A more extensive tabulation of these wavefunctions is given in Chap. 5 of Pauling and Wilson. These results can be extended to the case of a general single-electron hydrogenic atom, that is, a single electron orbiting a nucleus of charge $+Ze$, by replacing a_o by a_o/Z wherever it appears.

The probability of finding the electron in a volume of space $dV = r^2 \sin\theta \, dr \, d\theta \, d\phi$ is given by

$$\text{Probability of electron in } dV = \psi^*\psi dV = \psi^*\psi r^2 \sin\theta \, dr \, d\theta \, d\phi. \tag{7.81}$$

A central physical concept here is that of a *probability density*, literally, the probability per unit volume of space of finding the electron at a given position. We use the symbol P to designate probability density:

$$P = \frac{\text{Probability of finding electron in } dV}{dV} = \psi^*\psi. \tag{7.82}$$

Table 7.2 Hydrogen atom wavefunctions

$\psi_{100} = \frac{1}{\sqrt{\pi} a_o^{3/2}} e^{-r/a_o}$

$\psi_{200} = \frac{1}{4\sqrt{2\pi} a_o^{3/2}} (2 - r/a_o) e^{-r/2a_o}$

$\psi_{210} = \frac{1}{4\sqrt{2\pi} a_o^{3/2}} (r/a_o) e^{-r/2a_o} \cos\theta$

$\psi_{21\pm 1} = \frac{1}{8\sqrt{\pi} a_o^{3/2}} (r/a_o) e^{-r/2a_o} \sin\theta e^{\pm \iota\phi}$

$\psi_{300} = \frac{1}{81\sqrt{3\pi} a_o^{3/2}} \left[27 - 18\frac{r}{a_o} + 2\frac{r^2}{a_o^2} \right] e^{-r/3a_o}$

$\psi_{310} = \frac{\sqrt{2}}{81\sqrt{\pi} a_o^{3/2}} (r/a_o) (6 - r/a_o) e^{-r/3a_o} \cos\theta$

$\psi_{31\pm 1} = \frac{1}{81\sqrt{\pi} a_o^{3/2}} (r/a_o)(6 - r/a_o) e^{-r/3a_o} \sin\theta e^{\pm \iota\phi}$

$\psi_{320} = \frac{1}{81\sqrt{6\pi} a_o^{3/2}} (r/a_o)^2 e^{-r/3a_o} (3\cos^2\theta - 1)$

$\psi_{32\pm 1} = \frac{1}{81\sqrt{\pi} a_o^{3/2}} (r/a_o)^2 e^{-r/3a_o} \sin\theta \cos\theta e^{\pm \iota\phi}$

$\psi_{322} = \frac{1}{162\sqrt{\pi} a_o^{3/2}} (r/a_o)^2 e^{-r/3a_o} \sin^2\theta e^{\pm 2\iota\phi}$

To visualize these probability density distributions in three dimensions requires an imaginative ability bordering on clairvoyance. To illustrate some of the important points, we consider the (1, 0, 0) and (2, 0, 0) states in particular.

7.5.1   The (1, 0, 0) State of Hydrogen

The wavefunction for the (1, 0, 0) (= ground state) of hydrogen is

$$\psi_{100} = \frac{1}{\sqrt{\pi} a_o^{3/2}} e^{-r/a_o}. \tag{7.83}$$

The probability density for this state is given by

$$P_{100} = \psi_{100}^* \psi_{100} = \frac{1}{\pi a_o^3} e^{-2r/a_o}. \tag{7.84}$$

Note that P_{100} is independent of θ, that is, the ground-state electron cloud is spherically symmetric about the nucleus. The position of highest probability density for the electron is at the nucleus itself. However, the most likely radius is not $r = 0$, as the following argument shows. Imagine surrounding the nucleus with concentric spherical shells each of thickness Δr; the volume of each shell will be $4\pi r^2 \Delta r$. The probability of finding the electron within the shell at radius r will be

$$P(\text{electron in shell at radius } r) = P_{100}(4\pi r^2 \Delta r) = \frac{4}{a_o^3}(r^2 e^{-2r/a_o})\Delta r. \quad (7.85)$$

In what shell are we most likely to find the electron? This can be determined by maximizing P with respect to r to find the most probable radius r_{mp}:

$$\left(\frac{dP}{dr}\right)_{r=r_{mp}} = 0.$$

Therefore,

$$\frac{d}{dr}(r^2 e^{-2r/a_o}) = 2r e^{-2r/a_o} - (2/a_o)r^2 e^{-2r/a_o} = 0,$$

or, on cancelling a factor of re^{-2r/a_o},

$$(r_{mp})_{100} = a_o.$$

The most probable radius for the electron in the ground state of hydrogen is exactly one Bohr radius! This most probable radius, a_o, is different from the position of greatest probability density ($r = 0$) because there are more "points" near $r = a_o$ than there are near $r = 0$, albeit with the former each individually less probable than the latter. From this result one might hypothesize (in analogy with the Bohr model) that $(r_{mp})_{n\ell m} = n^2 a_o$, but we will see in the following subsection that this is in fact not the case.

The probability of finding an electron in the ground state at radius r is proportional to $r^2 e^{-2r/a_o}$. This function is shown graphically in Fig. 7.4, and is known as a radial probability distribution function. Specifically, $P(r) = 4\pi r^2 R_{n\ell}(r)$; the scale on the x-axis is in terms of Bohr radii. The units of $P(r)$ are reciprocal meters.

Fig. 7.4 Radial probability distribution for the ground state of hydrogen

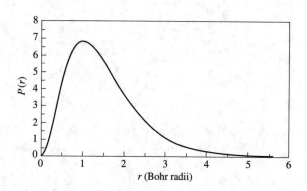

Example 7.3 The exponential tail in Fig. 7.4 indicates that the average radial position (expectation value) of the electron will be beyond a_o. Compute the average radial distance of the electron from the proton in the $(1, 0, 0)$ state of hydrogen.

From the concepts developed in Chap. 4, this is given by

$$\langle r_{100}\rangle = \langle \psi_{100}^*|r|\psi_{100}\rangle = \int_0^\infty \int_0^\pi \int_0^{2\pi} \psi_{100}^* \, r\psi_{100} r^2 \sin\theta \, dr \, d\theta \, d\phi$$

$$= \frac{1}{\pi a_o^3} \left(\int_0^\infty r^3 e^{-2r/a_o} dr\right) \left(\int_0^\pi \sin\theta \, d\theta\right) \left(\int_0^{2\pi} d\varphi\right)$$

$$= \frac{1}{\pi a_o^3} \left(\frac{3a_o^4}{8}\right) (2)(2\pi) = \frac{3}{2}a_o.$$

The interpretation of this result is that the average value of measurements of the radial distances of electrons in many ground-state hydrogen atoms would be $1.5a_o$ from the nucleus. You should be able to show that this result is consistent with the result of Example 4.2.

Example 7.4 What is the probability of finding the electron within distance r of the proton in the $(1, 0, 0)$ state of hydrogen?

The probability of finding the electron within any volume of space dV is given by $\psi^*\psi dv$. To answer this question, we need to integrate this over all volume out to radius r, that is, over all values of θ and ϕ for $r = 0$ to r:

$$P_{100}(\leq r) = \int_0^r \int_0^\pi \int_0^{2\pi} \psi_{100}^*\psi_{100} r^2 \sin\theta \, dr \, d\theta \, d\phi = \frac{4}{a_o^3} \left(\int_0^r r^2 e^{-2r/a_o} dr\right)$$

$$= 1 - e^{-2r/a_o}[2(r/a_o)^2 + 2(r/a_o) + 1].$$

From this result we find that the probability of finding the electron within one Bohr radius (that is, $r = a_o$) of the nucleus to be 0.323, and within two Bohr radii to be 0.762. The probability of finding the electron beyond 10 Bohr radii is only about 0.003. As $r \to \infty$, $P(\leq r) \to 1$. The run of this cumulative probability for the $(1, 0, 0)$ state of hydrogen is shown in Fig. 7.5.

We remark in passing that it is possible to derive an expression for the mean value of the s'th power of the electron-nucleus distance, $\langle r_{n\ell}^s\rangle$, for any (n, ℓ) state. It is not

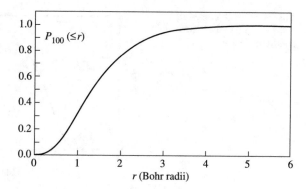

Fig. 7.5 Cumulative probability distribution for the (1, 0, 0) state of hydrogen

particularly neat or easy to work with, but has the advantage that s need not be an integer. This expression was developed by the author and is [4]

$$\langle r_{n\ell}^s \rangle = a_o^s \left(\frac{n^{s-1}}{2^{s+1}} \right) \frac{(n-\ell-1)!}{(n+\ell)!} \sum_{j=0}^{2(n-\ell-1)} c_j \Gamma(2\ell + s + 3 + j), \qquad (7.86)$$

where we must have $s > -2\ell - 3$. $\Gamma(x)$ denotes the Gamma function. The coefficients c_j are given by sums of products of coefficients appearing in the Laguerre polynomials:

$$c_j = \sum_{i=0}^{j} b_i b_{j-i}, \qquad (7.87)$$

where the b's are defined by (7.79),

$$L_{n-\ell-1}^{2\ell+1}(x) = \sum_{k=0}^{n-\ell-1} \frac{(-1)^k (n+\ell)!}{(n-\ell-1-k)!\,(2\ell+1+k)!\,k!} x^k = \sum_{k=0}^{n-\ell-1} b_k x^k. \qquad (7.88)$$

If you program the c-values of (7.87), be sure to set $b_i = 0$ for $i > (n - \ell - 1)$.

7.5.2 The (2, 0, 0) and Other States of Hydrogen

As a second example, consider the (2, 0, 0) or "first excited state" of hydrogen:

$$\psi_{200} = \frac{1}{\sqrt{32\pi a_o^{3/2}}} (2 - r/a_o) e^{-r/2a_o}. \qquad (7.89)$$

Fig. 7.6 Radial probability distribution for the $(2, 0, 0)$ state of hydrogen

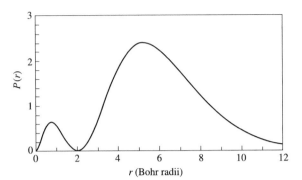

Let us test our previous hypothesis that the most probable radius for this state should be $4a_o$. To do this we find the value of r that satisfies

$$\frac{d}{dr}[r^2(2 - r/a_o)^2 e^{-r/a_o}] = 0. \tag{7.90}$$

On computing the derivative and canceling common factors of e^{-r/a_o} and $(2 - r/a_o)$, we are left with a quadratic equation in r/a_o:

$$(r/a_o)^2 - 6\,(r/a_o) + 4 = 0,$$

which has two real solutions,

$$(r_{mp})_{200} = 0.764a_o \quad or \quad 5.236a_o, \tag{7.91}$$

neither of which corresponds to the Bohr prediction. On back-substituting, it is easy to show that the second solution maximizes $(r R_{200})^2$. As shown in Fig. 7.6, the radial probability distribution for this state consists of an inner shell peaking at 0.76 Bohr radii, which is nested within a shell of higher probability density peaking at 5.236 Bohr radii. As a general rule, the most probable radius for energy level n is $n^2 a_o$ only when $\ell = n - 1$; this is shown in Appendix A.5.

Other Hydrogenic States

Figure 7.7 shows radial probability distributions for a number of (n, ℓ) states; the value of m does not affect these functions. For given (n, ℓ), there are a number of radii where the value of $P(r)$ is zero (nodes), that is, where we would never expect to find the electron. In general, the number of such nodes is given by $n - \ell - 1$; a proof of this is given in Appendix A.5. Also, examination of (7.77) shows that $R_{n\ell}(0) = 0$ for $\ell \neq 0$, always; similarly, $R_{n\ell}(0) \neq 0$ when $\ell = 0$.

If you look carefully at Fig. 7.7, it appears that the electron will on average reside closest to the nucleus when $\ell = n - 1$ for a given value of n. This is in fact generally true, as you will show in Problem 7.18. We can construct a classical analogy to this

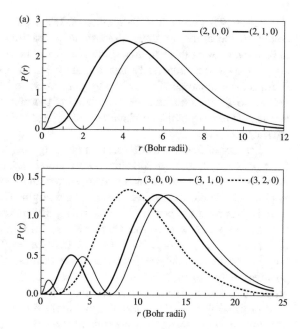

Fig. 7.7 Hydrogen radial probability distributions for **a** $n = 2$ and **b** $n = 3$

result by considering planets orbiting the Sun. If the orbits of two planets are of equal total energy, then the Keplerian ellipses that describe them will have identical semi-major axes. The precise shapes of the orbits, however, will depend on the planets' orbital angular momenta: lesser angular momentum means a more eccentric orbit because in a skinny, elongated orbit the r and L vectors in the angular momentum $L = r \times p$ are nearly parallel for most of the orbit. It turns out that the time-averaged distance of a planet from the Sun, $\langle r_{planet} \rangle$, is greater for a one in a high-eccentricity orbit than for one in a low-eccentricity orbit of the same energy; this is because the planet spends much of its time moving at low speeds far from the Sun. Conversely, a lower value of $\langle r_{planet} \rangle$ will be associated with greater L, and this is what Fig. 7.7 shows: $\langle r \rangle$ is least when ℓ is greatest for a fixed value of n (fixed total energy). Problem 7.19 explores this analogy further. This comparison should not be taken too literally, however, as the three-dimensional electron probability distributions (Fig. 7.10) look nothing like classical orbits.

The radial probability distributions do not tell the entire story. Because they have angular dependences as well through the spherical harmonics, the electron clouds are not in general spherically symmetric about the nucleus. Attempts to represent $\psi_{n,\ell,m}(r, \theta, \phi)$ graphically are always plagued by the difficulty that one is attempting to represent a function that associates an amplitude with each point in three-dimensional space on a two-dimensional surface. One approach is to pseudo-three-dimensionally plot the value of $|\psi^2|$ in a plane cutting through the nucleus; by convention, this is usually taken to be the $\phi = 0$ plane, that is, the xz plane. (Actually, the choice of plane is irrelevant as $|\psi|$ is rotationally symmetric about the z-axis, but choosing $\phi = 0$ is computationally simplest.) Such plots are shown in Fig. 7.8. To get an impression of the full three-dimensional probability distribution, it is necessary to imagine rotating these figures about the z-axis. In these figures, $|\psi|$ is plotted as opposed to the more physically meaningful $|\psi^2|$ in order to render smaller-amplitude features more apparent. The scales along the axes are in Bohr radii; the various figures are to different scales. For convenience of plotting, the vertical scales have been normalized to unity at the maximum value of $|\psi|$ for each case; what is important is the shapes of the surfaces.

A number of features can be seen in these figures. The single peak in Fig. 7.8a, $(n, \ell, m) = (1, 0, 0)$, represents the purely radial exponential function characteristic of the hydrogen ground state; when rotated about the z-axis, $|\psi|$ is spherically symmetric, and has its maximum value at the origin. The $(n, \ell, m) = (2, 0, 0)$ state likewise has a central peak and is spherically symmetric, but there is a radial node (a radius at which $\psi = 0$) located at $r = 2a_o$ (see Table 7.2), followed by a second maximum at $r = 4a_o$. On moving to $(n, \ell, m) = (2, 1, 1)$, no radial nodes are present, but there is an angular node at $\theta = 0$; recall that θ is measured from the positive z axis. For $(n, \ell, m) = (4, 1, 1)$ and $(5, 3, 1)$, both radial and angular nodes are present, with both having a central radial node at $r = 0$.

By following reasoning similar to that in the discussion of the radial probability distributions $P(r)$ above, it is possible to show that two cases obtain for the radial wavefunctions. If $\ell = 0$, then there will be no central radial node, and there will be a total of $(n - 1)$ radial nodes. On the other hand, if $\ell \neq 0$, then there will be a central radial node, and the total number of radial nodes (including the central one, but not that at $r = \infty$) will be $(n - \ell)$.

Given plots such as those in Fig. 7.8, these rules, when combined with those developed at the end of Sect. 6.5.3 for interpreting the Associated Legendre functions $\Theta_{\ell,m}$, can be used to make inferences as to what values of (n, ℓ, m) are involved. Figure 7.9 summarizes these rules in flowchart form. In some cases one can uniquely specify (n, ℓ, m) from the plot of $|\psi|$ alone. As an example, consider $(n, \ell, m) = (2, 0, 0)$. Here there is one (non-central) radial node and no angular nodes. The lack of a central radial node implies $\ell = 0$, which by default means $m = 0$. The presence of one radial node then demands $n = 2$. In contrast, $(n, \ell, m) = (5, 3, 1)$ is case where the (n, ℓ, m) values cannot be uniquely reconstructed from an examination of $|\psi|(x, z)$ alone. In this case we have two radial nodes (including a central one) and three angular nodes, including one along the z-axis. The presence of a central radial node indicates that $\ell \neq 0$ and $n - \ell = 2$. The presence of an angular node along the z-axis indicates $m \neq 0$ and $\ell - m = 2$. Beyond these constraints, we cannot further specify (n, ℓ, m). As a rule of thumb, however, the radial extent of a wavefunction is largely confined to $r < 2n^2 a_o$, so examination of the scales in the plots, if they are provided, can be helpful in such otherwise non-unique cases.

The shapes of the atomic orbitals one obtains on rotating such plots around the z-axis dictate the orientations in which atoms will bind to form molecules. The successful prediction of these bonding angles, the details of which lie beyond the scope of this book, represented a major success for the wave theory of quantum mechanics over the older semiclassical theory.

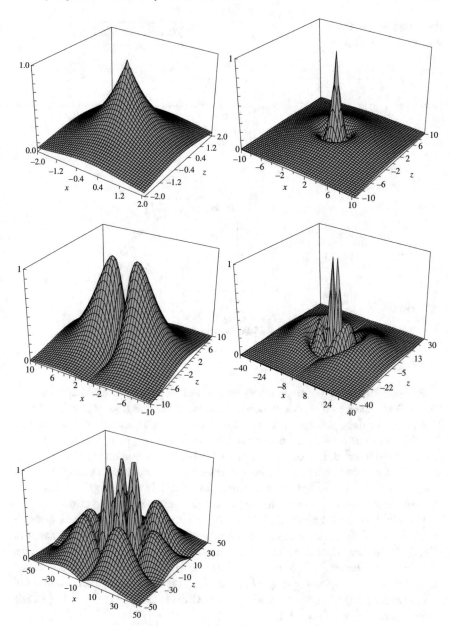

Fig. 7.8 $|\psi|$ surfaces for various (n, ℓ, m) states: Top row: $(1, 0, 0)$ and $(2, 0, 0)$; Middle row: $(2, 1, 1)$ and $(4, 1, 1)$; Bottom row: $(5, 3, 1)$

Fig. 7.9 Flowchart for
analysis of $|\psi|(x, z)$ plots

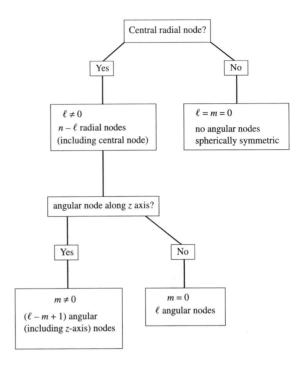

Figure 7.10 shows grayscale representations of all hydrogen states up to $n = 5$. These plots appear as if one were looking at those in Fig. 7.8 from above: the x-axes increase to the right, and the z-axes upwards, that is, we are viewing the $\phi = 0$ plane. The full width of each figure depends on the principle quantum number depicted, and is given by $4n^2$ Bohr radii. States with negative values of m are not shown, as the only effect of a negative value of m is to negate the Associated Legendre function when m is odd; this will have no effect in a plot of $|\psi|$. Negating m also conjugates the wavefunction through the Φ function, but this again will have no effect on a plot of $|\psi|$. Readers should check that the radial and angular node conditions given in Fig. 7.9 are satisfied in each case. Also, it is worthwhile comparing these images to Fig. 6.7 to see the effect of the spherical harmonics on the probability distributions.

In a classic paper, H.E. White described the construction of an ingenious set of models of the xz plane hydrogenic probability distributions, which, when set in motion and photographed, produced figures analogous to those in Fig. 7.10 [5]. His paper is still well worth reading.

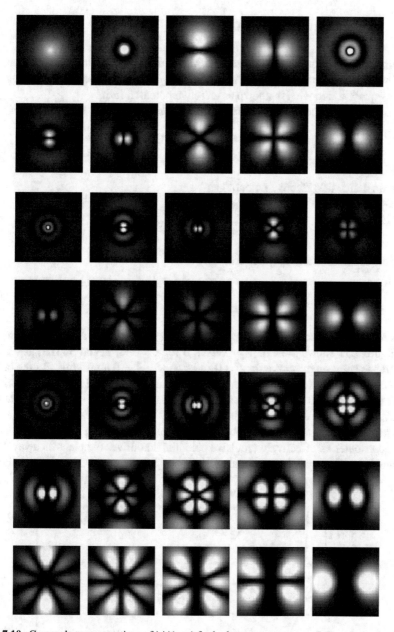

Fig. 7.10 Grayscale representations of $|\psi|(x, z)$ for hydrogen states to $n = 5$. Row 1: $(n, \ell, m) =$ $(1, 0, 0)$ $(2, 0, 0)$ $(2, 1, 0)$ $(2, 1, 1)$ $(3, 0, 0)$ Row 2: $(n, \ell, m) = (3, 1, 0)$ $(3, 1, 1)$ $(3, 2, 0)$ $(3, 2, 1)$ $(3, 2, 2)$ Row 3: $(n, \ell, m) = (4, 0, 0)$ $(4, 1, 0)$ $(4, 1, 1)$ $(4, 2, 0)$ $(4, 2, 1)$ Row 4: $(n, \ell, m) =$ $(4, 2, 2)$ $(4, 3, 0)$ $(4, 3, 1)$ $(4, 3, 2)$ $(4, 3, 3)$ Row 5: $(n, \ell, m) = (5, 0, 0)$ $(5, 1, 0)$ $(5, 1, 1)$ $(5, 2, 0)$ $(5, 2, 1)$ Row 6: $(n, \ell, m) = (5, 2, 2)$ $(5, 3, 0)$ $(5, 3, 1)$ $(5, 3, 2)$ $(5, 3, 3)$ Row 7: $(n, \ell, m) =$ $(5, 4, 0)$ $(5, 4, 1)$ $(5, 4, 2)$ $(5, 4, 3)$ $(5, 4, 4)$

7.6 The Effective Potential

In this section we briefly consider another way of thinking about the Coulomb potential that emphasizes some important qualitative points. This is done by casting the radial equation into what is known as effective potential form.

Consider the dimensionless radial equation for the Coulomb potential (7.53):

$$-\frac{d^2U}{d\rho^2} + \frac{\ell(\ell+1)}{\rho^2}U - \frac{2}{\rho}U = \epsilon\, U. \tag{7.92}$$

If we define the effective potential as

$$V_{eff} = -\frac{2}{\rho} + \frac{\ell(\ell+1)}{\rho^2}, \tag{7.93}$$

then the radial equation has the form

$$-\frac{d^2U}{d\rho^2} + V_{eff}U = \epsilon\, U. \tag{7.94}$$

This form looks exactly like the regular one-dimensional Schrödinger equation.

Figure 7.11 shows plots of V_{eff} versus ρ for $\ell = 0$, 1, 2, and 3. The qualitative difference between the curve for $\ell = 0$ and those for $\ell \neq 0$ is striking: For the latter, the combination of the two terms in (7.93) leads to potential wells with infinite walls as $\rho \to 0$; this is why there are central radial nodes ($R = 0$) in every panel of Fig. 7.10 when $\ell \neq 0$.

The $\ell(\ell + 1)/\rho^2$ term in the effective potential is known as a "centrifugal" term; this contributes an effectively repulsive potential that drives the electron away from the nucleus. The repulsion becomes stronger as ℓ increases; each curve reaches a

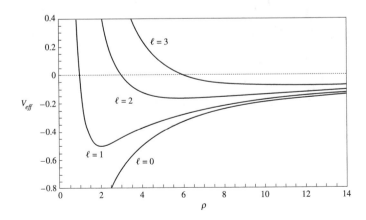

Fig. 7.11 Effective potential curves for Coulomb potential

minimum at $\rho = \ell(\ell + 1)$ where $V_{eff} = -1/\ell(\ell + 1)$. Thus, as ℓ increases, we can expect to find the electron residing further and further from the nucleus; this is consistent with the statement in the preceding section that the most probable radius is $n^2 a_o$ when $\ell = n - 1$. All of the curves approach $V_{eff} = 0$ from below as $\rho \to \infty$; the way in which they merge together conveys a sense of why the Bohr energy levels get closer and closer together as $n \to \infty$.

7.7 Some Philosophical Remarks

In solving Schrödinger's equation for hydrogen we have been rewarded with results of remarkable elegance and beauty. From a purely mathematical point of view, we are fortunate that Schrödinger's equation for a two-body problem can be separated into center-of-mass and relative-coordinate sub-equations, and that the latter can be further separated into three independent one-dimensional equations. The solutions of these equations required the imposition of various boundary conditions, requirements which in each case led to the introduction of a quantum number restricted to certain integral values. The purely-radial Coulomb potential function makes its influence known only in the radial equation and the subsequent quantization of energy levels. The quantization of L and L_z will be a feature of any central potential, independent of the precise form of the potential.

What is lost in this solution is the intuitively powerful concept of Bohr orbits. Instead of a planetary electron orbiting its nuclear Sun in some well-defined orbit, our view of the atom has evolved into amorphous clouds of probability density which peak here and there while extending to infinity.

Finally, beyond the evidence of spectra, how can we know whether or not this picture of probability distributions is fundamentally correct? Direct experimental evidence for probability orbitals can be obtained via a technique known as electron momentum spectroscopy. With this technique, it is possible to measure the distribution of electron orbitals in individual atoms as well as in molecules and solids; see [6]. When this technique is applied to hydrogen atoms, the measured orbitals prove to be in excellent agreement with what is deduced from solutions of Schrödinger's equation; see [7]. Any debate as to the meaning or interpretation of quantum mechanics pales before its success in yielding precise, verifiable, correct predictions of measurable quantities in the atomic domain.

Summary

The radial equation for a system of mass μ moving in a central potential $V(r)$ has the equivalent forms

$$-\frac{\hbar^2}{2\mu} \left\{ \frac{1}{R} \frac{\partial}{\partial r} \left(r^2 \frac{\partial R}{\partial r} \right) - \ell(\ell + 1) \right\} + V(r) r^2 = E r^2,$$

$$-\frac{\hbar^2}{2\mu}\left\{\frac{r}{R}\frac{\partial^2}{\partial r^2}(rR) - \ell(\ell+1)\right\} + V(r)r^2 = Er^2,$$

and

$$\frac{d^2U}{dr^2} + \frac{2\mu}{\hbar^2}\left\{E - V(r) - \frac{\ell(\ell+1)\hbar^2}{2\mu r^2}\right\}U(r) = 0,$$

where $U(r) = rR(r)$. The overall wavefunction ψ is of the form

$$\psi_{n\ell m}(r,\theta,\phi) = R_{n\ell}(r)Y_{\ell,m}(\theta,\phi) = \frac{U_n(r)}{r}Y_{\ell,m}(\theta,\phi).$$

For a zero angular momentum ($\ell = 0$) particle of mass μ in an infinite spherical well, the energy levels are analogous to those of an infinite rectangular well of width L, with L replaced by the radius a of the spherical well:

$$E(\ell = 0) = \frac{n^2\pi^2\hbar^2}{2\mu a^2}.$$

For a particle of mass μ with $\ell = 0$ moving in a spherical potential of finite depth V_o and radius a, the condition that the well possess n bound states is given by

$$V_o a^2 \geq \frac{(2n-1)^2 h^2}{32\mu} \quad (\ell = 0).$$

The solution of Schrödinger's equation for the Coulomb potential

$$V(r) = -\frac{e^2}{4\pi\epsilon_o r},$$

forms the cornerstone of our understanding of atomic structure. The solution of the radial part of Schrödinger's equation for this potential is

$$R_{n\ell}(r) = A_{n\ell}e^{-r/na_o}(2r/na_o)^\ell L_{n-\ell-1}^{2\ell+1}(2r/na_o),$$

where a_o is the Bohr radius and where $A_{n\ell}$ is a normalization constant given by

$$A_{n\ell} = \sqrt{\frac{4(n-\ell-1)!}{a_o^3 n^4(n+\ell)!}}.$$

$L_{n-\ell-1}^{2\ell+1}$ denotes an Associated Laguerre polynomial, which for a general argument x is given by

$$L_{n-\ell-1}^{2\ell+1}(x) = \sum_{k=0}^{n-\ell-1} (-1)^k \binom{n+\ell}{n-\ell-1-k} \frac{x^k}{k!}$$

$$= \sum_{k=0}^{n-\ell-1} \frac{(-1)^k (n+\ell)!}{(n-\ell-1-k)!(2\ell+1+k)!k!} x^k.$$

In the hydrogenic case, $x = 2r/na_o$. The energy levels of the hydrogen atom are given by

$$E_n = -\frac{\mu_e e^4}{32\pi^2 \varepsilon_o^2 \hbar^2 n^2} \quad (n \geq 1),$$

where μ_e is the reduced mass of the electron/proton system. Note that ℓ can take on integer values from 0 to $(n-1)$. The degeneracy of energy level n is n^2.

Although the Schrödinger and Bohr models of the hydrogen atom predict the same energies, any strict comparison ends there. In the Schrödinger picture, the orbital radius of the electron is not quantized. Instead, one has some probability (possibly zero) of finding the electron at any given position; for most states, this probability is a function of both r and θ. For a given principal quantum number n there are preferred regions where one is more likely but not guaranteed to find the electron.

Problems

7.1(E) Show that for an alpha-particle moving in an infinite spherical well with $\ell = 0$ and $a = 10^{-15}$ m, $E_\alpha = 51.6n^2$ MeV.

7.2(E) Derive an expression for $\langle r^2 \rangle$ for a particle in the $\ell = 0$ state of the infinite spherical well.

7.3(I) Normalize the wavefunction for the $\ell = 0$ finite spherical well. In terms of ξ and η, derive an expression for the probability of finding the particle outside the well. Evaluate your result numerically for the case discussed in Example 7.1.

7.4(I) Apply the results of Problem 6.21 to the infinite spherical well wavefunctions ($\ell = 0$) of Sect. 7.2. It is helpful to recall that for a mass m, $\langle p^2 \rangle = 2m\langle KE \rangle$, where $\langle KE \rangle$ is the kinetic energy. Is the uncertainty principle satisfied?

7.5(I) A particle trapped in a potential well of depth V_o at energy level energy level E is said to possess a binding energy $E_B = V_o - E$; this is the amount of energy one would have to supply to the particle to free it from the well. A deuteron is a nucleus formed by the bonding of a neutron and a proton under the action of nuclear forces. Assume that the corresponding potential can be modeled as a finite spherical well of radius $a = 2 \times 10^{-15}$ m and that the deuteron acts as a single particle of mass equivalent to the reduced mass of the neutron/proton pair. Assume further that the deuteron is in an $\ell = 0$ ground state. Experimentally, the deuteron has a binding energy of 2.23 MeV. What is the depth of the potential well?

7.6(I) Consider a radial analog of the harmonic potential, $V(r) = (k/2)r^2$ $(0 \le r \le \infty)$. Show that if we define α, ϵ, and λ as in Sect. 5.1 and put $\rho = \alpha r$, then the radial equation for $U(r)$ (7.11), reduces to

$$\frac{d^2U}{d\rho^2} + \left\{ \lambda - \rho^2 - \frac{\ell(\ell+1)}{\rho^2} \right\} U = 0.$$

Now, from the development in Sect. 7.4 it should be clear that for any central potential that does not diverge more strongly than $1/\rho^2$ as $\rho \to 0$, then $U(\rho) \sim \rho^{\ell+1}$ as $\rho \to 0$. This potential satisfies this requirement. Further, as $\rho \to \infty$, this differential equation has exactly the same form as that for the one-dimensional harmonic oscillator [(5.6) with ρ in place of ξ], so we must have $U \sim e^{-\rho^2/2}$ as $\rho \to \infty$. As in Sect. 7.4, then, we can posit that the overall solution must be of the form $U(\rho) \sim F(\rho)\rho^{\ell+1}e^{-\rho^2/2}$. Show that $F(\rho)$ must satisfy

$$\frac{d^2F}{d\rho^2} + 2 \left\{ \frac{(\ell+1)}{\rho} - \rho \right\} \frac{dF}{d\rho} + \{\lambda - (2\ell+3)\} F = 0.$$

This result is known as the Kummer-Laplace equation; its solution usually proceeds in terms of hypergeometric functions.

7.7(E) An electron is trapped in a finite spherical potential of radius 10^{-14} m. If it is in an $\ell = 0$ state, what must be the minimum depth of the potential (in MeV) to bind the electron? Compare to the results of Problem 4.17.

7.8(A) Consider a linear central potential of the form $V(r) = kr$ $(0 < r < \infty)$. Show that if we define $\rho = \alpha r$ and $\lambda = \epsilon E$ with $\alpha = (km/\hbar^2)^{1/3}$ and $\epsilon = (m/k^2\hbar^2)^{1/3}$ (see Problem 5.10; m = mass, not magnetic quantum number), then the radial equation can be expressed in the dimensionless form

$$\frac{d^2U}{d\rho^2} - \frac{\ell(\ell+1)}{\rho^2}U + 2(\lambda - \rho)U = 0.$$

Now show that as $\rho \to \infty$, U must have the asymptotic form $U(\rho) \sim e^{-\gamma\rho^{3/2}}$, where $\gamma = 2\sqrt{2}/3$. From the logic of Problem 7.6 we can then propose a general solution for this potential of the form $U(\rho) \sim F(\rho)\rho^{\ell+1}e^{-\gamma\rho^{3/2}}$. Hence show that $F(\rho)$ must satisfy

$$\rho\frac{d^2F}{d\rho^2} + 2\left(\ell + 1 - \sqrt{2}\rho^{3/2}\right)\frac{dF}{d\rho} + \left[2\lambda\rho - \sqrt{2}(2\ell+5/2)\rho^{1/2}\right]F = 0.$$

7.9(E) In (7.50) we redefined the Bohr radius to involve the reduced mass of the electron/proton system as opposed to just the electron mass as was used in Chap. 1. Consider a proton and an antiproton "orbiting" each other under the force of their Coulomb attraction. What is the Bohr radius of this system? Such atoms, known as "protonium" have been experimentally produced, but the proton and antiproton annihilate each other in about a millionth of a second; see [8].

7.10(I) Working directly from (7.66) and (7.76), determine the value of the expansion coefficient b_0 such that $R_{31}(r)$ will be properly normalized. Verify your result by consulting Table 7.1.

7.11(I) Determine $\langle r \rangle$ and $\langle 1/r \rangle$ for a hydrogen atom in the $(2, 1, 0)$ state. Verify that the most probable value of r for an electron in this state is $4a_o$. What is the probability of finding the electron between 3.9 and 4.1 Bohr radii from the nucleus?

7.12(I) Working directly from the expressions given in Chap. 6 for the spherical harmonics, verify by explicit calculation the expression given for $\psi_{3,1,-1}$ in Table 7.2. Use your result from Problem 7.10.

7.13(E) For a hydrogen atom in the $(1, 0, 0)$ state, what value of r corresponds to a cumulative probability of 50% of finding the electron within r?

7.14(E) Prove that the $(1, 0, 0)$ and $(2, 1, 0)$ states of hydrogen are orthogonal.

7.15(E) Verify the assertion made in Sect. 7.5.2 that the most probable radius for the electron in the $(2, 0, 0)$ state of hydrogen is $5.236a_o$.

7.16(I) Apply the results of Problem 6.21 to the $(1, 0, 0)$ state of hydrogen to compute $\Delta r \Delta p$ for this state; see also the hint in Problem 7.4. Is your result consistent with the uncertainty principle?

7.17(I) Write a computer program to generate the radial probability distribution function for any (n, ℓ) such as are shown in Figs. 7.4, 7.6, and 7.7. Check your results against those shown in the figures.

7.18(A) To compare the results of wave-mechanical calculations with those in the Bohr theory, it is useful to develop an expression for $\langle r \rangle$ for any hydrogenic state. Given that the Laguerre polynomials defined in (7.79) obey the following relations,

$$\int_0^\infty e^{-x} x^K L_N^K(x) L_M^K(x) \, dx = \frac{(N+K)!}{N!} \delta_N^M$$

and

$$x L_N^K(x) = (2N + K + 1) L_N^K(x) - (N + K) L_{N-1}^K(x) - (N + 1) L_{N+1}^K(x),$$

where $\delta_N^M = 1$ if $M = N$, (zero otherwise), show that

$$\langle r \rangle = (a_o/2)[3n^2 - \ell(\ell + 1)].$$

Show also that $\langle 1/r \rangle = 1/(n^2 a_o)$. HINT: Do not confuse N and M with the quantum numbers n and m; the upper-case indices used in the relations above are general.

7.19(I) Because Coulombic and gravitational forces are both inverse-square, analogies can be made between them. In Newtonian gravitational mechanics, the mean distance of the planet from the Sun is given by $\langle r_{planet} \rangle = a(1 + e^2/2)$, where a is the semi-major axis and e is the eccentricity of the orbit. Further, if the gravitational potential is written as $V(r) = -K/r$, then a and e are given by

$$a = -K/2E$$

and

$$e^2 = 1 - \frac{2\,|E|\,L^2}{mK^2},$$

where m is the mass of the planet, E is the total energy (< 0) of the orbit and L the planet's orbital angular momentum. Using the expression for K for the Coulomb potential (watch your signs) and those for the energies and angular momenta of the various states of the hydrogen atom in terms of the quantum numbers n and ℓ, show that $\langle r_{planet} \rangle$ reduces to that derived in the previous problem for the mean distance of the electron from the proton. (This analogy should not be taken too literally: $\langle r_{planet} \rangle$ is a *time* average!)

7.20(E) For an electron orbiting a nucleus of Z protons, the hydrogen atom wavefunctions are as given in Table 7.2 with $a_o \rightarrow a_o/Z$. Also, if the nucleus comprises a total of A nucleons, its radius is given approximately by $r \sim kA^{1/3}$, where $k \sim 1.2 \times 10^{-15}$ m. Hence show that the probability that an electron in the $(1, 0, 0)$ state will be found within the nucleus is given by $P \sim (4A/3)(kZ/a_o)^3$. What is P for a hydrogen atom? For a lone electron orbiting a uranium-238 nucleus? Nuclei can occasionally "swallow" electrons, a process known as electron capture. The electron combines with a nuclear proton to yield a neutron, thus lowering the atom by one place in the periodic table.

7.21(I) Using the angular momentum operators developed in Chap. 6, verify that $\langle L_x \rangle = 0$ for the $(2, 1, 1)$ state of hydrogen. Refer to Problem 6.7.

7.22(I) Verify the assertions made following Fig. 7.6 regarding the number of nodes in the radial wavefunctions $R_{n\ell}(r)$.

7.23(I) Using the rules developed in the text for interpreting plots of $|\psi|$, how many and what type of nodes would you expect to find in the case of $(n, \ell, m,) = (6, 2, 1)$? Check your conclusions against Fig. 7.12.

7.24(A) Consider a particle of mass μ trapped in an infinite cylindrical well of radius a and height L defined by

$$V(\rho, \phi, z) = \begin{cases} 0, & (0 \le \rho \le a,\ 0 \le z \le L,\ 0 \le \phi \le 2\pi) \\ \infty, & (\rho \ge a,\ z \le 0, z \ge L,\ 0 \le \phi \le 2\pi). \end{cases}$$

(a) Set up Schrödinger's equation for this potential. Use cylindrical coordinates.

Fig. 7.12 Problem 7.23

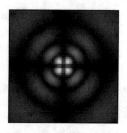

(b) Setting $\psi(\rho, \phi, z) = R(\rho)\Phi(\phi)Z(z)$, show that the z-dependence can be separated as in the one-dimensional infinite well, that is, if the separation constant is written as $-k_z^2$, then the z-dependence solution is

$$Z = \sqrt{\frac{2}{L}} \sin(k_z z), \quad k_z = \frac{n_z \pi}{L}, \quad n_z = 1, 2, 3, \ldots.$$

(c) Show that the ϕ-dependence then separates as in the hydrogen atom, that is, if the separation constant is written as $-m^2$, then the ϕ-dependence solution is

$$\Phi = \frac{1}{\sqrt{2\pi}} e^{im\phi}, \quad m = 0, \pm 1, \pm 2, \pm 3, \ldots.$$

(d) With your results from parts (b) and (c), show that Schrödinger's equation can be put in the form

$$\rho^2 \frac{d^2 R}{d\rho^2} + \rho \frac{dR}{d\rho} + \left[k^2 \rho^2 - m^2\right] R = 0,$$

where

$$k^2 = 2\mu E / \hbar^2 - k_z^2.$$

This differential equation is known as Bessel's equation; its solution is treated in any good text on mathematical physics. The solutions to this equation depend on the value of the azimuthal separation constant m and are known as Bessel functions of argument $k\rho$ and order m, usually designated $J_m(k\rho)$. In the present case of an infinite potential well of radius a, the energy eigenvalues are given by demanding that $J_m(ka) = 0$. A Bessel function of order m has an infinite number of zeros; they can be designated as Q_{pm}, read as "zero number p $(p = 1, 2, 3, \ldots)$ of order m." Hence, the eigenvalue condition becomes $ka = Q_{pm}$. For $m = 0$, the lowest zero is $Q_{10} \sim 2.4048$. For an electron trapped in such a well with $L = 50$ Å and $a = 10$ Å, what is the lowest energy state?

7.25(I) In Problem 6.19 it was shown that for a potential function of the form $V = V(r)$ and for wavefunctions of the separable form $\psi(r, \theta, \phi) = R(r)g(\theta, \phi)$, the Hamiltonian and L^2 operators commute, that is, $[H, L^2] = 0$. It is also true that the operators for L and L^2 commute: $[L^2, L] = 0$. For central potentials, L^2 can be written as $L^2 = g(r) - 2mr^2 H$; see (6.45). Use this expression in the identity $[L^2, L] = 0$ to prove that $[H, L] = 0$; assume that ψ is separable into the form $\psi = R\Theta\Phi$.

7.26(I) In the infancy of quantum mechanics there was considerable debate as to the probabilistic interpretation of ψ. Suppose that a friend proposes that the electron in a hydrogen atom in fact loses its identity as a particle, becoming instead a smeared-out cloud of electric charge whose density is distributed as $\psi^*\psi$. How would you respond to this argument?

References

1. E. Schrödinger, Ann. Phys. **384**, 361 (1926)
2. R. Shankar, *Principles of Quantum Mechanics*, 2nd edn. (Kluwer Academic/Plenum Publishers, New York, 1994), Sects. 12.6 and 13.1
3. L. Pauling, E.B. Wilson *Introduction to Quantum Mechanics With Applications to Chemistry* (Dover Publications, New York, 1985), Appendix VII
4. B.C. Reed, Eur. J. Phys. **42**(6), 065401 (2021)
5. H.E. White, Phys. Rev. **37**, 1416 (1931)
6. M. Vos, I. McCarthy, Am. J. Phys. **65**(6), 544 (1997)
7. B. Lohmann, E. Weingold, Phys. Lett. **86A**, 139 (1981)
8. N. Zurlo et al., Phys. Rev. Lett. **97**(15), 153401 (2006)

Chapter 8
Further Developments with Angular Momentum and Multiparticle Systems

Summary This chapter can be considered to be an optional supplement to Chap. 6, and is devoted to some more advanced aspects of the role of angular momentum in atomic systems. Topics addressed include angular momentum raising and lowering operators, the Stren-Gerlach experiment as evidence for quantized angular momentum, angular momentum quantization in diatomic molecules, and how the properties of wavefunctions relate to the famous Pauli Exclusion Principle.

This chapter takes up some more advanced aspects of angular momentum and multiparticle systems. This material should be considered optional; returning to it later or skipping it entirely will have no effect on your understanding of Chaps. 9–11. As each section of this chapter is independent of all others, problems are placed at the end of their respective sections as opposed to the usual procedure of placing them all at the end of the chapter. No summary is provided for this chapter.

In Sect. 6.3 it was remarked that it is possible to define angular momentum raising and lowering operators. These operators are the subject of Sect. 8.1. Section 8.2 is devoted to a discussion of the Stern-Gerlach experiment. This experiment, designed to test for quantization of angular momentum, revealed not only that such quantization exists but that, in addition, electrons possess a previously-unknown property which has come to be called *spin angular momentum*. Section 8.3 examines a simple application of quantized angular momentum, that of understanding the rotational states of diatomic molecules. Section 8.4 examines some consequences of the separability of Schrödinger's equation similar to the center-of-mass and relative-coordinates separation of Chap. 7.

8.1 Angular Momentum Raising and Lowering Operators

To begin here, it is helpful to summarize various properties of the angular momentum operators that were introduced in Chap. 6. We drop the "op" notation for convenience.

The operators for the (x, y, z) components of L can be written in Cartesian coordinates as

$$L_x = -\iota\hbar \left(y\frac{\partial}{\partial z} - z\frac{\partial}{\partial y} \right), \tag{8.1}$$

$$L_y = -\iota\hbar \left(z\frac{\partial}{\partial x} - x\frac{\partial}{\partial z} \right), \tag{8.2}$$

and

$$L_z = -\iota\hbar \left(x\frac{\partial}{\partial y} - y\frac{\partial}{\partial x} \right). \tag{8.3}$$

These operators satisfy the commutation relations (Problem 6.10)

$$[L_x, L_y] \equiv (\iota\hbar)L_z, \tag{8.4}$$

$$[L_y, L_z] \equiv (\iota\hbar)L_x, \tag{8.5}$$

and

$$[L_z, L_x] \equiv (\iota\hbar)L_y. \tag{8.6}$$

The operators for L^2 and L_z satisfy (Problem 6.18)

$$[L^2, L_z] \equiv 0. \tag{8.7}$$

Also, L^2 and L_z generate eigenvalues according as (Eqs. 6.85 and 6.87)

$$L^2 Y_{\ell,m} = \hbar^2 \ell(\ell + 1) Y_{\ell,m}, \tag{8.8}$$

and

$$L_z Y_{\ell,m} = m\hbar Y_{\ell,m}. \tag{8.9}$$

We reiterate that "m" here is the magnetic quantum number, not a mass; the latter makes no appearance in this section.

Before defining new raising and lowering operators, it is helpful to look at the commutator $[L^2, L_x]$. Again we let the operators act on some dummy wavefunction to help keep the order of operations straight:

$$[L^2, L_x]\psi = L^2(L_x\psi) - L_x(L^2\psi)$$

$$= (L_x^2 + L_y^2 + L_z^2)(L_x\psi) - L_x(L_x^2 + L_y^2 + L_z^2)\psi.$$

Both terms in this expression give rise to terms of the form $L_x^3\psi$ which cancel each other. The remaining terms give

$$[L^2, L_x]\psi = (L_y^2 + L_z^2)(L_x\psi) - L_x(L_y^2 + L_z^2)\psi$$

$$= (L_y^2 L_x - L_x L_y^2)\psi + (L_z^2 L_x - L_x L_z^2)\psi \qquad (8.10)$$

$$= [L_y^2, L_x]\psi + [L_z^2, L_x]\psi,$$

where square brackets again denote commutators.

We can rewrite the commutators in (8.10) with the help of (4.31):

$$[L_y^2, L_x] \equiv L_y[L_y, L_x] + [L_y, L_x]L_y, \qquad (8.11)$$

and

$$[L_z^2, L_x] \equiv L_z[L_z, L_x] + [L_z, L_x]L_z. \qquad (8.12)$$

Substituting (8.4)–(8.6) into (8.11) and (8.12) gives

$$[L_y^2, L_x] \equiv -(\iota\hbar)L_y L_z - (\iota\hbar)L_z L_y \qquad (8.13)$$

and

$$[L_z^2, L_x] \equiv +(\iota\hbar)L_z L_y + (\iota\hbar)L_y L_z. \qquad (8.14)$$

Equations (8.13) and (8.14), when back-substituted into (8.10), give

$$[L^2, L_x] \equiv 0. \qquad (8.15)$$

By an exactly analogous process, it can be shown that

$$[L^2, L_y] \equiv 0. \qquad (8.16)$$

(See also Problem 6.18)

As an aside, Eqs. (8.7), (8.15), and (8.16) together indicate that the operators for squared angular momentum and the total vector angular momentum itself commute:

$$[L^2, \mathbf{L}] \equiv 0. \qquad (8.17)$$

We are now in a position to introduce the *angular momentum raising and lowering operators*

$$L_+ \equiv L_x + \iota L_y$$

and

$$L_- \equiv L_x - \iota L_y,$$

which can be abbreviated as

$$L_\pm \equiv L_x \pm \iota L_y. \tag{8.18}$$

To discover what these operators raise and lower, consider the commutator formed by L_z and $L_\pm$:

$$\left[L_z, L_\pm\right] \equiv L_z L_\pm - L_\pm L_z$$

$$\equiv L_z(L_x \pm \iota L_y) - (L_x \pm \iota L_y)L_z$$

$$\equiv L_z L_x \pm \iota L_z L_y - L_x L_z \mp \iota L_y L_z.$$

The first and third, and second and fourth terms in this expression can be reduced to commutators:

$$[L_z, L_\pm] \equiv [L_z, L_x] \pm \iota [L_z, L_y].$$

From (8.5) and (8.6) this reduces to

$$[L_z, L_\pm] \equiv \iota\hbar L_y \mp \iota (\iota\hbar L_x) \equiv \pm \hbar L_\pm, \tag{8.19}$$

where you should verify the algebra for yourself. Now, circular as it may seem, expand out the commutator on the left side of this result; this gives

$$L_z L_\pm - L_\pm L_z \equiv \pm \hbar L_\pm,$$

or

$$L_z L_\pm \equiv L_\pm(L_z \pm \hbar). \tag{8.20}$$

Now apply this result to some general spherical harmonic $Y_{\ell,m}$:

$$L_z(L_\pm Y_{\ell,m}) = L_\pm\{(L_z \pm \hbar)Y_{\ell,m}\}.$$

From (8.9), $L_z Y_{\ell,m} = m\hbar\, Y_{\ell,m}$, hence

$$L_z(L_\pm Y_{\ell,m}) = \hbar(m \pm 1)\,(L_\pm Y_{\ell,m}). \tag{8.21}$$

Study this result carefully. In words, it tells us that $L_\pm$ operates on $Y_{\ell,m}$ so as to yield a new function, $L_\pm Y_{\ell,m}$, whose z-component of angular momentum is, in units of $\hbar$, exactly one unit more (in the case of L_+) or one unit less (in the case of L_-) than that possessed by $Y_{\ell,m}$. That is, L_+ and L_- respectively act to raise or lower the state of the z-component of the angular momentum of $Y_{\ell,m}$ by one unit in terms of $\hbar$.

You might wonder what effect the $L_\pm$ operators have on the ℓ-eigenvalue of $Y_{\ell,m}$. To see this, it is first helpful to compute the commutator of L^2 and $L_\pm$:

$$[L^2, L_\pm] \equiv L^2 L_\pm - L_\pm L^2$$

$$\equiv L^2(L_x \pm \iota L_y) - (L_x \pm \iota L_y)L^2$$

(8.22)

$$\equiv L^2 L_x \pm \iota L^2 L_y - L_x L^2 \mp \iota L_y L^2$$

$$\equiv [L^2, L_x] \pm \iota [L^2, L_y] \equiv 0,$$

where the last step follows from (8.15) and (8.16). Applying this result to some spherical harmonic $Y_{\ell,m}$ gives

$$L^2(L_\pm Y_{\ell,m}) = L_\pm(L^2 Y_{\ell,m}).$$

(8.23)

However, from (8.8) we know that $L^2 Y_{\ell,m} = \hbar^2 \ell(\ell + 1) Y_{\ell,m}$, hence

$$L^2(L_\pm Y_{\ell,m}) = \hbar^2 \ell(\ell + 1) (L_\pm Y_{\ell,m}).$$

(8.24)

Equation (8.24) indicates that when L^2 operates on $(L_\pm Y_{\ell,m})$, the result is $\hbar^2 \ell(\ell+1)$, just as it is when L^2 operates on $Y_{\ell,m}$ alone. Hence, L_+ and L_- have *no* effect on the ℓ-eigenvalue of $Y_{\ell,m}$.

In Chap. 5 we saw thet the harmonic oscillator raising and lowering operators do not return simply the next highest or lowest wavefunction. A like situation holds with L_+ and L_-, that is, $L_\pm Y_{\ell,m} \neq Y_{\ell,m\pm 1}$. In general,

$$L_\pm Y_{\ell,m} = K_{\ell,m}^\pm Y_{\ell,m\pm 1},$$

(8.25)

where

$$K_{\ell,m}^\pm = \hbar\sqrt{\ell(\ell + 1) - m(m \pm 1)} = \hbar\sqrt{(\ell \mp m)(\ell \pm m + 1)}.$$

(8.26)

To prove this, it is helpful to first consider the operators $L_+ L_-$ and $L_- L_+$. We do $L_+ L_-$ in detail:

$$L_+ L_- \equiv (L_x + \iota L_y)(L_x - \iota L_y) \equiv L_x^2 - \iota L_x L_y + \iota L_y L_x + L_y^2$$
$$\equiv L_x^2 + L_y^2 + \iota[L_y, L_x].$$

(8.27)

From Example 6.2, $[L_y, L_x] = -(\iota\hbar)L_z$, hence

$$L_+ L_- \equiv L_x^2 + L_y^2 + \hbar L_z.$$

(8.28)

If a factor of L_z^2 had also appeared on the right side of this expression, we could write $L_x^2 + L_y^2 + L_z^2 = L^2$. We can make this so by adding and subtracting factors of L_z^2 and writing

$$L_+ L_- \equiv L^2 + \hbar L_z - L_z^2. \tag{8.29}$$

By a similar development, you can show that

$$L_- L_+ \equiv L^2 - \hbar L_z - L_z^2. \tag{8.30}$$

Equation (8.29) makes it easy to compute the expectation value of $L_+ L_-$ as it acts on a spherical harmonic:

$$\langle Y_{\ell,m} | L_+ L_- | Y_{\ell,m} \rangle = \langle Y_{\ell,m} | L^2 + \hbar L_z - L_z^2 | Y_{\ell,m} \rangle.$$

Recalling that $L^2 Y_{\ell,m} = \hbar^2 \ell(\ell+1) Y_{\ell,m}$ and $L_z Y_{\ell,m} = m\hbar\, Y_{\ell,m}$, this evaluates as

$$\langle Y_{\ell,m} | L_+ L_- | Y_{\ell,m} \rangle = \hbar^2 \left\{ \ell\,(\ell+1) - m(m-1) \right\} \langle Y_{\ell,m} | Y_{\ell,m} \rangle$$

$$\tag{8.31}$$

$$= \hbar^2 \left\{ \ell\,(\ell+1) - m(m-1) \right\}.$$

Similarly, Eq. (8.30) gives

$$\langle Y_{\ell,m} | L_- L_+ | Y_{\ell,m} \rangle = \hbar^2 \left\{ \ell\,(\ell+1) - m(m+1) \right\} \langle Y_{\ell,m} | Y_{\ell,m} \rangle$$

$$\tag{8.32}$$

$$= \hbar^2 \left\{ \ell\,(\ell+1) - m(m+1) \right\}.$$

Now suppose that we desire to evaluate the expectation value of L_+^2 as it acts on some spherical harmonic $Y_{\ell,m}$, that is, $\langle Y_{\ell,m} | L_+^2 | Y_{\ell,m} \rangle$. Since L_+ acts to raise the m-state of $Y_{\ell,m}$ to $m+1$, this calculation, involving two successive applications of L_+, would appear to lead to (neglecting constants), a result of the form $\langle Y_{\ell,m} | Y_{\ell,m+2} \rangle$, which should be zero in view of the orthogonality of the spherical harmonics. However, this argument overlooks the fact that L_+ is a *complex* operator; we should really put $L_+^2 = L_+^* L_+$. Conveniently, L_+ and L_- are conjugates of each other, that is, $L_\pm^* = L_\mp$ (see 8.18), so we can write

$$\langle Y_{\ell,m} | L_+^2 | Y_{\ell,m} \rangle = \langle Y_{\ell,m} | L_+^* L_+ | Y_{\ell,m} \rangle = \langle Y_{\ell,m} | L_- L_+ | Y_{\ell,m} \rangle.$$

That is, $\langle L_+^2 \rangle$ would be the same as (8.32). But now, work through the calculation assuming that L_+ and L_- operate as posited in (8.25):

$$\langle Y_{\ell,m} | L_- L_+ | Y_{\ell,m} \rangle = K_{\ell,m}^+ \langle Y_{\ell,m} | L_- | Y_{\ell,m+1} \rangle$$

$$= K_{\ell,m}^+ K_{\ell,m+1}^- \langle Y_{\ell,m} | Y_{\ell,m} \rangle \tag{8.33}$$

$$= K_{\ell,m}^+ K_{\ell,m+1}^-.$$

Similarly, $\langle L_-^2 \rangle$ looks like (8.31):

$$\langle Y_{\ell,m} | L_-^2 | Y_{\ell,m} \rangle = \langle Y_{\ell,m} | L_-^* L_- | Y_{\ell,m} \rangle = \langle Y_{\ell,m} | L_+ L_- | Y_{\ell,m} \rangle$$

$$= K_{\ell,m}^- \langle Y_{\ell,m} | L_+ | Y_{\ell,m-1} \rangle$$

$$(8.34)$$

$$= K_{\ell,m}^- K_{\ell,m-1}^+ \langle Y_{\ell,m} | Y_{\ell,m} \rangle$$

$$= K_{\ell,m}^- K_{\ell,m-1}^+ .$$

Hence, Eqs. (8.32) and (8.33) tell us that

$$K_{\ell,m}^+ K_{\ell,m+1}^- = \hbar^2 \left\{ \ell \left(\ell + 1 \right) - m(m + 1) \right\}, \qquad (8.35)$$

whereas Eqs. (8.31) and (8.34) tells us that

$$K_{\ell,m}^- K_{\ell,m-1}^+ = \hbar^2 \left\{ \ell \left(\ell + 1 \right) - m(m - 1) \right\}. \qquad (8.36)$$

Actually, Eqs. (8.35) and (8.36) are not independent: Putting $m = m - 1$ in (8.35) leads to (8.36); equivalently, setting $m = m + 1$ in (8.36) reproduces (8.35). The only consistent way to satisfy these results is to set

$$K_{\ell,m}^\pm = \hbar \sqrt{\ell \left(\ell + 1 \right) - m(m \pm 1)} = \hbar \sqrt{(\ell \mp m) (\ell \pm m + 1)}, \qquad (8.37)$$

as claimed in (8.26).

Problems

8.1(E) Show that the angular momentum raising and lowering operators, when applied to the $Y_{2,1}$ spherical harmonic, give results in accordance with (8.26).

8.2(I) For a given spherical harmonic $Y_{\ell,m}$, we must have $|m| \le \ell$. Operating L_+ on $Y_{\ell,\ell}$ or L_- on $Y_{\ell,-\ell}$ should thus yield zero.

(a) Show that

$$Y_{\ell,\ell}(\theta, \phi) = \frac{(-1)^\ell}{2^\ell \ell!} \sqrt{\frac{(2\ell + 1)(2\ell)!}{4\pi}} \left(\sin^\ell \theta \right) e^{i\ell\phi}.$$

(b) Show that $L_+ Y_{\ell,\ell}(\theta, \phi) = 0$.

$$L_+ Y_{\ell,\ell}(\theta, \phi) = 0.$$

(c) Using the identity $Y_{\ell,-m} = (-1)^m (Y_{\ell,m}^*)$, show that $L_- Y_{\ell,-\ell}(\theta, \phi) = 0$.

8.2 The Stern-Gerlach Experiment: Evidence for Qunatized Angular Momentum and Electron Spin

The solution of the angular part of Schrödinger's equation for a central potential in Chap. 6 revealed that the z-component of the angular momentum of such systems is quantized in units of $\hbar$. To the extent of the development given in Chaps. 6 and 7, however, this result does not manifest itself in any way in the energy eigenvalues of the system; energy quantization is solely the purview of the radial equation. Inasmuch as quantization of angular momentum is a unique prediction of wave mechanics, experimental verification of such effects would constitute a powerful argument in favor of Schrödinger's equation. This verification was realized in 1921 as the Stern-Gerlach experiment; this was *before* the development of wave mechanics [1].

At the heart of the Stern-Gerlach experiment is the concept that atoms behave like tiny magnets. By virtue of its "orbital" motion around the nucleus, an electron creates an electric current i which gives rise to a magnetic field (see Problem 8.5). If such a circulating current encloses cross-sectional area A, then the atom acquires a vector magnetic moment of magnitude $\mu = iA$. μ is directed in the sense of the angular momentum L of the electron according as the right-hand rule. [Note that we now use μ to designate magnetic dipole moment as opposed to a mass, as was the case in Chaps. 6 and 7. In this section, masses will be represented by the symbol "m". No confusion with magnetic quantum number should arise.] Since a magnetic dipole immersed in an externally-supplied magnetic field B will experience a torque given by $\tau = \mu \times B$, an atom placed in a B-field will experience a force. Since i can be expressed in terms of L, the magnitude of this force consequently depends on the ℓ-state of the orbiting electron(s). Where there are forces, there will be accelerations; roughly speaking, a stream of atoms (a so-called atomic beam) passed through a suitable B-field can then be expected to split into a number of components depending on the values of ℓ present. As we shall see, Stern and Gerlach did observe this effect, but in a way that revealed new information about electrons and their interactions.

A full wave-mechanical treatment of this effect lies well beyond the scope of this text [2]; the intent here is to use a brief semiclassical argument to develop a basic understanding of the Stern-Gerlach experiment and establish what orders of magnitude are involved.

We first need an expression for the magnetic dipole moment created by an orbiting electron. Consider an electron orbiting the nucleus in a circular orbit of speed v and radius r as in the Bohr model. If the orbit is counterclockwise as shown in Fig. 8.1, then (classically) L is vertically upwards, as shown.

The orbital motion of the electron is equivalent to an electric current i:

$$i = \frac{\text{charge}}{\text{time}} = \frac{\text{charge}}{\text{orbital period}} = \frac{e}{(2\pi r /v)} = \frac{ev}{2\pi r}, \qquad (8.38)$$

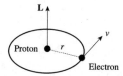

Fig. 8.1 Electron orbiting a proton. The angular momentum vector is perpendicular to the plane of the orbit

The area enclosed is πr^2. The dipole moment of this arrangement is then given by

$$\mu = i A = \left(\frac{ev}{2\pi r}\right)(\pi r^2) = \frac{evr}{2} = \frac{eL}{2m_e},\tag{8.39}$$

where $L = m_e vr$ is the classical orbital angular momentum of the electron (mass m_e). To be precise, both μ and L are vector quantities; we should write

$$\mu = \left(\frac{e}{2m_e}\right)L.\tag{8.40}$$

Now, if the magnitude of L is in fact quantized as $L = \hbar\sqrt{\ell(\ell+1)}$, we can write the magnitude of μ as

$$\mu = \left(\frac{e\hbar}{2m_e}\right)\sqrt{\ell(\ell+1)}.\tag{8.41}$$

On an intuitive level, μ can be thought of as quantifying the strength of a bar magnet, with μ conventionally represented as a vector from the south to the north pole of the magnet. The prefactor in (8.41) is known as the *Bohr magneton* μ_B, and serves as a natural unit of magnetic dipole moment for atomic systems:

$$\mu_B = \left(\frac{e\hbar}{2m_e}\right) = 9.2744 \times 10^{-24} \text{ A-m}^2.\tag{8.42}$$

A note on units: 1 A-m^2 is equivalent to 1 J/T. The Tesla is the MKS unit of magnetic field (which is also known as magnetic flux density); 1 T = 1 (N/A-m) = 1 (kg/C-s). A historical but still-used unit of magnetic field is the Gauss; 1 T = 10^4 Gs. The magnetic field of the Earth at its surface is about 0.5 Gs. A good laboratory electromagnetic can create a field of a few Tesla.

Now, it happens that if a stream of atoms possessing nonzero magnetic moments is directed to pass through a non-uniform magnetic field, the magnetic forces involved will split up the stream into a number of substreams, with one substream for each possible value of the z-component of μ, which is usually designated μ_z. Suppose for sake of simplicity that the atoms are all in the same ℓ state. If our interpretation of the angular part of Schrödinger's equation is correct, then for a given ℓ there can be $2\ell + 1$ possible values of the magnetic quantum number m (recall that $-\ell \le m \le +\ell$), hence $2\ell + 1$ possible values of μ_z. *Whatever* the value of ℓ, $2\ell + 1$ must be odd, so we should expect to see the beam split into an odd number of components on passing

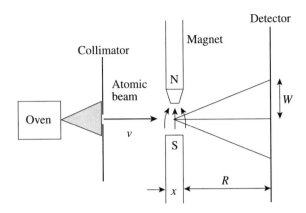

Fig. 8.2 Schematic illustration of the Stren-Gerlach experiment. A stream of fast-moving atoms passes through an inhomogeneous magnetic field, which splits the stream into multiple beams according as their ℓ-values

through the magnetic field. This is the essence of the Stern-Gerlach experiment, which is sketched in Fig. 8.2. The essential setup is that atoms of mass m are emitted with speed v from an oven and directed through a region of non-uniform magnetic field of linear extent x; the x-direction is to the right and the z-direction is upward in the figure. The entire apparatus is enclosed in a vacuum chamber to minimize collisions of the atoms with air molecules.

For atoms of mass m with z-components of μ given by μ_z, it can be shown that the vertical deflection W works out to

$$W = \frac{x}{mv^2}(R + x/2)\mu_z\left(\frac{dB}{dz}\right), \tag{8.43}$$

where the field gradient (dB/dz) is provided by machining suitably-shaped magnet pole-pieces.

The deflections predicted by (8.43) are not large, but are detectable; see Problem 8.4. On performing this now-historic experiment with a beam of silver atoms, Stern and Gerlach observed two components, in contradiction to the above prediction. The same result was found a few years later for hydrogen [3]. These results seemed to show that angular momentum is quantized, but not in the way predicted by the above analysis.

Resolution of this evidence with the theoretical expectation was provided in the form of a postulate by Uhlenbeck and Goudsmit [4, 5]. They suggested that in addition to their orbital angular momenta, electrons possess intrinsic spin angular momentum S. In analogy with the orbital angular momentum quantum number ℓ, Uhlenbeck and Goudsmit proposed the existence of an electron spin quantum number "s" restricted to magnitude $1/2$ (in units of $\hbar$, and, in analogy to the $2\ell + 1$ possible z-projections of L, that there can be only $2s + 1 = 2$ possible z-projections of S. The magnitude of the spin angular momentum is then

$$|S| = \sqrt{s(s+1)}\hbar = \sqrt{3/4}\,\hbar,$$

Table 8.1 Electron states

n	1	2		3		
ℓ	0	0	1	0	1	2
m	0	0	−1 0 1	0	−1 0 1	−2 −1 0 1 2
s	±1/2	±1/2	±1/2	±1/2	±1/2	±1/2
States	2	2	6	2	6	10
Subshell	1s	2s	2p	3s	3p	3d
Shell	K	L		M		
Total states	2	8		18		

and the possible z-projections are

$$s_z = s\hbar = \pm\hbar/2.$$

Uhlenbeck and Goudsmit's proposal was purely empirical: it worked. The classical analogy to electron spin is that of a planet spinning about its axis while orbiting the Sun. The word "spin" in the context of an electron is only a label, however: we cannot really conceive of a point particle spinning. The overall effect is to add a fourth quantum number $s = \pm1/2$ to the already existing quantum numbers (n, ℓ, m), thereby doubling the number of possible electronic states. Possible states up to $n = 3$ are enumerated in Table 8.1.

The situation demonstrated by Stern and Gerlach's experiment can now be roughly elucidated. A neutral silver atom has 47 electrons - an odd number. If it is assumed that electrons "pair off" with equal and opposite spin angular momenta as much as possible and that the remaining electron is in an $\ell = 0$ state, then the presence of two beams in the experiment can be explained since the lone unpaired electron will have $s_z = \pm1/2$. A full understanding of this result and of how electrons fill up available states requires an appreciation of Hund's Rule, the Pauli Exclusion Principle, and the symmetry properties of multi-electron atomic wavefunctions. Factors such as nuclear spin effects, non-Coulombic effects due to non-spherical nuclei, and mutual electron shielding complicate the situation further [6].

The existence of intrinsic electronic angular momentum was later found to be a natural consequence of reworking Schrödinger's equation to incorporate relativistic effects [7]. In some atoms, net non-zero orbital and spin angular momenta are present simultaneously, and the corresponding magnetic moments interact in such a way as to cause a slight splitting of energy levels: the energy of the system is slightly higher when they are parallel and slightly lower when antiparallel. This effect, known as spin-orbit coupling or L-S coupling, involves defining a total vector angular momentum J as $J = L + S$, and is fundamental to our understanding of the details of the structure and spectra of multi-electron atoms.

The splitting of spectral lines of atoms in the presence of a magnetic field is now known as the Zeeman effect; a similar effect for electric fields is known as the Stark effect; examples of these effects are discussed in Chap. 9. Quantitative measurements of line splittings can be used, for example, to determine the strength of such fields in the Sun and other stars.

Section 8.2 **Problems**

8.3(E) It was remarked that a magnetic dipole μ immersed in an externally-supplied magnetic field B will experience a torque given by $\tau = \mu \times B$. Use this to show that the unit of magnetic field must be N/(A-m).

8.4(I) Silver-107 atoms have a mass of 106.905 mass units. You are setting up a Stern-Gerlach experiment where the atoms leave an oven at 1000 m/s, travel through an inhomogeneous magnetic field of extent 1 centimeter, and then travel 2 meters before being detected. For simplicity, take $\ell = 1$, and assume $\mu_z = \mu$. If you want to achieve a deflection of $W = 0.5$ mm, what field gradient dB/dz will you need? If your magnet can provide a field of up to several tens of Teslas, is this experiment possible?

8.5(I) A circular electric current i of radius r creates a magnetic field at its center of magnitude $B = (\mu_o i / 2r)$. Consider an electron in a hydrogen atom. By using an argument like that at the start of this section, show that if the electron is orbiting in Bohr state n, then the magnetic field created by the electron at the location of the proton is given by

$$B = \frac{\mu_o m_e^2 e^7}{256 \pi^4 \varepsilon_o^3 \hbar^5 n^5}.$$

Evaluate this numerically for $n = 1$. In nuclear physics, the nuclear magneton μ_K serves as a natural unit of magnetic field and is given by $\mu_K = (e\hbar/2m_p)$, where m_p is the mass of the proton. The magnetic moment of the proton is $2.7893\mu_K$. A magnetic dipole moment μ placed in a magnetic field B acquires a potential energy of magnitude $V = \mu B$. What is the value of this energy (in eV) for a proton in the field of an $n = 1$ electron, and how does it compare to the orbital energy for the $n = 1$ state?

8.3   Diatomic Molecules and Angular Momentum

One of the important applications of the theory of angular momentum is to understanding of the rotational energy states of diatomic molecules as revealed by microwave spectroscopy. In this section we develop a straightforward semi-classical analysis of such molecules.

Our model of a diatomic molecule is shown in Fig. 8.3: Two electrically neutral atoms of masses m_A and m_B orbiting their common center of mass; r_A is the orbital radius of atom A and r_B that of atom B. To keep the situation simple, imagine the

Fig. 8.3 Diatomic molecule

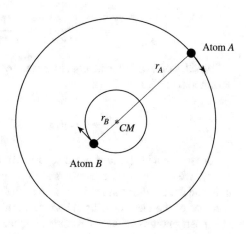

atoms to be on the ends of a massless, spinning rigid rod. If the atoms are uncharged, there will be no Coulomb potential to deal with. The electrical forces in molecules are often modeled as harmonic oscillator-type potentials such as were examined in Chap. 5; these lead to the equally-spaced vibrational energy levels of that chapter. Our concern here is with rotational energies.

To keep the center of mass in a fixed location, we must have

$$m_A r_A = m_B r_B,$$ (8.44)

a constraint we will make considerable use of below.

Imagine that the molecule is spinning about its center of mass with angular speed ω. Classically, the angular momentum of such an arrangement is $L = I\omega$, where I is the moment of inertia of the system as measured with respect to the rotational axis. If the atoms are uncharged, then the only energy in the system is the rotational kinetic energy $E = I\omega^2/2$. Putting $\omega = L/I$ gives

$$E = \frac{1}{2}I\omega^2 = \frac{L^2}{2I} = \frac{\hbar^2\ell(\ell+1)}{2I} = \frac{h^2\ell(\ell+1)}{8\pi^2 I},$$ (8.45)

where we have assumed that angular momentum is quantized in accordance with $L = \hbar\sqrt{\ell(\ell+1)}$.

Now consider such a system that undergoes a "rotational transition" $\ell \to \ell - 1$. The energy of the emitted photon will be

$$\Delta E = E_\ell - E_{\ell-1} = \frac{h^2}{4\pi^2 I}(\ell).$$ (8.46)

We can get an idea of the order of magnitude of ΔE via some approximations. Assume that the molecule consists of two equal-mass atoms each of mass m separated by distance r. Then $I = \Sigma m(r/2)^2$, and

$$\frac{h^2}{4\pi^2 I} = \frac{h^2}{2\pi^2 mr^2}.$$

For sake of explicitness, consider two hydrogen atoms ($m = 1.67 \times 10^{-27}$ kg) separated by $r = 1$ Å. Then

$$\frac{h^2}{2\pi^2 mr^2} = 1.33 \times 10^{-21} \text{ J} = 0.008 \text{ eV}.$$

For two heavier (and/or more separated) atoms we would find an even smaller energy. We can infer that the energy spacings of the rotational levels in a spinning molecule are smaller than either those of the vibrational levels of the molecule as a whole or of the electronic levels within atoms themselves. Smaller energy spacing will mean photons of lesser frequency, hence higher wavelength. Analysis of the rotational characteristics of molecules is thus typically a matter of far-infrared or microwave spectroscopy.

Most molecules involve atoms of unequal mass, however. We can modify (8.45) to accommodate this by looking in more detail at the moment of inertia of the system. From Fig. 8.3 this will be

$$I = m_A r_A^2 + m_B r_B^2. \tag{8.47}$$

Eliminating r_A in this expression via (8.44) leads to

$$I = m_B^2 r_B^2 \left[\frac{m_A + m_B}{m_A m_B} \right]. \tag{8.48}$$

Molecular spectroscopists prefer to speak in terms of the *bond length* $r = r_A + r_B$ of the system. Writing $r_B = r - r_A$ and again invoking (8.44) to eliminate r_A gives

$$r_B = r \left[\frac{m_A}{m_A + m_B} \right]. \tag{8.49}$$

Substituting (8.48) into (8.47) gives the moment of inertia in terms of the bond length,

$$I = \mu r^2, \tag{8.50}$$

where μ is the reduced mass of the system, defined in the same way as in Chap. 7:

$$\mu = \left[\frac{m_A m_B}{m_A + m_B} \right]. \tag{8.51}$$

With this modification, the rotational energy levels are given by

$$E = \frac{h^2 \ell(\ell + 1)}{8\pi^2 \mu r^2}. \tag{8.52}$$

In research practice, spectroscopists use measured photon frequencies for a given species (hence known μ) to infer bond lengths. The relevant information is often tabulated in handbooks by listing what is known as the equilibrium rotation constant B_e of a given species. This is defined as

$$B_e = \frac{h}{8\pi\,\mu r^2}. \tag{8.53}$$

B_e has units of reciprocal seconds; values are often tabulated in units of Hertz (Hz), and, in testament to the precision achievable in such spectroscopy, are known to a remarkable number of significant digits. B_e values and bond lengths can be found in a number of reference books; the *CRC Handbook of Chemistry and Physics* lists B_e values and bond lengths for dozens of diatomic species.

Example 8.1 The CRC Handbook lists the value of B_e for hydrogen chloride, HCl, as 3.175827×10^{11} Hz. Given that the isotopes involved are ${}^1_1\text{H}$ and ${}^{35}_{17}\text{Cl}$, what is the bond length of this molecule?

A table of isotopes gives the atomic weights of these species as 1.0078250 and 34.968853 mass units. These numbers give a reduced mass of 0.979593 mass units, or 1.626652×10^{-27} kg. The bond length then follows from (8.53) as

$$r = \sqrt{\frac{h}{8\pi^2\mu\,B_e}} = \sqrt{\frac{(6.62607 \times 10^{-34}\,\text{J-sec})}{8\pi^2\,(1.626652 \times 10^{-27}\,\text{kg})\,(3.175827 \times 10^{11}\,\text{Hz})}}$$

$$= 1.275 \times 10^{-10}\,\text{m} = 1.275\,\text{Å},$$

about the size one would expect for a molecule.

Section 8.3 **Problems**

8.6(E) Lithium fluoride (${}^7_3\text{Li}\,{}^{19}_9\text{F}$) has an equilibrium rotation constant of 4.032983×10^4 Hz. What is the bond length of this species?

8.7(E) The CRC Handbook lists the rotation constant and bond length of lithium oxide (${}^7_3\text{Li}\,{}^{16}_8\text{O}$) as 3.63597×10^{10} Hz and 1.688 Å, respectively. What is the reduced mass of this species as indicated by these values? How does your result compare to the reduced mass calculated from the isotopic masses, 7.016004 and 15.994915 mass units?

8.4 Identical Particles, Indistinguishability, and the Pauli Exclusion Principle

In this section we take up in a somewhat informal way a question similar to the center-of-mass and relative-coordinates separation of variables technique discussed in Chap. 7, but which turns out to have profound implications for the nature of particles and their wavefunctions.

The issue here concerns solutions of Schrödinger's equation for a system of two (or more) particles moving in the same potential, for example, two electrons in a helium atom. To keep the treatment simple, we shall assume that the electrons are non-interacting, that is, that they exert no forces on each other, or at least that inter-electronic forces can be neglected in comparison with those that they experience from the nucleus. Suppose that the particles have masses m_1 and m_1, and respectively experience potentials $V_1(\mathbf{r_1})$ and $V_2(\mathbf{r_2})$, where $\mathbf{r_1}$ and $\mathbf{r_2}$ are their coordinates (see Fig. 7.3). The potentials that they experience may be different even if they are moving within the same environment or even if $\mathbf{r_1} = \mathbf{r_2}$: they might, for example, carry opposite electrical charges. The total Hamiltonian operator for the system will be the sum of the kinetic and potential energies of each:

$$H = -\frac{\hbar}{2m_1}\nabla_1^2 - \frac{\hbar}{2m_2}\nabla_2^2 + V_1(\mathbf{r_1}) + V_2(\mathbf{r_2}). \qquad (8.54)$$

It is in writing the potential energy for each particle as a function of only that particle's position and not as a function involving the position of the other particle that we build in the fact that the particles do not interact with each other.

Now presume that there is some solution $\Psi(\mathbf{r_1}, \mathbf{r_2})$ of Schrödinger's equation for this system with some total energy E:

$$-\frac{\hbar}{2m_1}\nabla_1^2\Psi - \frac{\hbar}{2m_2}\nabla_2^2\Psi + V_1(\mathbf{r_1})\Psi + V_2(\mathbf{r_2})\Psi = E\Psi. \qquad (8.55)$$

If the particles do not interact, it would seem reasonable to assume that Ψ can be written as a product of two functions, one for each particle: $\Psi(\mathbf{r_1}, \mathbf{r_2}) = \psi_A(\mathbf{r_1})\psi_B(\mathbf{r_2})$; this is in analogy to the center-of-mass and relative-coordinates separation posited in Sect. 7.4. The interpretation of Ψ is that Ψ^2 gives the "joint probability" that particle 1 will be found in state A within volume $d\tau_1$ of position $\mathbf{r_1}$ while particle number 2 is simultaneously to be found in state B within volume $d\tau_2$ of position $\mathbf{r_2}$. The subscripts A and B serve as reminders that the two independent particles will in general be in different states. With this separability assumption, Schrödinger's equation becomes

$$-\frac{\hbar}{2m_1}\nabla_1^2\psi_A\psi_B - \frac{\hbar}{2m_2}\nabla_2^2\psi_A\psi_B + V_1\psi_A\psi_B + V_2\psi_A\psi_B = E\psi_A\psi_B, \qquad (8.56)$$

where we have dropped the understood coordinate dependences of the potentials. Now, the Laplacian operators ∇_1^2 and ∇_2^2 operate on only one set of coordinates each: $r_1 = (x_1, y_1, z_1)$ for the former and similarly for the latter. With this in mind, Eq. (8.56) can be factored into the form

$$\left\{ -\frac{\hbar}{2m_1}\nabla_1^2\psi_A + V_1\psi_A \right\}\psi_B + \left\{ -\frac{\hbar}{2m_2}\nabla_2^2\psi_B + V_2\psi_B \right\}\psi_A = E\psi_A\psi_B. \quad (8.57)$$

The form of (8.57) suggests that we write the total energy E as the sum of the energies for the individual-particle states: $E = E_A + E_A$. With this, as might be expected, Eq. (8.57) separates neatly into two independent expressions:

$$-\frac{\hbar}{2m_1}\nabla_1^2\psi_A + V_1\psi_A = E_A\psi_A, \quad (8.58)$$

and

$$-\frac{\hbar}{2m_2}\nabla_2^2\psi_B + V_2\psi_B = E_B\psi_B. \quad (8.59)$$

You should be able to convince yourself that this procedure can be extended to any number of non-interacting particles. In a three-dimensional situation, Eqs. (8.58) and (8.59) would each separate into three sub-equations if the potentials are of separable form.

We come now to an important point. Imagine that the above argument is applied to two *identical* particles, perhaps two electrons in the vicinity of a nucleus. We have been assuming that it is legitimate to imagine "labeling" the electrons as numbers 1 and 2, that is, that we can imagine being able to distinguish them from each other. However, it has been known since the early days of statistical mechanics that in reality nature admits no such notion: identical particles are in principle *indistinguishable*: There is no way to tell them apart. In classical mechanics the notion of distinguishability can be gotten away with because the probability distributions of particles do not appreciably overlap, but at the atomic level probability distributions intermingle, rendering it impossible to say that number 1 is *here* while number 2 is *there*. Indistinguishability is a fundamental physical fact and must be accounted for in our calculations.

The fact of indistinguishability presents an awkward situation for our presumed product wavefunction

$$\Psi(r_1, r_2) = \psi_A(r_1)\psi_B(r_2), \quad (8.60)$$

namely, that Ψ does not respect the fact of indistinguishability! If the particles were "swapped", that is, if r_1 is replaced by r_2 and vice-versa, Ψ would be altered since the states A and B will in general be different. Indistinguishability means that Ψ should remain unchanged if we swap particle labels. Since it is probabilities that are the ultimately measurable quantities, this means that we must demand that Ψ must satisfy

$$\Psi^2(r_1, r_2) = \Psi^2(r_2, r_1). \tag{8.61}$$

By trying various combinations of states and coordinates, it becomes apparent that only two forms for Ψ will satisfy (8.61):

$$\Psi_{symm} = C\left[\psi_A(r_1)\psi_B(r_2) + \psi_A(r_2)\psi_B(r_1)\right] \tag{8.62}$$

and

$$\Psi_{anti} = C\left[\psi_A(r_1)\psi_B(r_2) - \psi_A(r_2)\psi_B(r_1)\right], \tag{8.63}$$

where the C's are normalization constants. Equation (8.62) is said to be the *symmetric* form of Ψ, since swapping r_1 and r_2 leaves Ψ unchanged; correspondingly, Eq. (8.63) is called the *antisymmetric* form since the same swapping will result in negating the value of Ψ while still satisfying (8.61).

At this point it is instructive to consider an example of (8.62) and (8.63). For simplicity we consider a one-dimensional case: Two identical particles moving in the harmonic oscillator potential $V = kx^2/2$. We use the dimensionless ξ coordinates of Chap. 5, labeling them ξ_1 and ξ_2. Presume that the states involved are the lowest and first-excited harmonic-oscillator states

$$\psi_A = e^{-\xi^2/2} \tag{8.64}$$

and

$$\psi_B = \xi e^{-\xi^2/2}. \tag{8.65}$$

We can ignore normalization constants as they will not affect the following argument. Figures 8.4 and 8.5 respectively show the absolute values of the corresponding symmetric and antisymmetric wavefunctions.

$$\Psi_{symm}(\xi_1, \xi_2) = \psi_A(\xi_1)\psi_B(\xi_2) + \psi_A(\xi_2)\psi_B(\xi_1)$$

$$= (\xi_2 + \xi_1)\, \exp\left[-(\xi_1^2 + \xi_2^2)\right] \tag{8.66}$$

and

$$\Psi_{anti}(\xi_1, \xi_2) = \psi_A(\xi_1)\psi_B(\xi_2) - \psi_A(\xi_2)\psi_B(\xi_1)$$

$$= (\xi_2 - \xi_1)\, \exp\left[-(\xi_1^2 + \xi_2^2)\right]. \tag{8.67}$$

From Fig. 8.4 it should be clear that, in the case of the symmetric wavefunction, locations where the joint probability tends to be high are characterized by coordinates values such that $\xi_1 \sim \xi_2$. Qualitatively, we can describe this by saying that particles that are described by symmetric joint wavefunctions will preferentially be found at locations close to each other. Conversely, in the antisymmetric case (Fig. 8.5), locations where the joint probability tends to be high are characterized by coordinates

Fig. 8.4 $|\psi|$ for a two-particle symmetric wavefunction for $n = 0$ and $n = 1$ harmonic oscillator states

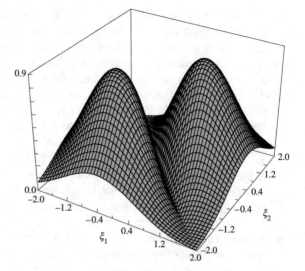

Fig. 8.5 $|\psi|$ for a two-particle antisymmetric wavefunction for $n = 0$ and $n = 1$ harmonic oscillator states

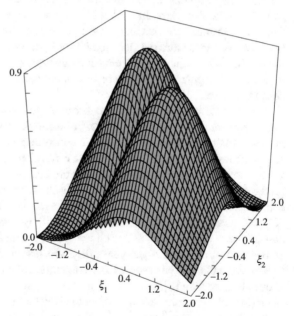

such that $\xi_1 \neq \xi_2$; in fact, it is highest when $\xi_1 \sim -\xi_2$. Such particles thus tend to avoid each other. These conclusions can also be seen directly by looking at the forms of (8.66) and (8.67). These "approach/avoidance" behaviors are general features of such two-particle joint wavefunctions.

It is important to emphasize that this behavior is not a consequence of any interaction between the particles involved; indeed, we have assumed all along that they do not interact. It is purely a consequence of the demand that Ψ must respect the fact that the particles are indistinguishable. It is remarkable that such a powerful conclusion can emerge from such a straightforward argument.

What dictates whether a given type of particle will be described by an overall symmetric or antisymmetric wavefunction? This issue cannot be resolved with the above development. A full understanding of this question is quite involved, but we can give a qualitative description of the essential points. In Sect. 8.2 we saw how experiments conducted in the early 1920's revealed that particles possess intrinsic angular momentum, an attribute now known as spin. These experiments further revealed that all particles can be classed as one of two types: those whose spin is an odd half-integer multiple of $\hbar$ ("spin-1/2" particles) and those whose spins are full-integer (including zero) multiples of $\hbar$ ("spin-one" particles). The former are now known as *fermions* in honor of Enrico Fermi; the latter are known as *bosons* in honor of Indian physicist Satyendra Bose. Fermions, which include electrons, protons, and neutrons, are described by antisymmetric overall wavefunctions; bosons, which include photons and other, more exotic particles, are of the symmetric kind. Hence, the presence of a boson in a particular state increases the probability that an identical particle will be found in the same state, whereas the presence of a fermion in a particular state decreases the probability that an identical particle will be found in that state. Curiously, the fact that photons possess intrinsic angular momentum could have been deduced from the Bohr model for hydrogen: when an electron transits to a lower (higher)-radius orbit by emitting (absorbing) a photon, the latter must carry off (bring in) angular momentum in an amount equal to an integer multiple of $\hbar$, as described in Sect. 1.3.

A composite structure such as an entire atom possesses the symmetry corresponding to the sum of its constituent spins. For example, normal helium $^{4}_{2}$He is a boson in view of it's even number of spin-1/2 constituents (2 each of electrons, protons, and neutrons), whereas helium-3 $^{3}_{2}$He is a fermion (2 protons, 1 neutron, 2 electrons). Bosons, being characterized by symmetric wavefunctions, will presumably prefer to be found in the proximity of each other. This prediction has been verified experimentally via creating what is known as a Bose-Einstein condensate (BEC). A liquid-phase BEC was first achieved in 1908 by cooling liquid helium-4 to near absolute-zero temperature; in this circumstance the helium acts as a superfluid with no resistance to motion. A gas-phase BEC was first achieved in 1995 with rubidium ($^{85}_{37}$Rb) atoms that had been cooled to a temperature of about 200 billionths of a degree above absolute zero [8]. In such circumstances, all of the atoms involved are in the same (lowest) quantum state and collectively act as a single atom.

One important point remains here. From (8.63) it should be clear that, in the antisymmetric (fermion) case, if the two particles should find themselves in the same state ($A = B$), then $\Psi = 0$ everywhere. However, this would be inconsistent with the presumed presence of the particles in the first place. Consequently, we are forced to conclude that no two electrons in the same atom can find themselves in the same quantum state. This is the *Pauli exclusion principle*.

Ultimately, four quantum number are necessary to specify the state of an electron: the principal quantum number n, the angular momentum quantum number ℓ, the magnetic quantum number m, and its spin-state quantum number, $\pm 1/2$. In view of the spin quantum number, the total number of states for principal quantum number n is $2n^2$; see (7.80). For $n = 1, 2, 3, \ldots$, this amounts to $2, 8, 18, \ldots$ states; these values correspond to the noble gases helium, neon, argon, and so on, which are said to possess "closed" (completely filled) electron shells. Elements with $3, 9, 19, \ldots$ electrons will thus have their "last" electron being the lone occupant of $n = 2, 3, 4, \ldots$ states. Such elements are those in the first column of the periodic table. As one might expect, their lone electrons can be dissociated relatively easily, making these elements particularly reactive. A full treatment of the chemical nature of elements involves analysis of the order in which energy states are filled in view of the interaction between electron wavefunctions as well as relativistic effects. Suffice it to say that the important concepts of symmetry and antisymmetry survive these issues.

Problems

8.8(E) Classify the following neutral atoms as to whether they will be fermions or bosons: $^{29}_{14}Si$, $^{40}_{20}Ca$, $^{69}_{31}Ga$, $^{105}_{46}Pd$, $^{137}_{55}Cs$, $^{235}_{92}U$.

8.9(E) In Problem 4.17 it was remarked that early in the history of nuclear physics, the electrically neutral mass of nuclei now attributed to neutrons was considered to arise from neutral particles comprising combinations of protons and electrons as opposed neutrons being fundamental particles in their own right. That problem explored the implications of the uncertainty principle for that theory. Here we look at the spin statistics of the situation. Consider a nitrogen-14 nucleus ($^{14}_{7}N$). If the "protons + electrons" model were correct, would you predict N-14 to be a spin-1/2 or spin-1 system? What about in the case of the "protons + neutrons" model? Spectroscopic evidence indicates that N-14 is a spin-1 system. Which model does this support?

References

1. W. Stern, O. Gerlach, Z. Phys. **8**, 110 (1921)
2. L.D. Landau, E.M. Lifshitz, *Quantum Mechanics (Non-relativistic Theory)*, revised 3rd ed. (Pergamon, Oxford, 1977), p. 461
3. T.E. Phipps, J.B. Taylor, Phys. Rev. **29**, 309 (1927)
4. G.E. Uhlenbeck, S.A. Goudsmit, Naturwiss. **13**, 953 (1925)
5. G.E. Uhlenbeck, S.A. Goudsmit, Nature **117**, 264 (1926)
6. A.P. French, E.F. Taylor, *An Introduction to Quantum Physics* (W. W. Norton, New York, 1978), Section 13.9
7. P.A.M. Dirac, Proc. Roy. Soc. **A117**, 610 (1928)
8. G.P. Collins, Phys. Today **48**(8), 17 (1995)

Chapter 9
Approximation Methods

Summary The number of potentials for which Schrödinger's equation can be solved exactly is very limited. In view of this, methods for developing approximate analytic and numerical solutions take on great importance. Numerical solutions are the topic of Chap. 10; this chapter deals with approximate analytic techniques. Three methods are explored. These are (1) The Wentzel-Kramers-Brillouin (WKB) method for estimating the energy of the n'th energy level of a system via a straightforward integral involving the potential function; (2) Perturbation theory, a technique which allows one to get an approximate solution for a potential which can be expressed as one which can be solved exactly plus a perturbing effect; and (3) The variational theorem, a method for estimating the lowest possible energy of a system via an educated guess at the shape of the corresponding wavefunction. Some of these techniques are applied to exactly-solvable problems, an approach which helps in understanding how successful they are in practice.

In previous chapters we explored solutions to Schrödinger's equation for a variety of potentials. In most cases these solutions were expressible in closed analytic forms. As you might imagine, however, the number of potentials for which Schrödinger's equation is soluble in closed form is very limited. For many potentials, even some simple ones such as the linear potential of Problem 4.10, it is impossible to develop closed-form analytic solutions.

How can we analyze such potentials within the framework of the knowledge of solutions of Schrödinger's equation that we have built up? The purpose of this chapter is to explore analytic methods of obtaining approximate solutions; numerical methods are discussed in Chap. 10. We consider three methods in particular. In Sect. 9.1 we examine the Wentzel-Kramers-Brillouin or *WKB method*, a scheme for determining approximate energy eigenvalues [1–3]. This method is also known as the JWKB method in honor of Harold Jeffreys, who had established the basic mathematical approach in 1923 [4]. In Sects. 9.2 and 9.3 we develop *perturbation theory*, a powerful, quite general method for establishing approximate wavefunctions and energies for potentials that can be expressed as variants of potentials for which Schrödinger's equation can be solved exactly. In Sects. 9.4 and 9.5 we examine how the *variational method* can be employed to generate approximate ground-state

© The Author(s), under exclusive license to Springer Nature Switzerland AG 2022   287
B. C. Reed, *Quantum Mechanics*, https://doi.org/10.1007/978-3-031-14020-4_9

energies based on an educated guess at the form of the ground-state wavefunction. By applying these techniques to problems known to possess exact solutions, it is possible to get a sense of how accurate they can be.

9.1 The WKB Method

We have seen that for regions of space where the total energy of a particle exceeds the local potential energy, that is, where $E > V(x)$, the de Broglie wavelength characterizing a particle of mass m is given by

$$\lambda(x) = \frac{h}{\sqrt{2m\,[\,E - V(x)\,]}}. \tag{9.1}$$

We revert to the use of m to designate mass in this chapter. This symbol will also sometimes be used as a summation index, but no confusion with mass or magnetic quantum number should arise in such cases.

In cases where $V(x)$ is constant (such as inside a finite or infinite rectangular well), λ will be constant. In such a situation, the wavefunctions are mathematically identical in form to the standing-wave patterns of a vibrating string clamped at both ends, as sketched in Fig. 9.1. From elementary mechanics you may recall that such standing-wave patterns are defined by the requirement that an integral number of half-wavelengths fit between the clamped ends. The essence of the WKB method is to similarly assume that if ψ is characterized by a wavelength λ, then the particle's energy E must be such that an approximately half-integral number of de Broglie wavelengths fit between the classical turning points of the motion. The approximation in this method comes from neglecting the fact that, quantum mechanically, the particle can penetrate into regions of space where $E < V(x)$. In effect, this method is predicated on the idea that if penetration of the wave into the potential barrier is slight in comparison to the extent of the domain for which $E > V(x)$, then $\psi(x)$ goes approximately to zero where $E = V(x)$.

In general, however, $V(x)$ will not be constant and λ will be a function of position; consequently, it is not justifiable to simply equate $n(\lambda/2)$ to the distance between the turning points. We can, however, build up an integral representation of (9.1). A short interval of distance dx over which the potential $V(x)$ is effectively constant will contain a number of wavelengths given by

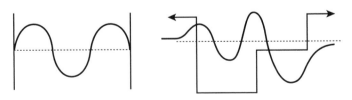

Fig. 9.1 Left: A string vibrating in the $n = 3$ mode. Right: A quantum wavefunction with $n = 4$

$$\frac{dx}{\lambda} = \frac{\sqrt{2m\,[\,E - V(x)\,]}}{h}dx. \tag{9.2}$$

The total number of wavelengths between the turning points is given by the integral of (9.2) between those limits; if an integral number of half-wavelengths are to approximately fit between the turning points, then

$$\int \frac{dx}{\lambda} \sim \frac{n}{2}, \tag{9.3}$$

or, on combining (9.2) and (9.3),

$$2\sqrt{2m} \int \sqrt{E - V(x)}dx \sim nh. \tag{9.4}$$

This is the WKB approximation. A more sophisticated derivation renders the right side (9.4) as $(n + 1/2)h$, an expression which gives better results for low-n states, but this is a refinement with which we will not concern ourselves [5].

With this result, we can estimate the energy corresponding to quantum level n for a particle of mass m under the influence of any potential function $V(x)$. In practice, the limits of integration are given by the positions at which some general energy E cuts the $V(x)$ curve; the result is an expression for $E(n)$.

Before illustrating use of the WKB method, it is instructive to look a little more closely at the conditions of its validity. The accuracy of (9.3) depends on the quantum wavelength λ not varying too rapidly over short distances. What we are really demanding is that any change $d\lambda$ in λ over distance dx be small:

$$\frac{d\lambda}{dx} \ll 1. \tag{9.5}$$

To compute actual numbers, it is more useful to express (9.5) in terms of the energy eigenevalues and potential function $V(x)$. From the de Broglie relation, the quantum-mechanical wavelength is given in terms of momentum by $\lambda = h/p$. Classically, we can write momentum as

$$p(x) = \sqrt{2m[E - V(x)]}. \tag{9.6}$$

Taking the derivative gives

$$\frac{d\lambda}{dx} = -\frac{h}{p^2}\frac{dp}{dx} = \frac{mh}{2m\,[\,E - V(x)\,]^{3/2}}\left(\frac{dV}{dx}\right).$$

That is, Eq. (9.5) is equivalent to

$$\frac{mh}{\{2m[E - V(x)]\}^{3/2}}\left(\frac{dV}{dx}\right) \ll 1. \tag{9.7}$$

This is sometimes written with the absolute value of the gradient of the potential, $|dV/dx|$, to accommodate the fact that λ can either increase (or decrease) as $V(x)$ decreases (increases).

If dV/dx is small or the momentum is large (or both), Eq. (9.7) is likely to be satisfied. The phrase "momentum is large" has a macroscopic ring to it, so (9.7) is referred to as the *classical approximation* in this context. At locations where an energy level E cuts the $V(x)$ curve or where $E \sim V(x)$, that is, at or near the classical turning points of the motion, Eq. (9.7) will clearly not be satisfied. If such regimes form a substantial part of the domain of the problem at hand, then the WKB approximation cannot be expected to be terribly accurate. An illustration of using (9.4) and (9.7) is given in the following example.

Example 9.1 Use the WKB approximation to determine approximate energy eigenvalues for a particle of mass m moving in the one-dimensional linear potential

$$V(x) = \begin{cases} \infty, & x \leq 0 \\ \alpha x, & 0 < x < \infty. \end{cases}$$

Consider some general energy E as shown in Fig. 9.2. If energy E cuts the $V(x)$ line at $x = a$, then $E = \alpha a$. From (9.4) we have

$$2\sqrt{2m} \int_0^a \sqrt{E - \alpha x}\, dx \sim nh.$$

Evaluating the integral gives

$$nh \sim 2\sqrt{2m} \left[-\frac{2}{3\alpha} \sqrt{(E - \alpha x)^3} \right]_0^a \sim -\frac{4\sqrt{2m}}{3\alpha} [(E - \alpha a)^{3/2} - E^{3/2}].$$

With $E = \alpha a$ we find

$$E_n \sim \left(\frac{3\alpha h}{4\sqrt{2m}} n \right)^{2/3} \sim \left(\frac{6\pi\alpha\hbar}{4\sqrt{2m}} n \right)^{2/3} \sim 2.811 \left(\frac{\alpha\hbar}{\sqrt{2m}} n \right)^{2/3}.$$

This result is consistent with that in Problem 4.10; for an electron in such a potential with $\alpha = 1$ eV/Å, $E_1 \sim 4.39$ eV. An analytic solution to this problem involves an infinite polynomial known as an Airy function, the roots of which are related to the energy eigenvalues. The exact ground-state energy is of the form above but with a numerical factor of 2.338 as opposed to 2.811; that is, the WKB method overestimates the ground-state energy by about 20%. In Sect. 9.4 we will explore an alternate analytic method for estimating the ground-state energy corresponding to this potential, and a direct numerical solution of Schrödinger's equation for this problem is considered in Chap. 10.

To what extent is the classical approximation satisfied for this problem? To get a sense of this, we evaluate the left side of (9.7) at an x-position halfway between $x = 0$ and where a general energy level E_n cuts the potential line $V(x)$. For convenience, write the above result as $E_n = \beta n^{2/3}$, where $\beta = (3\pi\alpha\hbar/2\sqrt{2m})^{2/3}$. A given energy level E_n will cut the potential line $V(x) = \alpha x$ at the turning point $x_{turn} = \beta n^{2/3}/\alpha$. Evaluating the left side of (9.7) at $x_{turn}/2$ gives

$$\frac{mh}{\{2m[E - V(x)]\}^{3/2}} \left(\frac{dV}{dx}\right) = \frac{mh\alpha}{\{2m[E_n - \alpha x_{turn}/2]\}^{3/2}}$$
$$= \frac{h\alpha}{\sqrt{m}\,\beta^{3/2}\,n} = \frac{4\sqrt{2}}{3n}.$$

Equation (9.7) then demands

$$\frac{4\sqrt{2}}{3} \ll n \Rightarrow n \gg 1.886.$$

There is no hard-and-fast answer to the question of "For what values of n can this be considered to be satisfied?" It depends on your definition of "much greater than"; also, the numbers can be made to change by evaluating (9.7) at a different value of x. But "n less than at least a few" is likely to be unsatisfactory by any reasonable definition. It is perhaps somewhat surprising that the WKB method is in error by only 20% for the ground state.

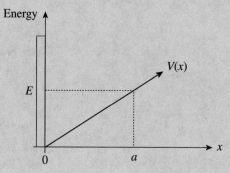

Fig. 9.2 Linear potential

Example 9.2 Treat the hydrogen atom as a one-dimensional system in the radial coordinate r and use the WKB approximation to estimate the energy levels.

The situation is shown schematically in Fig. 9.3, where a bound-state energy $-E$ cuts the Coulomb potential at $r = a$.

The WKB approximation gives

$$2\sqrt{2m} \int_0^a \sqrt{E - V(r)}\, dr = 2\sqrt{2m} \int_0^a \sqrt{E + (e^2/4\pi\varepsilon_o r)}\, dr \sim nh.$$

The upper limit of integration is given by

$$E = -\frac{e^2}{4\pi\varepsilon_o a} \Rightarrow aE = -\frac{e^2}{4\pi\varepsilon_o}.$$

Substituting this expression into the integral gives

$$2\sqrt{2m} \int_0^a \sqrt{E - \frac{aE}{r}}\, dr \sim nh.$$

To simplify this integral, it is helpful to remember that we are seeking bound states, that is, $E < 0$. Writing $E = -|E|$, we have

$$2\sqrt{2m\,|E|} \int_0^a \sqrt{\frac{a}{r} - 1}\, dr \sim nh.$$

This integral is most easily solved by invoking the change of variable $r = a\sin^2\beta$ to yield

$$4a\sqrt{2m\,|E|} \int_0^{\pi/2} \cos^2\beta\, d\beta \sim nh.$$

This integral evaluates to $\pi/4$. The energy levels are then

$$|E| \sim \frac{n^2 h^2}{2m\pi^2 a^2} \sim \frac{n^2 h^2}{2m\pi^2}\frac{16\pi^2\varepsilon_o^2\, E^2}{e^4},$$

or, on canceling a common factor of E from both sides,

$$\left|E_n^{WKB}\right| \sim \frac{me^4}{8\varepsilon_o^2 h^2 n^2}.$$

Remarkably, the WKB method yields *exactly* the energy levels as given by the Bohr and Schrödinger models!

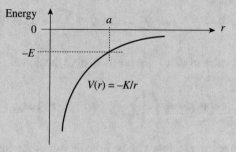

Fig. 9.3 Coulomb potential

9.2 The Superposition Theorem Revisited

In Chap. 4, we encountered the idea of constructing a superposition state by forming a linear sum of solutions to Schrödinger's equation for some potential $V(x)$. In this section we examine a closely related mathematical technique that lies at the heart of both the perturbation and variational methods developed later in this chapter.

An essential ingredient in developing the superposition theorem was the orthogonality theorem: That the family of solutions $\psi_n(x)$ to Schrödinger's equation for a given potential satisfies the relationship

$$\langle \psi_j | \psi_i \rangle = \delta_j^i, \tag{9.8}$$

where ψ_j and ψ_i are any of the ψ_n and where $\delta_j^i = 1$ if $i = j$, zero otherwise (review Sect. 4.2 for the angle-bracket notation and Sect. 4.6 for the δ_j^i notation.) This relationship will prove useful in what follows.

In developing the superposition theorem, we had in mind the idea of a system in a sort of indeterminate state comprising contributions from individual states ψ_n, with the contribution of any particular state dictated by a corresponding "expansion coefficient" a_n. In the present development we reverse the sense of the argument to ask: Suppose that $\phi(x)$ is some continuous but otherwise arbitrary function valid over the same domain as the $\psi_n(x)$. What values must the expansion coefficients take if we desire to write $\phi(x)$ as a linear sum of the $\psi_n(x)$, that is, if we wish to have

$$\phi(x) = \sum_n a_n \psi_n(x)? \tag{9.9}$$

With both $\phi(x)$ and the $\psi_n(x)$ known, this seems a curious demand of no practical value. We shall see, however, that this sort of construction can be very useful.

To establish the a_n it is helpful to use the orthogonality properties of the $\psi_n(x)$. To do this, begin by multiplying through (9.9) by any one of the $\psi(x)$, say $\psi_j(x)$. This gives

$$\psi_j(x)\phi(x) = \psi_j(x) \sum_n a_n \psi_n(x) = \sum_n a_n \psi_j(x)\psi_n(x). \qquad (9.10)$$

Integrating both sides of (9.10) [in practice, over the domain of the $\psi(x)$] gives

$$\langle \psi_j | \phi \rangle = \int \left\{ \sum_n a_n \psi_j(x)\psi_n(x) \right\} dx.$$

Since the a_n are constants, they can be brought out of the integral provided that their positions within the sum are properly maintained:

$$\langle \psi_j | \phi \rangle = \sum_n a_n \langle \psi_j | \psi_n \rangle.$$

From the orthogonality condition of (9.8), the integral on the right side reduces to δ_j^n, hence,

$$\langle \psi_j | \phi \rangle = \sum_n a_n \delta_j^n. \qquad (9.11)$$

The sum in (9.11) runs over all possible values of n. But since $\delta_j^n = 0$ when $n \neq j$, the only surviving term is that for which $n = j$, that is, a_j. The result is a recipe for the j-th expansion coefficient in terms of $\phi(x)$ and $\psi_j(x)$:

$$a_j = \langle \psi_j | \phi \rangle. \qquad (9.12)$$

The critical step in this derivation was simplifying the integral in the equation preceding (9.11): Had the $\psi_n(x)$ not obeyed the orthogonality relation, Eq. (9.12) would not have followed. It is worth emphasizing that $\phi(x)$ need not (and in general will not) be a solution to Schrödinger's equation.

The resulting expression for $\phi(x)$ will be valid only over the domain of the $\psi(x)$. If we desire in addition that $\phi(x)$ be normalized over this range, a constraint on the a_n arises. Normalization demands

$$\langle \phi^* | \phi \rangle = \int \left\{ \left(\sum_n a_n^* \psi_n^*(x) \right) \left(\sum_m a_m \psi_m(x) \right) \right\} dx = 1. \qquad (9.13)$$

Note that two different dummy summation indices are used here to maintain generality. This expresssion can be rearranged to read

$$\sum_n \sum_m a_n^* a_m \langle \psi_n^* | \psi_m \rangle = 1. \tag{9.14}$$

The integral here reduces to δ_n^m via (9.8), leaving

$$\sum_n \sum_m a_n^* a_m \delta_n^m = 1. \tag{9.15}$$

Now consider the inner sum in this result: For some particular value of n, it runs over all possible values of m. The index n is then incremented and all possible values of m are run through again. This process is continued until all of the possible values of n are exhausted. In each case, however, $\delta_n^m = 0$ unless $m = n$. The result is that in each run of the inner sum only one term, that for $m = n$, is nonzero:

$$\sum_n \sum_m a_n^* a_m \delta_n^m = \sum_n a_n^* a_n = \sum_n |a_n|^2 = 1. \tag{9.16}$$

The condition that $\phi(x)$ be normalized means that the sum of the squares of the absolute values of the expansion coefficients must be unity. Conversely, if $\phi(x)$ is initially chosen such that it is normalized over the domain of the $\psi(x)$, this must happen automatically when the a_n are computed from (9.12) since (9.9) was assumed to represent an exact expression for $\phi(x)$.

Example 9.3 Express the function $\phi(x) = x$ as a linear sum of infinite rectangular well wavefunctions.

The relevant wavefunctions are

$$\psi_n(x) = \sqrt{\frac{2}{L}} \sin\left(\frac{n\pi x}{L}\right).$$

From (9.12) the expansion coefficients are given by

$$a_n = \langle \psi_n | \phi \rangle = \sqrt{\frac{2}{L}} \int_0^L x \sin\left(\frac{n\pi x}{L}\right) dx.$$

Defining $c = n\pi/L$ and evaluating the integral gives

$$a_n = \sqrt{\frac{2}{L}} \left[\frac{1}{c^2} \sin(cx) - \frac{x}{c} \cos(cx) \right]_0^L$$

$$= \sqrt{\frac{2}{L}} \left[-\frac{L^2}{n\pi} \cos(n\pi) \right] = (-1)^{n+1} \frac{\sqrt{2} L^{3/2}}{n\pi},$$

where we used the fact that $\cos(n\pi) = (-1)^n$. Note that the expansion coefficients are functions of n; this is a common occurrence. Hence our expression for $\phi(x)$ is

$$\phi(x) = x = \sum_n a_n \psi_n(x) = \frac{2L}{\pi} \sum_n \frac{(-1)^{n+1}}{n} \sin\left(\frac{n\pi x}{L}\right).$$

Readers familiar with Fourier series will recognize the nature of this example: that any function can be reproduced by adding together an infinite number of sinusoidal functions. For computational purposes it is not practical to sum an infinite number of terms; however, the "n" in the denominator of the sum ensures that the high-n terms in the sum will make relatively little contribution. This is demonstrated in Fig. 9.4, where $\phi(x)$ is plotted for the case of $L = 1.0$. The two curves correspond to including terms up to $n = 5$ and $n = 20$. Both curves collapse to zero at $x = 1.0$ because the infinite-well wavefunctions vanish there.

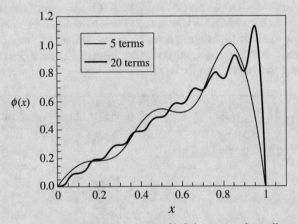

Fig. 9.4 Approximating a straight line as a sum of infinite rectangular well wavefunctions

9.3 Perturbation Theory

Sometimes we encounter problems that we have not solved *per se* but which bear a strong resemblance to others we have previously met. Almost intuitively, accumulated wisdom of past experience begins to suggest strategies for attacking the new problem. Perturbation theory systematizes this strategy by specifying a recipe for constructing an approximate solution to Schrödinger's equation for a potential that cannot be solved exactly by expressing the solution in terms of the eigenfunctions

Fig. 9.5 Perturbed infinite
rectangular well

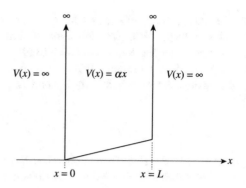

and eigenvalues of a similar problem that can. The "modification" required to cast
the exactly-soluble problem into to the desired one is known as a "perturbation."

Figure 9.5 shows an example: an infinite one-dimensional potential well with
a tilted floor described by $V(x) = \alpha x$. Schrödinger's equation for this potential
cannot be solved in closed form (see Problem 5.10); however, we know that if $\alpha = 0$,
the solution is trivial. Hypothesizing a sort of continuity of nature, it would seem
reasonable to presume that if α is small (say, such that αL is small in comparison
with the lowest energy level of the unperturbed well), then the wavefunctions and
energy levels of the perturbed well will probably not differ much from their $\alpha = 0$
counterparts. The essence of perturbation theory is to work out correction factors to
be added to the unperturbed solutions to approximate the solution of the perturbed
problem.

Developing perturbation theory is a somewhat lengthy task that hinges on the
superposition theorem developed in the preceding section. While the physical concept
underlying it – that the solution to Schrödinger's equation for the perturbed potential
can be expressed as a sum over solutions to a closely similar "unperturbed" potential
– is intuitively appealing, the mathematics can be daunting at first glance and so are
worked out in some detail.

As in the previous section, an important aspect of what follows is the orthogonality theorem of Chap. 4. Here, however, we invoke it in its full form: That for
two wavefunctions ψ_k and ψ_n which satisfy the same potential, the energies and
wavefunctions must satisfy

$$(E_n - E_k)\langle \psi_k^* | \psi_n \rangle = 0.$$

If ψ_k and ψ_n should happen to be degenerate, then $E_n = E_k$ and no conclusion can
be made about the orthogonality of ψ_k and ψ_n: $\langle \psi_k^* | \psi_n \rangle$ might be zero, but this cannot
be assumed. In what follows, no assumptions are made regarding orthogonality (or
not) of wavefunctions unless it is clear that they refer to the different energies; the
derivation is consequently equally valid for both non-degenerate and degenerate
systems. We do, however, restrict the discussion to bound-state, time-independent
problems.

In developing perturbation theory, it is convenient to introduce some simplifying notation. Suppose that we wish to solve Schrödinger's equation for the wavefunctions $\psi(r)$ and energies E corresponding to some potential $V(r)$; we put r-dependences for ψ and V as a reminder that we may be dealing with a system of more than one dimension. We write Schrödinger's equation in the usual form

$$H\psi(r) = E\psi(r),\; H = -\frac{\hbar^2}{2m}\nabla^2 + V(r), \tag{9.17}$$

where H is the Hamiltonian operator. $V(r)$ is presumed to be such that a closed analytic solution to Schrödinger's equation is difficult or impossible to obtain. Suppose further that there is some similar potential $V^0(r)$ for which we can solve Schrödinger's equation exactly. Designate the wavefunctions and energies corresponding to this potential as $\psi_n^0(r)$ and E_n^0. Superscript 0 refers to the unperturbed problem. In this "unperturbed" case, Schrödinger's equation is

$$H^0\psi_n^0(r) = E_n^0\psi_n^0(r),\; H^0 = -\frac{\hbar^2}{2m}\nabla^2 + V^0(r). \tag{9.18}$$

These unperturbed states may be infinite in number. Irrespective of whether or not some of them may be degenerate, the numbering scheme $n = 0, 1, 2, 3, \ldots$ can proceed arbitrarily; for practical computational purposes we will have to assume some upper limit N on the index n. (There is an exception to this as discussed at the end of the subsection on first-order perturbation theory following (9.31) below, but don't worry about this for now.) The boundary conditions for the perturbed and unperturbed potentials are assumed to be the same.

Comparing (9.17) and (9.18) shows that we can write the perturbed Hamiltonian in terms of the unperturbed Hamiltonian as

$$H = H^0 + [V(r) - V^0(r)] = H^0 + V'(r). \tag{9.19}$$

$V'(r)$ is defined as the difference between the perturbed and unperturbed potentials, in the sense (perturbed minus unperturbed). In the spirit of the superposition process of Sect. 4.7, we assume that a solution for the perturbed potential can be written as a linear sum over all of the solutions of the unperturbed potential:

$$\psi(r) = \sum C_m \psi_m^0(r), \tag{9.20}$$

where the C's are dimensionless expansion coefficients and the index of summation $m = 1, 2, 3, \ldots N$.

Now substitute (9.19) and (9.20) into Schrödinger's equation for the perturbed problem (9.17). Dropping the understood dependences of the potential and wavefunctions on r, this gives

$$\left(H^0 + V'\right) \sum C_m \psi_m^0 = E \sum C_m \psi_m^0.$$

On expanding out the left side of this expression, the unperturbed Hamiltonian can be taken within the summation, yielding

$$\sum C_m E_m^0 \, \psi_m^0 + \sum C_m (V' \psi_m^0) = E \sum C_m \psi_m^0, \qquad (9.21)$$

where we have used $H \psi_m^0 = E_m^0 \psi_m^0$.

Now multiply through (9.21) by one of the unperturbed wavefunctions, say ψ_j^0 (or, if ψ_j^0 is complex, by its complex conjugate) and integrate over the range of the potential. This gives

$$\sum C_m E_m^0 \, \langle \psi_j^0 | \psi_m^0 \rangle + \sum C_m \, \langle \psi_j^0 | V' | \psi_m^0 \rangle = E \sum C_m \, \langle \psi_j^0 | \psi_m^0 \rangle,$$

or

$$\sum C_m \left\{ (E_m^0 - E) \, \langle \psi_j^0 | \psi_m^0 \rangle + \langle \psi_j^0 | V' | \psi_m^0 \rangle \right\} = 0. \qquad (9.22)$$

We can cast this into the more compact form

$$\sum C_m V_{jm} = 0 \qquad (9.23)$$

by defining matrix elements

$$V_{jm} = \left\{ (E_m^0 - E) \, \langle \psi_j^0 | \psi_m^0 \rangle + \langle \psi_j^0 | V' | \psi_m^0 \rangle \right\}. \qquad (9.24)$$

If the summation in (9.20) runs from $m = 1$ to N, then (9.23) and (9.24) can be expressed as

$$\begin{bmatrix} V_{11} & \cdots & V_{1k} & \cdots & V_{1r} & \cdots & V_{1N} \\ \vdots & \ddots & \vdots & \vdots & \vdots & \vdots & \vdots \\ V_{k1} & \cdots & V_{kk} & \cdots & V_{kr} & \cdots & V_{kN} \\ \vdots & \vdots & \vdots & \ddots & \vdots & \vdots & \vdots \\ V_{r1} & \cdots & V_{rk} & \cdots & V_{rr} & \cdots & V_{rN} \\ \vdots & \vdots & \vdots & \vdots & \vdots & \ddots & \vdots \\ V_{N1} & \cdots & V_{Nk} & \cdots & V_{Nr} & \cdots & V_{NN} \end{bmatrix} \begin{bmatrix} C_1 \\ \vdots \\ C_k \\ \vdots \\ C_r \\ \vdots \\ C_N \end{bmatrix} = 0. \qquad (9.25)$$

The only way to satisfy this expression is to require that the determinant of the large square matrix, hereafter abbreviated $[V]$, be zero. To determine the eigenvalues of the perturbed potential, one must search, usually by trial and error, for those values of E that yield $det[V] = 0$. Since we have an N by N matrix, there will be N independent although not necessarily different solutions for E.

Once the perturbed energies have been established, we can go back to solve for the expansion coefficients corresponding to each of them. It is important to bear in mind that there will be a perturbed wavefunction $\psi(E)$ corresponding to each value of E, so each E must be treated separately. To get the C's, split up one of the elements of $[V]$ with the help of (9.24), say the (r, k)-th element (mnemonic: *r*ow, *k*olumn), and re-cast (9.25) as

$$
\begin{bmatrix}
V_{11} & \cdots & V_{1k} & \cdots & V_{1r} & \cdots & V_{1N} \\
\vdots & \ddots & \vdots & & \vdots & \vdots & \vdots \\
V_{k1} & \cdots & V_{kk} & \cdots & V_{kr} & \cdots & V_{kN} \\
\vdots & \vdots & \vdots & \ddots & \vdots & \vdots & \vdots \\
V_{r1} & \cdots & \langle\psi_r^0|V'|\psi_k^0\rangle & \cdots & V_{rr} & \cdots & V_{rN} \\
\vdots & \vdots & \vdots & & \vdots & \ddots & \vdots \\
V_{N1} & \cdots & V_{Nk} & \cdots & V_{Nr} & \cdots & V_{NN}
\end{bmatrix}
\begin{bmatrix}
C_1 \\ \vdots \\ C_k \\ \vdots \\ C_r \\ \vdots \\ C_N
\end{bmatrix}
=
\begin{bmatrix}
0 \\ \vdots \\ C_k(E - E_k^0)\,\langle\psi_r^0|\psi_k^0\rangle \\ \vdots \\ 0 \\ \vdots \\ 0
\end{bmatrix}.
$$

$$(9.26)$$

In carrying out this rearrangement, it is crucial to choose an element (r, k) such that $(E - E_k^0)\langle\psi_r^0|V'|\psi_k^0\rangle$ is nonzero; otherwise, the large matrix of (9.26) will not be different from that of (9.25) and the determinant would still be zero, rendering it impossible to proceed from this point.

A general expansion coefficient C_i can be obtained by inverting (9.26),

$$
C_i = C_k \left\{ V_{ik}^* (E - E_k^0) \, \langle\psi_r^0|\psi_k^0\rangle \right\}, \quad (i = 1, N), \tag{9.27}
$$

where $V_{i,k}^*$ designates the (i, k)-th element of the inverse of the square matrix in (9.26), not the complex conjugate of V_{ik}. Coefficient C_k itself is established by demanding that the perturbed wavefunction be normalized:

$$
\langle\psi|\psi\rangle = 1 \Rightarrow \left\langle \sum C_n \psi_n^0 \Big| \sum C_m \psi_m^0 \right\rangle = 1
$$

$$
\Rightarrow \sum_n \sum_m C_n C_m \langle\psi_n^0|\psi_m^0\rangle = 1. \tag{9.28}
$$

With (9.27), this reduces to

$$
C_k^2 (E - E_k^0)^2 \, \langle\psi_r^0|\psi_k^0\rangle^2 \sum_n \sum_m \left\{ (V_{nk}^*)(V_{mk}^*) \, \langle\psi_n^0|\psi_m^0\rangle \right\} = 1,
$$

or

$$
C_k = \frac{1}{(E - E_k^0) \, \langle\psi_r^0|\psi_k^0\rangle \, \sqrt{\sum_n \sum_m \left\{ (V_{nk}^*)(V_{mk}^*) \, \langle\psi_n^0|\psi_m^0\rangle \right\}}}. \tag{9.29}
$$

This expression, in combination with (9.27), gives, finally,

$$C_i = \frac{V_{ik}^*}{\sqrt{\sum_n \sum_m \left\{ (V_{nk}^*) (V_{mk}^*) \left\langle \psi_n^0 | \psi_m^0 \right\rangle \right\}}}. \tag{9.30}$$

If the unperturbed wavefunctions are non-degenerate, then $\left\langle \psi_n^0 | \psi_m^0 \right\rangle = \delta_n^m$, and only $m = n$ survives the innermost sum in the denominator:

$$C_i = \frac{V_{ik}^*}{\sqrt{\sum_n (V_{nk}^*)^2}} \quad \text{(non-degenerate only)}. \tag{9.31}$$

Non-degenerate First-Order Perturbation Theory
The development above indicates that the calculations involved in a perturbation problem could rapidly escalate to a daunting degree. In the event that the unperturbed states are non-degenerate, however, considerable simplification can be had.

Consider the matrix elements of (9.24):

$$V_{jm} = \left\{ (E_m^0 - E) \left\langle \psi_j^0 | \psi_m^0 \right\rangle + \left\langle \psi_j^0 | V' | \psi_m^0 \right\rangle \right\}.$$

If the unperturbed states are non-degenerate, we must have $\left\langle \psi_j^0 | \psi_m^0 \right\rangle = 0$ from the orthogonality theorem, that is, the first term within the curly brackets will vanish unless $m = j$. As to the remaining term, $\left\langle \psi_j^0 | V' | \psi_m^0 \right\rangle$, it will likely be small except perhaps when $j = m$ if the perturbing potential V' is small—with "small" having to be justified after the fact. If we accept this approximation, then the only elements of V which would be significantly different from zero are the diagonal ones:

$$V_{mm} \sim \left\{ (E_m^0 - E) + \left\langle \psi_m^0 | V' | \psi_m^0 \right\rangle \right\}. \tag{9.32}$$

In this case the determinant of V reduces to

$$det\,[\mathbf{V}] \sim \prod_m \{ (E_m^0 - E) + \left\langle \psi_m^0 | V' | \psi_m^0 \right\rangle \}, \tag{9.33}$$

which means that the only way (9.25) can be satisfied is if

$$E \sim E_m^0 + \left\langle \psi_m^0 | V' | \psi_m^0 \right\rangle \quad (m = 1, 2, 3, ...). \tag{9.34}$$

This result indicates that every state of the unperturbed system will give rise to a perturbed state whose energy is approximately that of the unperturbed state plus the expectation value of the perturbing potential acting on that unperturbed state. This result applies only when the unperturbed states are non-degenerate.

To establish the expansion coefficients in this case it is easiest to return to (9.21):

$$\sum C_m E_m^0 \, \psi_m^0 + \sum C_m \, (V' \psi_m^0) = E \sum C_m \, \psi_m^0.$$

There will be a set of expansion coefficients corresponding to each possible perturbed energy of (9.34). Let us find those corresponding to the perturbed state which develops from some general unperturbed state, say the n'th one:

$$E \sim E_n^0 + \langle \psi_n^0 | V' | \psi_n^0 \rangle. \tag{9.35}$$

Substitute (9.35) into (9.21), breaking out the $m = n$ terms in the sums explicitly. The result is, after some rearranging,

$$- C_n \, \psi_n^0 \, \langle \psi_n^0 | V' | \psi_n^0 \rangle + \sum_{m \neq n} C_m \, \left\{ E_m^0 - E_n^0 - \langle \psi_n^0 | V' | \psi_n^0 \rangle \right\} \psi_m^0 \tag{9.36}$$

$$+ \sum_{m \neq n} C_m \, (V' \psi_m^0) + C_n \, (V' \psi_n^0) = 0.$$

Now multiply through this expression by one of the unperturbed wavefunctions, say ψ_k^0 ($k \neq n$), and integrate over the domain of the potential, invoking the orthogonality theorem where appropriate. The result is

$$C_k \left\{ E_k^0 - E_n^0 - \langle \psi_n^0 | V' | \psi_n^0 \rangle \right\} + \sum_{m \neq n} C_m \langle \psi_k^0 | V' | \psi_m^0 \rangle + C_n \langle \psi_k^0 | V' | \psi_n^0 \rangle = 0. \tag{9.37}$$

Study this expression carefully. If V' is small and the unperturbed wavefunctions are orthogonal, all of the terms in the summation are likely to be negligible with the possible exception of the case where $m = k$, that is, we can approximate this expression as

$$C_k \left\{ E_k^0 - E_n^0 - \langle \psi_n^0 | V' | \psi_n^0 \rangle + \langle \psi_k^0 | V' | \psi_k^0 \rangle \right\} + C_n \, \langle \psi_k^0 | V' | \psi_n^0 \rangle \sim 0,$$

or

$$C_k \sim \frac{-C_n \, \langle \psi_k^0 | V' | \psi_n^0 \rangle}{\left\{ E_k^0 - E_n^0 - \langle \psi_n^0 | V' | \psi_n^0 \rangle + \langle \psi_k^0 | V' | \psi_k^0 \rangle \right\}}. \tag{9.38}$$

In (9.38), $\langle \psi_n^0 | V' | \psi_n^0 \rangle$ and $\langle \psi_n^0 | V' | \psi_k^0 \rangle$ are likely to be small in comparison to E_k^0 and E_n^0. Further, since ψ_n^0 can probably be expected to make the dominant contribution to the perturbed wavefunction which arises from the unperturbed energy E_n^0, we are likely safe in setting $C_n \sim 1$. Hence we have

$$C_k \sim - \frac{\langle \psi_k^0 | V' | \psi_n^0 \rangle}{\left(E_k^0 - E_n^0 \right)} \quad (k \neq n), \tag{9.39}$$

which, in combination with (9.20), gives us an approximate expression for the perturbed wavefunction:

$$\psi_n \sim \psi_n^0 - \sum_{k \neq n} \frac{\psi_k^0 \langle \psi_k^0 | V' | \psi_n^0 \rangle}{\left(E_k^0 - E_n^0 \right)}. \tag{9.40}$$

The notation ψ_n serves as a reminder that this result has evolved from a perturbation to the unperturbed energy corresponding to ψ_n^0. In principle, the sum in (9.40) runs over an infinite number of states. However, only those values of E_k^0 that are close to E_n^0 will yield significant contributions, so there is some hope that it will converge quickly.

Equations (9.35) and (9.40) comprise time-independent, non-degenerate first-order perturbation theory; if one is dealing with a non-degenerate system and if the perturbations $\langle \psi_n^0 | V' | \psi_n^0 \rangle$ caused by V to the unperturbed energies E_n^0 are small in comparison to E_n^0, this first-order approach is usually adequate. Otherwise, the full matrix approach must be invoked. In dealing with the perturbed energy states in first-order perturbation theory, one can (in theory at least) handle an infinite number of states since only the m'th-state eigenfunction and eigenvalue appears in the correction term for the m-th energy state in (9.34). The perturbed eigenfunction for state m calculated from (9.40) still, however, requires in principal a summation over an infinite number of states.

Example 9.4 To a first order approximation, what are the energy levels and wavefunctions for the perturbed infinite rectangular well depicted in Fig. 9.5?
This is a one-dimensional, non-degenerate problem, so we can use (9.35) and (9.40). The unperturbed energies and wavefunctions are given by

$$E_n^0 = \frac{n^2 h^2}{8 m L^2},$$

and

$$\psi_n^0(x) = \sqrt{\frac{2}{L}} \sin \left(\frac{n \pi x}{L} \right).$$

The perturbing potential is $V' = \alpha x$. We first compute the perturbed energy levels. From (9.35),

$$\langle \psi_n^0 | V' | \psi_n^0 \rangle = \frac{2\alpha}{L} \int_0^L x \sin^2 \left(\frac{n \pi x}{L} \right) dx.$$

The integral is of the form

$$\int\limits_{0}^{L} x \sin^2(ax)\, dx = \left[\frac{x^2}{4} - \frac{x \sin(2ax)}{4a} - \frac{\cos(2ax)}{8a^2} \right]_0^L,$$

where $a = (n\pi/L)$. Substituting the limits gives

$$\langle \psi_n^0 |V'| \psi_n^0 \rangle = \frac{2\alpha}{L} \left\{ \left[\frac{L^2}{4} - \frac{L \sin(2n\pi)}{4a} - \frac{\cos(2n\pi)}{8a^2} \right] - \left[0 - 0 - \frac{\cos(0)}{8a^2} \right] \right\}$$

$$= \frac{2\alpha}{L} \left\{ \left[\frac{L^2}{4} - \frac{1}{8a^2} \right] + \frac{1}{8a^2} \right\} = \frac{\alpha L}{2}.$$

Therefore, to first order, the energy levels of the perturbed well are

$$E_n \sim E_n^0 + \alpha L/2.$$

Each level is perturbed upward by the average value of the perturbing potential across the well. Note that if $\alpha \to 0$, the result reduces to the energy levels of the unperturbed well; this is always a useful way of checking a perturbation calculation.

The perturbed wavefunction corresponding to any perturbed energy E_n is more involved. From (9.40) we have

$$\langle \psi_k^0 |V'| \psi_n^0 \rangle = \frac{2\alpha}{L} \int\limits_0^L \sin\left(\frac{k\pi x}{L} \right) x \sin\left(\frac{n\pi x}{L} \right) dx.$$

Using the trigonometric identity

$$\sin(ax) \sin(bx) = (1/2)[\cos[(a - b)x] - \cos[(a + b)x],$$

this integral can be simplified to

$$\langle \psi_k^0 |V'| \psi_n^0 \rangle = \frac{\alpha}{L} \left\{ \int\limits_0^L x \cos\left[\frac{(k - n)\pi x}{L} \right] dx - \int\limits_0^L x \cos\left[\frac{(k + n)\pi x}{L} \right] dx \right\}.$$

Designate the first integral as $I1$ and the second as $I2$. Both are of the form

$$\int x \cos(ax)\, dx = \frac{\cos(ax)}{a^2} + \frac{x \sin(ax)}{a}.$$

Consider $I1$. Defining $m = k - n$, we have

$$I1 = \left[\frac{\cos(m\pi x/L)}{(m\pi/L)^2} + \frac{x\sin(m\pi x/L)}{(m\pi/L)} \right]_0^L.$$

Clearly, m must be an integer; since $\sin(m\pi) = 0$ for all m, the second term vanishes at both limits. The remaining term yields

$$I1 = \frac{L^2}{m^2\pi^2}[\cos(m\pi) - 1].$$

Now, if m is even, $\cos(m\pi) = 1$, and $I1$ vanishes. If m is odd, $\cos(m\pi) = -1$, and the term in square brackets evaluates to -2. Therefore,

$$I1 = \begin{cases} 0, & (k-n) \text{ even} \\ -\frac{2L^2}{(k-n)^2\pi^2}, & (k-n) \text{ odd}. \end{cases}$$

Similarly,

$$I2 = \begin{cases} 0, & (k+n) \text{ even} \\ -\frac{2L^2}{(k+n)^2\pi^2}, & (k+n) \text{ odd}. \end{cases}$$

Now, if $(k-n)$ is even, $(k+n)$ will be as well; similarly, $(k-n)$ odd implies $(k+n)$ odd. Therefore,

$$\langle\psi_k^0|V'|\psi_n^0\rangle = \frac{\alpha}{L}(I1 + I2) = \frac{2\alpha L}{\pi^2}\left[\frac{1}{(k+n)^2} - \frac{1}{(k-n)^2}\right], \quad (k-n) \text{ odd}.$$

This result indicates that in the sum over unperturbed states in (9.40), only those terms when $(k-n)$ is odd will contribute. Bringing the above expression to a common denominator gives

$$\langle\psi_k^0|V'|\psi_n^0\rangle = -\frac{8\alpha L}{\pi^2}\left[\frac{nk}{(k+n)^2(k-n)^2}\right], \quad (k-n) \text{ odd}.$$

In the denominator of (9.40) we have

$$E_k^0 - E_n^0 = \frac{\pi^2\hbar^2}{2mL^2}(k^2 - n^2),$$

hence

$$\psi_n(x) \sim \psi_n^0(x) - \sum_{\substack{k \neq n \\ (k-n)\,odd}} \frac{\left[\sqrt{\frac{2}{L}}\sin\left(\frac{k\pi x}{L}\right)\right]\left[-\frac{8\alpha L}{\pi^2}\left(\frac{nk}{(k+n)^2(k-n)^2}\right)\right]}{\frac{\pi^2\hbar^2}{2mL^2}(k^2 - n^2)},$$

or, after some simplification,

$$\psi_n(x) \sim \psi_n^0(x) + \frac{16\sqrt{2}\alpha m L^{5/2}}{\hbar^2 \pi^4} \sum_{\substack{k \neq n \\ (k-n)\,odd}} \frac{nk \sin(k\pi x/L)}{(k^2 - n^2)^3}.$$

Note again that if $\alpha \to 0$, we recover the unperturbed rectangular–well wavefunctions. This result is plotted in Fig. 9.6 for the case of an electron in the $n = 1$ state of an infinite well with $L = 1$ Å and $\alpha = 100$ eV/Å; terms up to $k = 25$ were included, and $\psi/10^5$ is plotted for convenience of scale. The perturbation "pushes" the particle preferentially to the left side of the well, a result consistent with the precepts for sketching wavefunctions developed in Sect. 3.6.

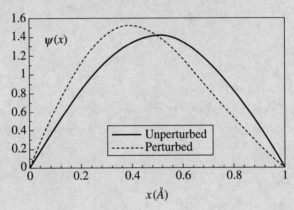

Fig. 9.6 Unperturbed and perturbed $n = 1$ state of an infinite rectangular well

Example 9.5 Determine first-order expressions for the ground-state energy and wavefunction of a particle of mass m moving in the perturbed harmonic-oscillator potential

$$V(x) = kx^2/2 + \beta x \quad (\beta > 0, -\infty \le x \le \infty).$$

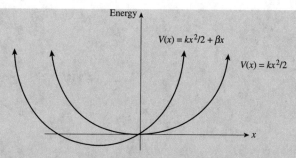

Fig. 9.7 Sketches of unperturbed and perturbed harmonic oscillator potential. Not to scale

Again we have a non-degenerate problem; the perturbing potential is $V' = \beta x$. As sketched in Fig. 9.7, the effect of this perturbation is to lower the potential well for $x < 0$ while raising it for $x > 0$. This has the effect of destroying the symmetry of the well; we can expect that the perturbed wavefunction will be asymmetric.

The ground-state of a harmonic oscillator is

$$\psi_0^0(x) = A_0 \exp(-\alpha^2 x^2/2),$$

where

$$A_0 = \sqrt{\alpha}/\pi^{1/4}$$

and

$$\alpha^4 = mk/\hbar^2.$$

The perturbed ground state energy is then

$$\langle \psi_0^0 | V' | \psi_0^0 \rangle = \beta A_0^2 \int\limits_{-\infty}^{\infty} x \exp(-\alpha^2 x^2) dx.$$

Since the integrand is an antisymmetric function, this integral evaluates to zero. To first order, a linear potential perturbing a harmonic oscillator causes *no change* to the ground state energy. The same result will clearly hold for any odd-powered perturbation (why?).

As to determining the perturbed ground-state wavefunction, we need to recall from Chap. 5 the general expression for the harmonic-oscillator wavefunction for any level k:

$$\psi_k^0(x) = A_k H_k(\alpha x) \exp(-\alpha^2 x^2/2), \quad k = 0, 1, 2, ...,$$

with

$$A_k = \sqrt{\alpha / \sqrt{\pi} 2^k k!}.$$

Therefore,

$$\langle \psi_k^0 | V' | \psi_0^0 \rangle = \beta A_k A_0 \int_{-\infty}^{\infty} x \exp(-\alpha^2 x^2) H_k(\alpha x) H_0(\alpha x) dx,$$

or, on defining $\xi = \alpha x$,

$$\langle \psi_k^0 | V' | \psi_0^0 \rangle = \frac{\beta A_k A_0}{\alpha^2} \int_{-\infty}^{\infty} \xi \exp(-\xi^2) H_k(\xi) H_0(\xi) d\xi.$$

Now, it is a property of Hermite polynomials that

$$\int_{-\infty}^{\infty} \xi \exp(-\xi^2) H_n(\xi) H_m(\xi) d\xi = \sqrt{\pi} \left[2^{n-1} n! \delta_m^{n-1} + 2^n (n+1)! \delta_m^{n+1} \right].$$

This property of the Hermite polynomials has an interesting consequence for the perturbed ground-state wavefunction. We have, with $n = k$ and $m = 0$ to correspond to what we need,

$$\psi_0 \sim \psi_0^0 - \frac{\beta \sqrt{\pi} A_0}{\alpha^2} \sum_{k \neq 0} \frac{A_k \psi_k^0 \left[2^{k-1} k! \delta_0^{k-1} + 2^k (k+1)! \delta_0^{k+1} \right]}{(E_k^0 - E_0^0)}.$$

The sum in this expression runs over all values of k, excluding $k = 0$. The first Kronecker delta, δ_0^{k-1}, is zero except when $k - 1 = 0$, that is, except when $k = 1$. The second will be zero except when $k + 1 = 0$, that is, except when $k = -1$. But the series runs from $k = 1$ upward, so the second Kronecker delta makes no contribution at all to the sum. The only surviving term is that for $k = 1$, which gives

$$\psi_0 \sim \psi_0^0 - \frac{\beta \sqrt{\pi} A_0 A_1}{\alpha^2} \frac{\psi_1^0}{(E_1^0 - E_0^0)}.$$

On substituting explicit expressions for A_1, ψ_1^0, E_1^0, and E_0^0, we get

$$\psi_0 \sim A_0 e^{-\alpha^2 x^2 / 2} \left(1 - \frac{\beta}{\hbar \omega} x \right).$$

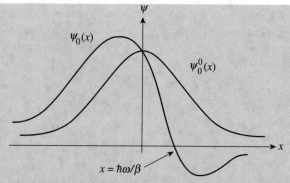

Fig. 9.8 Sketches of unperturbed (ψ_0^0) and approximate perturbed (ψ_0) ground-state harmonic oscillator potential wavefunctions. Not to scale

If β vanishes, we recover the unperturbed ground-state solution. To a first-order approximation, the perturbation causes *no* change to the ground-state energy but it does destroy the symmetry of the wavefunction. The forms of the perturbed and unperturbed ground-state wavefunctions are sketched in Fig. 9.8. First-order perturbation theory predicts that at $x = 0$, the perturbed and unperturbed wavefunctions will have the same value.

A closer look at the results of this example reveals a disturbing feature, however: That the perturbed wavefunction possesses *two* extrema. How can this be for a ground state? The resolution of this question serves as a cautionary tale for users of perturbation theory.

It is possible to solve this example exactly, and so glean some insight as to the limits of perturbation theory. First, on completing the square, the perturbed potential can be written as

$$V(x) = \left(\sqrt{\frac{k}{2}} x + \frac{\beta}{\sqrt{2k}} \right)^2 - \frac{\beta^2}{2k}.$$

On defining a new coordinate $\xi = \gamma(Ax + B)$ where $\gamma^4 = 2m/\hbar^2 A^2$, $A = \sqrt{k/2}$, and $B = \beta/\sqrt{2k}$, Schrödinger's equation for this potential can be written in the form

$$\frac{d^2\psi}{dx^2} + (\lambda - \xi^2)\psi = 0,$$

where $\lambda = \gamma^2(E + \beta^2/2k)$. This is identical to the form of Schrödinger's equation for the unperturbed harmonic oscillator (5.5). The ground state energy corresponds to $\lambda = 1$ [(5.34) with $n = 0$],

$$E_{ground}^{exact} = \frac{\hbar\omega}{2} - \frac{\beta^2}{2k},$$

where $\omega = \sqrt{k/m}$ as usual. Contrary to the conclusion of first-order perturbation theory, the exact solution shows that the ground state energy is *less* than that for the unperturbed case. The essential behavior of the exact ground-state wavefunction is then

$$\psi_{ground}^{exact} \sim e^{-\xi^2/2} \sim e^{-[\gamma(Ax+B)^2]}$$

$$\sim [e^{-\alpha^2 x^2/2}][e^{-(\beta/\hbar\omega)x}][e^{-\beta^2/2\hbar\omega k}],$$

where α is as we defined in Chap. 5, $\alpha = (mk/\hbar^2)^{1/4}$. As expected for a ground state, this solution has no nodes other than at $x = 0$ and $x = \infty$. The first factor on the right side is the normal harmonic oscillator ground state, whereas the third factor is a constant. The middle factor is a new x-dependent term; if $\beta \to 0$, we recover the unperturbed case. The interesting thing about this new term is that if we expand it to first order we get

$$e^{-(\beta/\hbar\omega)x} \approx 1 - \frac{\beta}{\hbar\omega}x + \cdots,$$

that is, we arrive at a correction expression identical to that given by first-order perturbation theory! The result yielded by first-order perturbation theory is thus but the first term of an exponential function. The exact solution is asymmetric, and has only one extremum.

The point of this example should be clear: perturbation theory yields only an *approximate* solution, and one must be careful when interpreting the results.

Example 9.6 A collection of hydrogen atoms is subject to a uniform external electric field in the z-direction of strength λ Volt/meter: $\boldsymbol{E} = \lambda\hat{z}$. Determine the perturbations to the electronic energy levels.

The effect of an external electric field on the energy levels of an atom is known as the *Stark effect* after German physicist Johannes Stark (1874-1957), who won the 1919 Nobel prize for Physics for its discovery.

Here we must use the full machinery of degenerate perturbation theory. The electrical potential corresponding to this field is $V = -\lambda z = -\lambda r \cos\theta$ [recall $\boldsymbol{E} = -\nabla V$]. An electron of charge $-e$ will then experience a perturbing potential energy $V' = qe$:

$$V' = (-e)(-\lambda r \cos\theta) = +e\lambda r \cos\theta.$$

To keep the calculations involved with the matrix of (9.25) tractable, we limit the number of states considered to those of just the first two hydrogenic energy levels, that is, to those of fundamental quantum numbers $n = 1$ and $n =$

2. In writing the matrix elements (9.24) we label these states from 1 to 5, with states (1,2,3,4,5) corresponding to hydrogenic states $(n, \ell, m) = (1,0,0), (2,0,0)$, $(2,1,0)$, $(2,1,1)$, and $(2,1,-1)$, respectively. Also, we shall concern ourselves only with the perturbed energies as opposed to the expansion coefficients. Table 7.2 gives the relevant unperturbed wavefunctions. Since some of these functions are explicitly complex, it must be remembered to use the complex conjugates of the ψ_j^0 in computing the matrix elements.

Tedious if straightforward calculations (which the reader should verify) show that most of the elements of the matrix in (9.25) vanish. The matrix emerges as

$$
\begin{bmatrix}
(E_1 - E) & 0 & 256\lambda a_o e/243\sqrt{2} & 0 & 0 \\
0 & (E_1/4 - E) & 3\lambda a_o e & 0 & 0 \\
256\,\lambda a_o e/243\sqrt{2} & 3\lambda a_o e & (E_1/4 - E) & 0 & 0 \\
0 & 0 & 0 & (E_1/4 - E) & 0 \\
0 & 0 & 0 & 0 & (E_1/4 - E)
\end{bmatrix},
$$

where a_o is the Bohr radius and E_1 is the hydrogen ground-state energy. After further algebra, the determinant of this matrix emerges as

$$
B^2 \left(AB^2 - AD^2 - C^2 B \right) = 0,
$$

where $A = E_1 - E$, $B = E_1/4 - E$, $C = 256\lambda a_o e/243\sqrt{2}$, and $D = 3\lambda a_o e$. Writing this out explicitly gives

$$
(E_1/4 - E)^2 \left\{ E^3 - (3E_1/2)\, E^2 + (9\, E_1^2/16 - D^2 - C^2)\, E \right.
$$
$$
\left. + (E_1 D^2 + E_1 C^2/4 - E_1^3/16) \right\} = 0.
$$

From the prefactor $(E_1/4 - E)^2$, we can infer that there is a double root of $E = E_1/4$. This means that, to this level of approximation, two of the four $n = 2$ states remain unperturbed. The energies of the other three solutions are obtained by solving the cubic equation in E that resides within the curly brackets. Because this is algebraically messy, we content ourselves with a numerical example.

To aid in examining this cubic equation, it is helpful to cast it into a dimensionless form. This can be accomplished by expressing the electric field strength λ as a multiple (say, λ^*) of some field strength that could be regarded as natural for the problem at hand. To this end, we use the strength of the electric field of the proton at a distance of one Bohr radius:

$$
\lambda = \lambda^* \left(\frac{e}{4\pi\varepsilon_o a_o^2} \right).
$$

Hence

$$\lambda e a_o = \lambda^* \left(\frac{e^2}{4\pi \varepsilon_o a_o} \right) = \lambda^* \left(\frac{e^2}{4\pi \varepsilon_o} \right) \left(\frac{\pi m_e e^2}{\varepsilon_o h^2} \right) = -2\lambda^* E_1.$$

With this we can put

$$C = -\gamma E_1, \gamma = \lambda^* (256/243\sqrt{2})$$

and

$$D = -\delta E_1, \delta = 6\lambda^*.$$

With these results, the cubic in E becomes

$$(E/E_1)^3 - (3/2)(E/E_1)^2 + (9/16 - \delta^2 - \gamma^2)(E/E_1) + (\delta^2 + \gamma^2/4 - 1/16) = 0.$$

Parenthetically, the strength of the proton's electric field at a_o is about 5×10^{11} N/C; a pulsed laser system can briefly create fields of about this strength.

Taking $\lambda^* = 0.01$ as an example, three solutions emerge for (E/E_1): $\sim$ 0.1899, 0.3099, and 1.0002. The first two of these represent not-quite symmetrically perturbed $n = 2$ hydrogenic states normally at $(E/E_1) = 0.25$, and the last a very slightly perturbed $n = 1$ state. This example illustrates a situation where the original degeneracy is said to be *partially lifted* since two degenerate states at energy $E_1/4$ remain.

A word of warning is appropriate here. If λ^* is made too large (0.25, for example) you will find that one of the roots for E/E_1 emerges as a negative number, which would have to be interpreted as an unbound state; recall that all of the hydrogenic states have $E_n < 0$. This is because the perturbing potential energy that an $n = 2$ electron would experience in such a situation, on the order of $-2\lambda^* E_1$, is greater in magnitude than its initial energy, $E_1/4$. The level of approximation used here would thus not be adequate in such circumstances; more unperturbed states would have to be allowed to contribute to the solution. Perturbation theory is intended for use when the perturbing effect is small compared to the initial condition of what is being perturbed.

9.4 The Variational Method

The variational method is a technique for estimating ground-state energies of potentials where one can make an educated guess at the form of the corresponding wavefunction. The only requirement is a normalized trial wavefunction that satisfies the boundary conditions of the problem; it need not satisfy Schrödinger's equation. We will first show how an estimate on the ground-state energy can be obtained, then address some strategies for establishing trial wavefunctions.

If $\phi(x)$ is the trial ground-state wavefunction, then the predicted ground-state energy E is

$$E = \langle \phi^* | H | \phi \rangle, \tag{9.41}$$

where H is the Hamiltonian operator

$$H = -\frac{\hbar^2}{2m} \frac{d^2}{dx^2} + V(x). \tag{9.42}$$

Now, Schrödinger's equation for the potential under consideration will in principle possess some family of exact solutions; call these $\psi_n(x)$ and label their corresponding energies E_n. If we knew the $\psi_n(x)$ we could expand $\phi(x)$ in terms of them along the lines of the development of the superposition theorem of Sect. 9.2:

$$\phi(x) = \sum_n a_n \psi_n(x), \tag{9.43}$$

where

$$a_j = \langle \psi_j | \phi \rangle. \tag{9.44}$$

In reality, the idea is that we do not know the $\psi_n(x)$: if we did, there would be no sense in trying to establish an approximate solution! The important point for the moment is that if we did know the $\psi_n(x)$'s, we could in principle expand out (9.43) explicitly.

If $\phi(x)$ is to be plausible as a trial wavefunction, it must satisfy two physical criteria: (i) It must satisfy the boundary conditions of the problem at hand, and (ii) It must be normalized. In practice, condition (i) is ensured by a judicious choice of $\phi(x)$ as discussed below. Condition (ii) was explored in Sect. 9.2, where we found that the constraint

$$\sum |a_n|^2 = 1 \tag{9.45}$$

will ensure normalization. While this result plays a central role in the development that follows, we will see that the a_n need not be computed in practice.

Substitute (9.43) into (9.41). This gives

$$\langle \phi^* | H | \phi \rangle = \int \left(\sum_n a_n^* \psi_n^* \right) H \left(\sum_m a_m \psi_m \right) dx$$

$$= \sum_n \sum_m a_n^* a_m \int \psi_n^* (H \, \psi_m) dx$$

$$= \sum_n \sum_m a_n^* a_m \int \psi_n^* E_m \, \psi_m dx \qquad (9.46)$$

$$= \sum_n \sum_m a_n^* a_m E_m \int \psi_n^* \psi_m dx$$

$$= \sum_n \sum_m a_n^* a_m E_m \, \delta_n^m.$$

Because of the δ_n^m, the inner sum over m vanishes for all terms except when $m = n$, leaving

$$E = \sum_n a_n^* a_n E_n = \sum_n |a_n|^2 E_n. \qquad (9.47)$$

Consider the right side of (9.47). Designate the true ground state energy as E_0; since all of the other states must have $E_n > E_0$, then

$$\sum_n |a_n|^2 E_n \geq \sum_n |a_n|^2 E_0,$$

or, since E_0 is a constant,

$$\sum_n |a_n|^2 E_n \geq E_0 \sum_n |a_n|^2.$$

The sum on the right side is equal to 1 from (9.45), so we have

$$\sum_n |a_n|^2 E_n \geq E_0. \qquad (9.48)$$

Comparing (9.47) and (9.48) shows that

$$E \geq E_0. \qquad (9.49)$$

The essential result here is that if $\phi(x)$ is a normalized trial wavefunction for the ground state of a system, $\langle \phi^* | H | \phi \rangle$ sets an upper limit to the ground-state energy. In practice, one starts with a trial wavefunction which obeys the relevant boundary conditions and which contains some parameter(s), and then varies the parameter(s) to minimize $\langle \phi^* | H | \phi \rangle$. In the following section we will examine how this scheme can be modified to give a lower limit on E_0 as well.

How does one make a "judicious" choice for the trial wavefunction? A potential that supports bound states can ultimately be conceived of as some sort of potential well, in which case we know that $\psi \to 0$ as $x \to \infty$. From prior experience (finite rectangular well, harmonic oscillator, hydrogen atom) we also know that ground states are often characterized by exponential-decay functions. If the domain of the potential runs over $-\infty \le x \le +\infty$, a function of the form $\phi(x) = Ae^{-\beta x^2}$ will both satisfy the boundary conditions and be normalizable over the domain. Note that ϕ contains two constants, A and β. A is a normalization constant; on computing $\int \phi^2(x)dx$, it will emerge in terms of β. When the energy estimate E is computed from (9.41) and (9.42), it too will emerge in terms of β; the resulting expression can then be differentiated with respect to β and minimized in order to yield the lowest possible estimate of E. β is the *variational* parameter. [Parenthetically, if the domain is $-\infty \le x \le +\infty$, any even-powered exponential would do – why? Similarly, if the domain is restricted to $0 \le x \le +\infty$, then an odd-powered exponential such as $Ae^{-\beta x}$ or $Ae^{+\beta x}$ could be used - why?] If your potential contains an infinite wall at, say, $x = 0$ but otherwise runs over $0 \le x \le +\infty$, such an exponential function will be acceptable as $x \to \infty$, but will be unsatisfactory at $x = 0$ as it does not respect the boundary condition there. In such a case we could modify $\phi(x)$ to, say, the form $\phi(x) = Axe^{-\beta x^2}$. If you are ambitious, you could try a function of the form $\phi(x) = Af(x)e^{-\beta x^2}$, where $f(x)$ is a function containing yet more variational parameters; you would then face a multi-parameter minimization problem. An approach for a two-parameter function might be $\phi(x) = Af(x)e^{-\beta x^n}$, where the power "$n$" in the exponential is considered to be a variational parameter along with β. This sort of extreme measure is usually not required for the level of problems intended here, however.

Although it has been emphasized that this method yields an upper limit on the ground-state energy, there is no explicit requirement in the derivation that $\phi(x)$ actually look anything like a ground-state wavefunction; indeed, *all* states, ground and excited alike, have to satisfy the boundary conditions and be normalizable. The variational theorem tells you only that the energy corresponding to $\phi(x)$ must be greater than or equal to that of the ground-state. If you choose $\phi(x) = Af(x)e^{-\beta x^2}$ with $f(x)$ as a polynomial, for example, you might actually be choosing a function that does a better job of mimicking an excited state rather than the ground-state, in which case the resulting value of E will likely be a very poor estimate of the ground state energy. Simplicity is a virtue when choosing trial wavefunctions.

Example 9.7 Use the variational method to set an upper limit on the ground-state energy of the linear potential

$$V(x) = \begin{cases} \infty & (x < 0) \\ \alpha x & (x \ge 0). \end{cases}$$

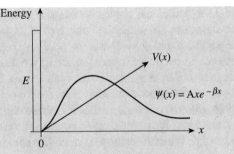

Fig. 9.9 Sketch of hypothetical linear potential ground state wavefunction

This problem was addressed with the uncertainty principle in Problem 4.10 and with the WKB method in Example 9.1; it will be interesting to compare the merits of these approximation schemes.

First we need a plausible trial wavefunction $\phi(x)$. From Chap. 3, we know that the ground state wavefunction will have one maximum. We also know that it must go to zero at $x = 0$ and decay exponentially at large x. These features are sketched in Fig. 9.9.

As suggested by the foregoing, a simple function that satisfies these requirements is

$$\phi(x) = Axe^{-\beta x},$$

where A is the normalization constant and β the variational parameter. Normalization demands

$$A^2 \int\limits_0^\infty x^2 e^{-2\beta x}\, \mathrm{d}x = 1,$$

that is, $2A^2/(2\beta)^3 = 1$, or $A^2 = 4\beta^3$.

It is always necessary to compute $d^2\phi/d\phi^2$ in order to compute $H\phi$:

$$\frac{\mathrm{d}^2\phi}{\mathrm{d}x^2} = A(\beta^2 x - 2\beta)e^{-\beta x},$$

hence

$$H\phi = -\frac{\hbar^2}{2m}\frac{\mathrm{d}^2\phi}{\mathrm{d}x^2} + V(x)\,\phi = Ae^{-\beta x}(2\beta\varepsilon - \beta^2\varepsilon x + \alpha x^2),$$

where

$$\varepsilon = \hbar^2/2m.$$

The expectation value of the Hamiltonian is then

$$\langle \phi \,|\, \boldsymbol{H} \,|\, \phi \rangle = A^2 \left\{ 2\beta\varepsilon \int\limits_0^\infty x e^{-2\beta x} \mathrm{d}x - \beta^2 \varepsilon \int\limits_0^\infty x^2 e^{-2\beta x} \mathrm{d}x + \alpha \int\limits_0^\infty x^3 e^{-2\beta x} \mathrm{d}x \right\}.$$

These integrals evaluate to $1/(4\beta)^2$, $1/(4\beta)^3$, and $3/(8\beta)^4$, respectively, which in combination with the expression for A give

$$E = \varepsilon\beta^2 + (3\alpha/2\beta).$$

To set the strictest possible limit on E_0 we need to minimize E with respect to β. Computing $\mathrm{d}E/\mathrm{d}\beta$, setting the result to zero, and solving for β gives

$$\beta_{\min} = (3\alpha/4\varepsilon)^{1/3}.$$

Back-substituting this result into the expression for E gives

$$E_{\min} = \varepsilon(3\alpha/4\varepsilon)^{1/3} + (3\alpha/2)(3\alpha/4\varepsilon)^{-1/3} = 2.476\,\alpha^{2/3}\varepsilon^{1/3}.$$

Therefore, the ground state energy must satisfy $E_0 \le 2.476\alpha^{2/3}\varepsilon^{1/3}$. This is consistent with the results yielded by the uncertainty principle and the WKB method: $E_0/(\alpha^{2/3}\varepsilon^{1/3}) > 1.191$ and ~ 2.811, respectively. The exact solution is $2.338(\alpha^{2/3}\varepsilon^{1/3})$. Our variational upper limit exceeds the correct result by only about 6%.

The power of the variational method is that any trial wavefunction will do, so long as it respects the boundary conditions of the problem. It need not and in general will not satisfy Schrödinger's equation for the potential at hand. The closer one's guess for the trial wavefunction is to the actual ground-state wavefunction, the more accurate will be the result.

Example 9.8 This example is somewhat abstract; you might want to come back to it after first working a few of the end-of-chapter problems on the variational theorem.

In this example we apply the variational theorem in spherical coordinates to determine an upper limit for the ground-state energy for a particle of mass m moving in the attractive central potential (ACP)

$$V(r) = -KE_1 \left(\frac{a_o}{r}\right)^n \quad (0 \le r \le \infty),$$

where K is a dimensionless strength parameter, a_o is the Bohr radius, and E_1 is the absolute value of the ground-state energy of the hydrogen atom. Both K and n are assumed to be positive; bound states will have $E < 0$.

As a trial wavefunction, we use a function of the form of the hydrogenic ground-state:

$$\phi(r) = A e^{-\beta r/2} \quad (0 \le r \le \infty),$$

where β is the variational parameter. Normalizing this demands

$$4\pi A^2 \int_0^\infty r^2 e^{-\beta r}\, dr = 1,$$

where the factor of 4π arises from integrating over θ and ϕ. The integral reduces to $2/\beta^3$, so our trial wavefunction is

$$\phi(r) = \sqrt{\frac{\beta^3}{8\pi}}\, e^{-\beta r/2} \, (0 \le r \le \infty).$$

The strategy in using this particular trial wavefunction is that when $n = 1$ and $K = 2$, we should recover the exact ground-state energy for hydrogen, thus providing a check on the calculations.

Since both the potential and the trial wavefunction have only radial dependencies, we can write the Hamiltonian as

$$H = -\frac{\hbar^2}{2m}\nabla^2 + V(r) \equiv -\frac{\hbar^2}{2m}\frac{1}{r^2}\frac{d}{dr}\left(r^2\frac{d}{dr}\right) + V(r).$$

The expectation value for the energy is

$$E = 4\pi \int_0^\infty \phi(H\phi)r^2 dr,$$

where the factor of 4π arises as above.

The Laplacian of the trial wavefunction is

$$\nabla^2\phi = -\frac{\beta^{5/2}}{2\sqrt{8\pi}}\left(\frac{2}{r} - \frac{\beta}{2}\right)e^{-\beta r/2},$$

leading to

$$E = \frac{\hbar^2 \beta^4}{8m} \int\limits_0^\infty (2r - \beta r^2/2)\, e^{-\beta r}\, dr - \frac{K E_1 a_o^n \beta^3}{2} \int\limits_0^\infty r^{2-n} e^{-\beta r} dr.$$

Evaluating the integrals gives, after some algebra,

$$E = \frac{\hbar^2 \beta^2}{8m} - \frac{K E_1 a_o^n \beta^n \Gamma(3 - n)}{2}, \quad (n < 3)$$

where $\Gamma(x)$ denotes a gamma function (see Appendix C). The solution of the second integral in the expression for E leads to a constraint on the value of n. Minimizing E with respect to β gives

$$\beta = \left\{ \frac{\hbar^2}{2 m n K E_1 a_o^n \Gamma(3 - n)} \right\}^{\frac{1}{n-2}}, \quad (n < 3)$$

The reader should verify that for $n = 1$, $K = 2$, and $m = m_e$, this expression gives $\beta = 2/a_o$ and that the trial wavefunction reduces to exactly the hydrogenic ground-state wavefunction. Back-substituting this result into the expression for E yields

$$\frac{E}{(E_1/\mu)} = \frac{1}{4[(K\mu)n\Gamma(3 - n)]^{2/(n-2)}} \left\{ 1 - \frac{2}{n} \right\},$$

where μ is the ratio of the mass of the particle to that of the electron: $\mu = (m/m_e)$.

A key point here is that the variational method is predicated on minimizing E; on computing the second derivative of E with respect to β, one finds that the above expression for E yields a minimum only for $n < 2$.

Clearly, the ground-state energies of ACP's depend on n and the combination $(K\mu)$ of the dimensionless parameters K and μ. Figure 9.10 illustrates ground-state upper limits for ACP's of various $(K\mu)$ as a function of n; the quantity plotted on the vertical axis is the dimensionless energy $W = E/(E_1/\mu)$. A number of conclusions follow directly: (i) Negative-energy bound state are possible for ACP's with $0 < n < 2$. (ii) The prediction of the expression for $E/(E_1/\mu)$ for the Coulomb potential, $(K, n, \mu) = (2, 1, 1)$, is $W = -1$, the exact value. This is to be expected on the basis that the trial wavefunction is the exact solution of Schrödinger's equation in this case. (iii) The most striking feature of Fig. 9.10 is that for $(K\mu) > 0.5$ (see below), there exists a least-bound ground state of an ACP; as $(K\mu)$ increases, both the value of n at which the maximum energy occurs and the energy of the ground state both decrease, that is, the ground state becomes more tightly bound. This is reflecetd in the maxima in the curves in the figure. Interestingly, the ground state of the Coulomb potential is about as unbound as it could possibly be. Numerical solutions of the radial Schrödinger equation (Chap. 10) verify the presence of the maxima seen in the figure. The expression for $E/(E_1/\mu)$ actually proves to yield fairly stringent upper limits: for $(K\mu) = 1$, for example, it predicts $-W = (0.435, 0.250, 0.260)$ for $n = (0.5, 1.0, 1.5)$, whereas the true solutions are $-W \sim (0.438, 0.250, 0.292)$; the variational calculation overestimates the true energies by about $(0.7, 0, 11)$ percent.

How does the value $K = 0.5$ come to be critical for determining whether or not an ACP will exhibit a least-bound ground state? To answer this, it is instructive to study the behavior of E as $n \to 2$ from below. In this limit, $\Gamma(3 - n)$ approaches unity from above, in which case we have

$$\left[\frac{E}{(E_1/\mu)}\right]_{n\to2} \Rightarrow \frac{1}{4}(2K\mu)^\infty(1 - 2/n).$$

Two fates are possible for E when $0 < n < 2$: if $(K\mu) < 0.5$, then $E \to 0$ from below, whereas if $(K\mu) > 0.5$, $E \to -\infty$. An exact solution of Schrödinger's equation for this potential reveals that the critical value of $(K\mu)$ that discriminates between these two behaviors is $(K\mu) = 1/4$. That the present result for the critical value of $(K\mu)$ differs from this is not surprising in view of the fact that we have utilized a trial wavefunction that is approximate except when $n = 1$.

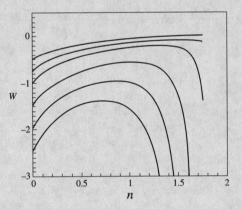

Fig. 9.10 Upper limits for ground-state energies of attractive central potentials. From top to bottom, the curves correspond to $K\mu = 0.5, 0.75, 1, 1.5, 2$, and 2.5

9.5 Improving the Variational Method

The variational method is somewhat limited in that it yields no idea how far the true ground state energy E_0 may be from the upper limit that it provides. In this section we take up a modification of the variational technique that gives both upper and lower limits on E_0.

Again, let $\phi(x)$ be a normalized trial ground-state wavefunction, and let E be calculated as before by

$$E = \langle \phi^* | H | \phi \rangle. \tag{9.50}$$

Now define a new quantity D given by

$$D = \int (H \phi)^* (H \phi) dx. \tag{9.51}$$

D will have units of E^2. Expanding $\phi(x)$ in terms of the exact solutions $\psi_n(x)$ to the problem, we can write

$$D = \int \left(H \sum_n a_n \psi_n \right)^* \left(H \sum_m a_m \psi_m \right) dx.$$

Recalling that $H \psi_j = E_j \psi_j$, we have

$$D = \int \left(H \sum_n a_n \psi_n \right)^* \left(H \sum_m a_m \psi_m \right) dx$$

$$= \sum_n \sum_m \left\{ a_n^* a_m E_n E_m \int \psi_n^* \psi_m \, dx \right\}$$

$$= \sum_n \sum_m \left\{ a_n^* a_m E_n E_m \delta_m^n \right\} = \sum_n a_n^* a_n E_n^2,$$

or

$$D = \sum_n |a_n|^2 E_n^2. \tag{9.52}$$

Now consider the quantity Δ defined by

$$\Delta = \sum_n |a_n|^2 (E_n - E)^2, \tag{9.53}$$

where E is as defined in (9.50). Expanding the binomial in this definition gives

$$\Delta = \sum_n |a_n|^2 (E_n^2 - 2E_n E + E^2)$$

$$= \sum_n |a_n|^2 E_n^2 - 2E \left(\sum_n |a_n|^2 E_n \right) + E^2 \sum_n |a_n|^2.$$

The first term in this expression is D itself (9.52), the bracketed term is E from (9.47), and the sum in the last term is equal to 1 via (9.45). Hence

$$\Delta = D - 2E(E) + E^2 = D - E^2. \tag{9.54}$$

Again, the ground-state energy E_0 must be less than (or at most equal to) all of the higher energy levels: $E_0 \leq E_n$. Therefore,

$$(E_0 - E)^2 \leq (E_n - E)^2,$$

or

$$\sum_n |a_n|^2 (E_0 - E)^2 \leq \sum_n |a_n|^2 (E_n - E)^2.$$

The right side of this last expression is just Δ; in the left side, $(E_0 - E)^2$ is a constant and can be brought in front of the sum to give

$$(E_0 - E)^2 \sum_n |a_n|^2 \leq \Delta.$$

But the sum of the $|a_n|^2$ is just 1, so

$$\Delta \geq (E_0 - E)^2, \tag{9.55}$$

or

$$\sqrt{\Delta} \geq \pm(E_0 - E). \tag{9.56}$$

Now, Eq. (9.49) tells us that $E \geq E_0$, or $(E_0 - E) \leq 0$. Considering the lower-sign case in (9.56), we have

$$\sqrt{\Delta} \geq (E_0 - E),$$

or

$$E_0 \geq E - \sqrt{\Delta}.$$

Recalling from (9.54) that $\Delta = D - E^2$, we can combine this result with (9.49) to give a revised statement of the variational method that provides both upper and lower limits on E_0:

$$E \geq E_0 \geq E - \sqrt{D - E^2}. \tag{9.57}$$

In practice, one again sets up a trial function with a variational parameter and varies the parameters until $\Delta - E^2$ is minimized; this ensures the most stringent possible limits on E_0 consistent with how well the trial wavefunction matches the true ground-state wavefunction. Selection of the upper sign in (9.56) leads to a weaker upper limit on E_0 than is already available from (9.49).

Example 9.9 Rework Example 9.7 with the improved variational method.
From Example 9.7 we had

$$H\phi = -\varepsilon\frac{d^2\phi}{dx^2} + V(x)\phi = Ae^{-\beta x}(2\beta\varepsilon - \beta^2\varepsilon x + \alpha x^2),$$

where

$$A^2 = 4\beta^3$$

and

$$E = \varepsilon\beta^2 + (3\alpha/2\beta).$$

From (9.51) and the result for $(H\phi)$, D is given by

$$D = \int (H\phi)^*(H\phi)dx = 4A^2\beta^2\varepsilon^2\int_0^\infty e^{-2\beta x}dx - 4A^2\beta^3\varepsilon^2\int_0^\infty xe^{-2\beta x}dx$$

$$+ (4A^2\beta\varepsilon\alpha + A^2\beta^4\varepsilon^2)\int_0^\infty x^2e^{-2\beta x}dx - 2A^2\beta^2\varepsilon\alpha\int_0^\infty x^3e^{-2\beta x}dx$$

$$+ A^2\alpha^2\int_0^\infty x^4e^{-2\beta x}dx.$$

These integrals are all standard; after some algebra one finds

$$D = 5\beta^4\varepsilon^2 + \beta\varepsilon\alpha + 3\alpha^2/\beta^2.$$

Combining this result with our previous result for E gives

$$(D - E^2) = 4\beta^4\varepsilon^2 - 2\beta\varepsilon\alpha + 3\alpha^2/4\beta^2.$$

Minimizing $(D - E^2)$ with respect to β results in a quadratic equation in β^3:

$$16\varepsilon^2(\beta^3)^2 - 2\varepsilon\alpha(\beta^3) - (3\alpha^2/2) = 0.$$

Solving this gives

$$\beta_{min}^3 = \frac{3\alpha}{8\varepsilon} \text{ or } -\frac{\alpha}{4\varepsilon}.$$

The second root implies an imaginary value for β. This would lead to an imaginary estimate for E, which we reject as unphysical. The remaining root, $\beta_{min}^3 = 3\alpha/8\varepsilon$, is different from the result obtained in Example 9.7 ($\beta_{min}^3 = 3\alpha/4\varepsilon$), where E alone was minimized. Back-substituting this root into the expression for $(D - E^2)$ results in

$$(D - E^2)_{min} = [4(3/8)^{4/3}](\alpha^{4/3}\varepsilon^{2/3}),$$

or

$$\sqrt{(D - E^2)_{min}} = 1.040\,(\alpha^{2/3}\varepsilon^{1/3}).$$

In the expression for E alone, this root gives

$$E_{min} = [5(3/8)^{2/3}](\alpha^{2/3}\varepsilon^{2/3}) = 2.600(\alpha^{2/3}\varepsilon^{2/3}),$$

and hence

$$E - \sqrt{(D - E^2)_{min}} = 1.560(\alpha^{2/3}\varepsilon^{1/3}).$$

Consistent with the remarks following (9.57) above, this result for E_{min} is slightly greater than that which was obtained by minimizing E alone in Example 9.7; combining the results from both methods gives

$$2.476 \geq (E_0/\alpha^{2/3}\varepsilon^{1/3}) \geq 1.560.$$

These limits bracket the exact value of 2.338.

An improved version of the trial wavefunction for this problem is

$$\phi(x) = Ax \exp\left(-\beta x^n\right),$$

where n is taken to be a second variational parameter which is held constant through the variational calculation. The result is an expression for $\beta(n)$ which sets maximum and minimum limits on the ground state energy according as (9.57); n can then be varied until the most restrictive limits are found, which can be shown to be $2.298 < (E_0/\alpha^{2/3}\varepsilon^{1/3}) < 2.378$ when $n = 1.773$. These bracket the exact result to better than 2% [6].

Summary

This chapter explores three methods of obtaining approximate energies and/or wavefunctions corresponding to potentials for which Schrödinger's equation cannot be solved exactly.

The WKB approximation says that the energy E corresponding to principal quantum number n of a particle of mass m moving in a potential $V(x)$ satisfies

$$2\sqrt{2m} \int \sqrt{E - V(x)}\,dx \sim nh.$$

This approximation is valid where

$$\frac{mh}{\{2m[E - V(x)]\}^{3/2}} \left(\frac{dV}{dx}\right) \ll 1.$$

The development of perturbation theory hinges on the superposition theorem: that any function $\phi(x)$ can be represented as a linear sum over all of the $\psi_n(x)$ that satisfy Schrödinger's equation for some potential:

$$\phi(x) = \sum_n a_n \psi_n(x),$$

where the expansion coefficients are in given by

$$a_n = \langle \psi_j(x) | \phi(x) \rangle.$$

In notation where superscript 0 designates the energies and wavefunctions corresponding to a potential for which one can solve Schrödinger's equation exactly, the energies and wavefunctions corresponding to a perturbation are given by those values of E that render the determinant

$$\begin{vmatrix} V_{11} & \cdots & V_{1k} & \cdots & V_{1r} & \cdots & V_{1N} \\ \vdots & \ddots & \vdots & \vdots & \vdots & \vdots & \vdots \\ V_{k1} & \cdots & V_{kk} & \cdots & V_{kr} & \cdots & V_{kN} \\ \vdots & \vdots & \vdots & \ddots & \vdots & \vdots & \vdots \\ V_{r1} & \cdots & V_{rk} & \cdots & V_{rr} & \cdots & V_{rN} \\ \vdots & \vdots & \vdots & \vdots & \vdots & \ddots & \vdots \\ V_{N1} & \cdots & V_{Nk} & \cdots & V_{Nr} & \cdots & V_{NN} \end{vmatrix}$$

equal to zero, where the elements are given by

$$V_{jm} = \left\{ (E_m^0 - E) \langle \psi_j^0 | \psi_m^0 \rangle + \langle \psi_j^0 | V' | \psi_m^0 \rangle \right\}.$$

If a solution for the perturbed potential is written as a linear sum of N of the solutions of the unperturbed potential,

$$\psi(r) = \sum_{m=1}^{N} C_m \psi_m^0(r),$$

the expansion coefficients are given by

$$C_i = \frac{V_{ik}^*}{\sqrt{\sum_n \sum_m \left\{ (V_{nk}^*)(V_{mk}^*) \langle \psi_n^0 | \psi_m^0 \rangle \right\}}} \quad (i = 1, N),$$

where V_{ik}^* designates the (i, k)-th element of the inverse of the square matrix

$$
\begin{bmatrix}
V_{11} & \cdots & V_{1k} & \cdots & V_{1r} & \cdots & V_{1N} \\
\vdots & \ddots & \vdots & & \vdots & \vdots & \vdots & \vdots \\
V_{k1} & \cdots & V_{kk} & \cdots & V_{kr} & \cdots & V_{kN} \\
\vdots & \vdots & \vdots & \ddots & \vdots & \vdots & \vdots \\
V_{r1} & \cdots & \langle \psi_r^0 | V' | \psi_k^0 \rangle & \cdots & V_{rr} & \cdots & V_{rN} \\
\vdots & \vdots & \vdots & & \vdots & \ddots & \vdots \\
V_{N1} & \cdots & V_{Nk} & \cdots & V_{Nr} & \cdots & V_{NN}
\end{bmatrix},
$$

not the complex conjugate of V_{ik}. First-order expressions for the energies and wavefunctions corresponding to a perturbation are much easier to deal with, being given by

$$E \sim E_m^0 + \langle \psi_m^0 | V' | \psi_m^0 \rangle$$

and

$$\psi_n \sim \psi_n^0 - \sum_{k \neq n} \frac{\psi_k^0 \langle \psi_k^0 | V' | \psi_n^0 \rangle}{\left(E_k^0 - E_n^0 \right)}.$$

The variational method is a powerful technique for estimating ground state energies when one can formulate a trial ground-state wavefunction. The trial function need not satisfy Schrödinger's equation for the potential at hand, only the boundary conditions of the problem. In the simplest form of this method, if $\phi(x)$ is a normalized trial function, then the ground-state energy E_0 is constrained by

$$E_0 \leq \langle \phi^* | H | \phi \rangle$$

where H is usual the Hamiltonian operator.

Problems

9.1(I) Expand the function $\phi(x) = x^2$ in terms of the infinite rectangular well wave-functions. Take $L = 1$; plot your result for terms in the sum up to $n = 5$ and $n = 20$.

9.2(I) Expand the function $\phi(x) = e^{ax}$ in terms of the infinite rectangular well wave-functions. Take $L = 1$ and $a = 3$; plot your result for terms in the sum up to $n = 5$ and $n = 20$.

9.3(I) Using the WKB approximation, determine the energy eigenvalues for a particle of mass m moving in a potential given by $V(x) = \alpha|x|$, $(-\infty \le x \le \infty)$.

9.4(I) In Example 9.2 it was shown that the WKB method predicts exactly the energy levels of the hydrogen atom. Following the approach of Example 9.1, investigate the application of the classical approximation to this system at the point $r_{turn}/2$ for a general energy level E_n. What constraint on n emerges?

9.5(I) Show that application of the WKB method to the harmonic-oscillator potential $V(x) = kx^2/2$ leads to $E_n \sim n\hbar\omega$. Investigate the application of the classical approximation to this system at the point $x_{turn}/2$ for a general energy level E_n. What constraint on n emerges?

9.6(I) Set up the WKB expression for the energy levels of a potential given by

$$V(x) = \begin{cases} \infty & x \le 0 \\ A\sin^2(x/\alpha) & 0 \le (x/\alpha) \le \pi/2 \\ 0 & \text{otherwise.} \end{cases}$$

If an electron were trapped in such a potential with $A = 24$ eV and $\alpha = 6$ Å at an energy of 12 eV, approximately what quantum state would it be in?

$$Note: \int_0^{\pi/4} \sqrt{1 - 2\sin^2 y}\, dy = 0.5991.$$

9.7(I) Set up the WKB integral for the potential

$$V(x) = \begin{cases} \infty & (x < 0) \\ V_o(e^{\alpha x} - 1) & (0 \le x \le \ln 2/\alpha) \\ V_o & (x > \ln 2/\alpha). \end{cases}$$

If there is an energy level with $n = 4$ at $E = 7$ eV when $V_0 = 28$ eV, determine the value of α.

9.8(A) Consider the family of potential wells defined by

$$V(x) = \alpha|x|^q, -\infty \le x \le \infty; q > 0.$$

(a) Show that the WKB integral for this potential can be put in the form

$$\frac{4\sqrt{2m}}{q\alpha^{1/q}} E^{(q+2)/2q} \int_0^1 y^{1/q-1}\sqrt{1-y}\,dy \sim nh,$$

where $y = (\alpha/E)x^q$.

(b) An integral of this form is related to a so-called Beta function, which can be expressed in terms of Gamma functions:

$$\int_0^1 x^{m-1}(1-x)^{n-1}\,dx = \frac{\Gamma(m)\,\Gamma(n)}{\Gamma(m+n)}.$$

Two properties of the Gamma function, $\Gamma(x+1) = x! = x\Gamma(x)$, and $\Gamma(1/2) = \sqrt{\pi}$ are helpful here. Hence show that the energy levels are given approximately by

$$E_n \sim \left\{ \frac{nh\alpha^{1/q}\Gamma(3/2+1/q)}{2\sqrt{2\pi m}\Gamma(1+1/q)} \right\}^{2q/(q+2)}.$$

Check this result in the case of the harmonic oscillator: $q = 2$ and $\alpha = k/2$. For further treatment of this problem, see [7].

9.9(I) An infinite rectangular well is subject to a perturbing potential $V(x) = \alpha x^2$, $(0 \le x \le L)$. Determine a first order approximation for the energy for any state $\psi_n(x)$.

9.10(A) A infinite rectangular well is subject to a perturbing potential $V(x) - V^0(x) = -V_0$, $(0 \le x \le L/2)$; see Fig. 9.11. Determine, to a first order approximation, the energy and wavefunction for any state $\psi_n(x)$.

9.11(A) A particle of mass m with $\ell = 0$ is trapped in an infinite spherical well of radius a (see Chap. 7). The well is subjected to a perturbing potential $V(r) = \beta r^2$, $(0 \le r \le a)$. Derive expressions for the first order corrections to the energy and wavefunction for any state n.

9.12(I) As the previous problem but with $V(r) = Ar^2 \sin\theta$. Derive an expression for the first order correction to the energy for state n.

9.13(A) A ground state $(n = 0)$ harmonic oscillator is subject to the perturbing potential $V' = \beta x^4$.

(a) Determine the first-order perturbation to the ground-state energy.

(b) Derive an expression for the perturbed ground-state wavefunction. Note that Hermite polynomials satisfy the identity

Fig. 9.11 Problem 9.10: Perturbed infinite well

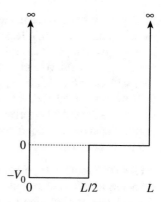

$$\int_{-\infty}^{\infty} \xi^r e^{-\xi^2} H_n(\xi) H_{n+p}(\xi)\, d\xi = \begin{cases} 0, & (p > r) \\ 2^n (n+r)! \sqrt{\pi}, & (p = r). \end{cases}$$

HINT: The parity properties of Hermite polynomials are useful in eliminating some of the integrals not covered by this identity. Leave your answer in terms of $H_2(\xi)$ and $H_4(\xi)$.

9.14(A) As a consequence of some process within its nuclear proton, the electron in a hydrogen atom finds itself subject to a perturbing potential $V' = \lambda r$. Using Example 9.6 as a model, derive an expression for the determinant to be solved for the perturbed energies. Include contributions form the $n = 1$ and $n = 2$ hydrogenic states. In computing the matrix elements, the following identity for spherical harmonics is helpful:

$$\int \int Y_{\ell_1 m_1} Y_{\ell_2 m_2} \sin\theta\, d\theta\, d\phi = \delta_{\ell_1}^{\ell_2} \delta_{m_1}^{m_2}.$$

Also, the result of Problem 7.18 for $\langle r \rangle$ may be helpful.

9.15(I) A hydrogen atom in the ground state is subject to the perturbing potential $V'(r) = \beta r$. Use non-degenerate first-order perturbation theory to determine the perturbation to the ground-state energy. If it is desired that the perturbation amount to no more than $E_1/10$ where E_1 is the ground-state energy, what is the maximum tolerable value of β? Express your results in units of eV and Å.

9.16(E) In Chap. 8 it was shown that an electron orbiting a nucleus with orbital angular momentum L gives rise to a vector magnetic dipole moment $\mu = (e/2m_e)L$. A magnetic dipole moment placed in a magnetic field B acquires a potential energy given by $V = -\mu \cdot B$. Consider hydrogen atoms in general (n, ℓ, m) states suddenly subjected to a magnetic field $B = B\hat{z}$, where $\hat{z}$ denotes the usual Cartesian-coordinate unit vector in the z-direction. Use non-degenerate first-order perturbation theory to show that the hydrogenic states will be perturbed by an amount $\Delta E = -\mu_B m B$, where μ_B is the Bohr magneton, $\mu_B = (e\hbar/2m_e)$. What does this result imply for the normally fourfold-degenerate $n = 2$ states? Within sunspots, magnetic fields can be as strong as $B = 0.3$ Tesla. In the unperturbed Bohr model, $3 \rightarrow 2$ transitions usually give rise to photons of wavelength 6564 Å. What alteration in the photon wavelength would you expect to observe for hydrogen in the vicinity of such a sunspot? The perturbation of atomic energy levels by an external magnetic field is known as the Zeeman effect after Dutch physicist Pieter Zeeman (1865–1943), who shared the 1902 Nobel prize for physics with Hendrik Lorentz for its discovery. A fuller analysis of this effect takes into account both the orbital and spin angular momenta of electrons.

9.17(E) In Example 9.6, the term $256\lambda a_o e / 243\sqrt{2}$ actually plays a fairly small role in the solution of the cubic equation for the perturbed energy levels. If it is assumed that $C \rightarrow 0$, then the solution for the perturbed energies reduces to $AB^2(B + D)(B - D) = 0$ in the notation of that example. Hence, the result is an unperturbed $n = 1$ level at energy E_1 (the A term), two unperturbed $n = 2$ levels at energy $E_1/4$ (the B^2 term), and two perturbed $n = 2$ levels at $E = E_1/4 \pm 3\lambda a_o e$ (the $B \pm D$ terms). For a field of strength $\lambda^* = 0.01$, what is the value of the perturbation term $3\lambda a_o e$ expressed as a wavelength, that is, what would be the wavelength of a photon arising from an electron transiting from a perturbed $n = 2$ level to an unperturbed $n = 2$ level? Stark-effect spectroscopy is usually carried out in the microwave region of the electromagnetic spectrum.

9.18(A) A one-dimensional potential is given by

$$V(x) = \begin{cases} \infty & (x < 0) \\ -A\sqrt{x}\,e^{-\alpha x^2} & (0 \le x \le \infty). \end{cases} \tag{9.58}$$

Analyze this potential by the variational method of Sect. 9.4 with the trial wavefunction $\phi(x) = Bxe^{-\beta x^2}$, where B is a normalization constant and β is the variational parameter. Develop an expression for the value of β which minimizes the system energy for a particle of mass m; you will not be able to solve this expression analytically for β. For the case of an electron moving in this potential with $A = 100$ eV Å$^{-1/2}$ and $\alpha = 1.75$ Å^2, solve numerically for the value of β that minimizes the energy, and so determine an upper-limit on the ground-state energy.

9.19(A) Analyze the potential $V(x) = -Axe^{-\alpha x^2}$ ($-\infty \le x \le \infty$) by the variational method with the trial wavefunction $\phi(x) = Be^{-\beta x^2}$ ($-\infty \le x \le \infty$), where B is the normalization constant and β is the variational parameter. You will not be able to solve the energy-minimization condition analytically for β. In the case of an electron moving in such a potential with $A = 10$ eV and $\alpha = 5$ Å^{-2}, what is the resulting upper-limit on the ground-state energy?

9.20(A) In the theory of nuclear forces, a modification of the Coulomb potential known as the Yukawa potential is sometimes utilized. This is a purely radial potential given by

$$V(r) = -\left(\frac{K}{r}\right)e^{-(\alpha/a_o)r}, \tag{9.59}$$

where a_o is the Bohr radius, $\kappa = e^2/4\pi\varepsilon_o$ (as in the Coulomb potential), and where $\alpha > 0$. Carry out a variational analysis for an electron moving in this potential using as the trial wavefunction $\phi(r) = Ce^{-(\beta/a_o)r}$, where C is the normalization constant and β is the variational parameter; show that the ground-state energy satisfies

$$E_0 \leq -E_1 \left\{ \beta^2 - \frac{8\beta^3}{(\alpha + 2\beta)^2} \right\}, \tag{9.60}$$

where E_1 is the hydrogenic ground-state energy. For $\alpha = 0$, the Yukawa potential reduces to the standard hydrogen-atom potential. Since the trial wavefunction is of the form of that for the hydrogenic ground state, then for $\alpha = 0$ we should recover the exact solution for the hydrogen ground state; verify that this is so. For $\alpha = 0.5$, determine the value of β that minimizes the expression for E_0 and hence determine an upper limit on the ground state energy in this case.

9.21(I) Consider the following potential, a one-dimensional analog of the hydrogen atom:

$$V(x) = \begin{cases} -\frac{\kappa}{x} & x \geq 0 \\ \infty & x \leq 0. \end{cases} \tag{9.61}$$

Carry out a variational analysis for a particle of mass m moving in this potential, taking as the trial wavefunction

$$\phi(x) = Cxe^{-\beta x} \quad (x \geq 0; \ \phi(x) = 0 \text{ otherwise}), \tag{9.62}$$

where C is the normalization constant and β is the variational parameter. If $\kappa = e^2/4\pi\varepsilon_o$ as in the Coulomb potential, how does your estimate of the ground-state energy for an electron compare to that for the usual Coulomb potential?

9.22(I) Use the trial wavefunction $\phi(x) = A(\lambda^2 - x^2)$, $|x| \leq \lambda$ ($\phi(x) = 0$ otherwise) to estimate the ground-state energy of the harmonic-oscillator potential $V(x) = kx^2/2$ via the variational method. A is the normalization constant; λ is the variational parameter. Compare your result to the true ground-state energy.

9.23(A) Consider a potential well defined by

$$V(x) = \begin{cases} \infty & x \leq 0 \\ \alpha x^n & 0 \leq x \leq \infty. \end{cases} \tag{9.63}$$

Using as a trial wavefunction $\phi(x) = Axe^{-\beta x}$ where A is the normalization constant and β the variational parameter, use the variational method to derive an expression for the ground-state energy for $V(x)$. Check your result against Example 9.7 for $n = 1$.

9.24(A) Use the trial wavefunction $\phi(r) = Ce^{-\beta r^2}$ to estimate the ground-state energy of the hydrogen atom. Compare your result to the true ground-state energy.

9.25(I) Consider two potentials $V_1(x)$ and $V_2(x)$ which run over the same domain. For a mass m moving in each, you are given that their ground-state solutions are (ψ_1, E_1) and (ψ_2, E_2). Now form a new potential by summing them: $V = V_1 + V_2$. By using the trial wavefunction $\psi(x) = A\psi_1(x)\psi_2(x)$ where A is the normalization constant, show by using the variational theorem that the ground state of the new potential must satisfy

$$E \leq E_1 + E_2 - \frac{A^2 \hbar^2}{m} \int \psi_1 \psi_2 (\nabla \psi_1 \nabla \psi_2) dx. \qquad (9.64)$$

The fact that the trial wavefunction has no variational parameter does not render it invalid for use in the variational theorem; there is no requirement that a trial function has to be a particularly good one. If $V_1 = V_2$, what does this result indicate for E in comparison to $E_1 + E_2$? Can you make any general statement about how E compares to $E_1 + E_2$?

9.26(E) In the analysis of the attractive central potential in Example 9.8, it was remarked that $K = 2$ and $n = 1$ should cause that example to reduce to the Coulomb potential. Verify this statement.

References

1. G. Wentzel, Z. Phys. **38**, 518 (1926)
2. H.A. Kramers, Z. Phys. **39**, 828 (1926)
3. L. Brillouin, J. Phys. Radium **7**, 353 (1926)
4. H. Jeffreys, Proc. London Math. Soc. **23**, 428 (1923)
5. L. Pauling, E.B. Wilson *Introduction to Quantum Mechanics with Applications to Chemistry* (Dover Publications, New York, 1985)
6. B.C. Reed, Am. J. Phys. **58**(4), 407 (1990)
7. U.P. Sukhatme, Am. J. Phys. **41**(8), 1015 (1973)

Chapter 10
Numerical Solution of Schrödinger's Equation

Summary This chapter develops a straightforward spreadsheet-based method of numerically solving Schrödinger's equation that is suitable for use on a personal computer. By exploring a few examples, techniques for deciding upon appropriate integration stepsizes and boundary conditions are illustrated. Once an operating spreadsheet is in hand, it can be applied to a wide variety of problems.

All of the solutions to Schrödinger's equation that we considered so far, whether exact or approximate, have been expressible analytically. In many cases, however, the algebra attending these solutions was extremely tedious. Based on this experience, it is not hard to imagine that even a slight change to an otherwise soluble potential could result in an intractable problem, rendering Schrödinger's equation difficult if not impossible to solve.

Through the use of computers, however, it is possible to solve Schrödinger's equation for complicated potentials via numerical integration. For most modern researchers this is the method of choice. Numerical solution not only shifts much drudgery from human to computer, but has the additional advantage that once a working program has been written, investigating a different potential becomes as easy as changing a few statements in a subroutine or spreadsheet cell. The drawback of a numerical approach is that the eigenstates of a system have to be found one-at-a-time, as opposed to establishing an exact analytic expression which comprises the full spectrum of solutions. But sometimes one has no choice in the matter.

Numerical solution of differential equations is an active, complex subdiscipline of mathematics and computing science. In this chapter we discuss a straightforward method for numerically solving the one-dimensional Schrödinger equation. It is suitable for use on any personal computer. The method described here is by no means the most elegant or efficient one available, but it is straightforward and reasonably accurate. The point here is not so much computational elegance as to explore how physical insight and computational power can be used to address problems.

© The Author(s), under exclusive license to Springer Nature Switzerland AG 2022
B. C. Reed, *Quantum Mechanics*, https://doi.org/10.1007/978-3-031-14020-4_10

10.1 Atomic Units

Before plunging into computational details, some practical considerations need to be addressed. The one-dimensional Schrödinger equation can be written as

$$\frac{d^2\psi}{dx^2} = -\beta[E - V(x)]\psi(x), \tag{10.1}$$

where

$$\beta = 2m/\hbar^2. \tag{10.2}$$

For an electron, $\beta \sim 1.6 \times 10^{38}$ in MKS units. Given that numerical integration of a differential equation is likely to involve thousands of repetitive calculations, each accompanied by possible overflows, underflows, and accumulating round-off errors, numbers of such magnitude are inconvenient at best and potentially catastrophic at worst. In computer work, one is best to keep all quantities of order unity to maintain maximum accuracy.

There are two ways of circumventing the magnitude of β: Either (i) cast Schrödinger's equation into a dimensionless form (as we have often done) and worry later about restoring proper units and conversion factors into the results, or (ii) work from the outset in a system of units appropriate to atomic-scale problems. Here I adopt the latter approach.

In setting up any system of units, one must adopt definitions for three fundamental quantities, usually chosen to be length, mass, and time. If you are dealing with problems involving electromagnetism, a fourth fundamental quantity has to be added to this list: electric charge, for which the charge on the electron is a useful practical standard. Units for all other quantities can then be *derived* from the fundamental units, for example, [momentum] = [mass][length]/[time], or [energy] = [mass][length]^2/[time]^2, where square brackets denote "dimensions of". From previous experience, we know that atomic-scale problems are characterized by lengths on the order of Ångstroms and energies on the order of eV. Let us construct a system of units from these quantities and the mass of the electron as the defined units:

$$[\text{length}] = 1\,\text{Å} = 1.0000 \times 10^{-10}\,\text{m}. \tag{10.3}$$

$$[\text{mass}] = 1\,m_e = 9.1094 \times 10^{-31}\,\text{kg}. \tag{10.4}$$

$$[\text{energy}] = 1\,\text{eV} = 1.6022 \times 10^{-19}\,\text{J}. \tag{10.5}$$

While it is of no direct use here, the unit of time in this system is a derived quantity:

$$[\text{time}] = \sqrt{\frac{[\text{mass}]\,[\text{length}]^2}{[\text{energy}]}} = 2.3844 \times 10^{-16}\,\text{s}.$$

Now,

$$\hbar = 1.0546 \times 10^{-34} \,\text{J s}.$$

The units of β in (10.2) are $(\text{J m}^2)^{-1}$. With the definitions above, we can construct a conversion factor to atomic units:

$$1 \,(\text{J m}^2)^{-1} = 1.6022 \times 10^{-39} \,(\text{eV Å}^2)^{-1}. \tag{10.6}$$

For an electron,

$$\beta_e = 2m_e/\hbar^2 = 1.6831 \times 10^{38} \,(\text{J m}^2)^{-1} = 0.26246 \,(\text{eV Å}^2)^{-1}. \tag{10.7}$$

If we agree to specify mass (m) in terms of electron masses, lengths in Å, and energies in eV, we can then write Schrödinger's equation as

$$\frac{d^2\psi}{dx^2} = -0.26246m[E - V(x)]\psi(x). \tag{10.8}$$

This system of units is in no way special. One could, for example, choose the unit of length to be the Bohr radius and the unit of energy to be the ionization energy of hydrogen from the $n = 1$ orbit. In a problem involving nuclear physics one might choose [length] $= 10^{-15}$ m and [energy] $= 1$ MeV. The point is to build a system such that any numerical factors appearing in Schrödinger's equation for the problem(s) you want to solve are of order unity.

10.2 A Straightforward Numerical Integration Method

In essence, all methods of numerically integrating differential equations boil down to the following recipe. First, the domain involved (x, in our case) is divided up into a number of (usually) equally-spaced discrete points separated by a stepsize Δx. Then, starting from a specified value of the dependent variable $y(x)$ [here, $\psi(x)$] at some initial point, say x_o, one bootstraps along increasing x by adding small increments to $y(x)$ based on derivatives multiplied by stepsizes. We will utilize a very straightforward implementation of this approach. For readers interested in exploring more sophisticated techniques, Sect. 8.7 of Arfken and Weber is a good starting place; see also Chap. 15 of Press et al.

To motivate this approach, we need three expressions. The first is Schrödinger's equation written in atomic units:

$$\frac{d^2\psi}{dx^2} = 0.26246m[V(x) - E]\psi(x). \tag{10.9}$$

The second expression is the definition of the second derivative of ψ at some position, say x_o,

$$\left(\frac{d^2\psi}{dx^2}\right)_{x_o} = \frac{\left[\left(\frac{d\psi}{dx}\right)_{x_o+\Delta x} - \left(\frac{d\psi}{dx}\right)_{x_o}\right]}{\Delta x}.$$

Formally, we should define the second derivative as the limit of this expression as $\Delta x \to 0$, but in practice Δx will be finite. We rearrange this to the form

$$\left(\frac{d\psi}{dx}\right)_{x_o+\Delta x} = \left(\frac{d^2\psi}{dx^2}\right)_{x_o} \Delta x + \left(\frac{d\psi}{dx}\right)_{x_o}. \tag{10.10}$$

Finally, we need the (truncated) expression for the Taylor-series expansion of $\psi(x)$:

$$\psi(x_o + \Delta x) = \psi(x_o) + \left(\frac{d\psi}{dx}\right)_{x_o} \Delta x + \left(\frac{d^2\psi}{dx^2}\right)_{x_o} \frac{\Delta x^2}{2} + \cdots . \tag{10.11}$$

The procedure for numerically integrating $\psi(x)$ can be summarized as follows.

(i) Specify $V(x)$, the stepsize Δx, the initial position x_o, and the values of ψ and $(d\psi/dx)$ at x_o.

(ii) Make a trial guess at an energy eigenvalue E.

(iii) A new integration "cycle" begins by using (10.9) to compute $(d^2\psi/dx^2)_{x_o}$.

(iv) From the values for ψ and $(d\psi/dx)$ at x_o and the result from step (iii), use (10.11) to compute $\psi(x_o + \Delta x)$.

(v) From the value of $(d\psi/dx)$ at x_o and the result from step (iii), use (10.10) to compute $(d\psi/dx)_{x_o+\Delta x}$.

(vi) Now loop back to step (iii) and start a new cycle beginning with computing the second derivative of ψ at $(x_o + \Delta x)$. Step (iv) will then give the value of ψ at $(x_o + 2\Delta x)$, step (v) the value of $(d\psi/dx)_{x_o+2\Delta x}$, and so forth. Proceed until x reaches some appropriate upper limit, plotting ψ as you go. A correct eigenvalue will be one that makes $\psi \to 0$ as $x \to \infty$. If this does not happen, begin again at step (ii).

A crucial qualification here is that even if you are numerically analyzing a problem known to possess an exact analytic solution, you will never be able to make your solution behave as $\psi \to 0$ as $x \to \infty$ even if E is chosen to be one of the exactly-known eigenvalues. This is because all computers operate with finite numerical accuracy; accumulation of small round-off errors is ultimately unavoidable, with the result that the computed run of ψ will eventually diverge to $\pm\infty$. As we will see below, the practical issue becomes one of isolating limits on guesses for E between which ψ "wags" between positive and negative divergences.

This procedure is ideally suited for a spreadsheet: Successive rows hold successive values of x, while columns hold the values of V, ψ, and derivatives of the latter. One advantage of using a spreadsheet is that you can plot $\psi(x)$ as the computation proceeds and so directly see the effect of changing your estimate for E.

Some operational questions remain. Where do initial values for ψ and $(d\psi/dx)$ come from? What is an appropriate stepsize? How far should the calculation be carried in x? These questions require some physical judgment on the part of the operator. For example, if $V(x)$ possesses an infinite wall at $x = 0$, then we immediately know that $\psi = 0$ there. If $V(x)$ is symmetric about $x = 0$ and you seek an even-parity wavefunction, then set $\psi(0) \neq 0$ and $(d\psi/dx)_{x_o} = 0$. If you already have an odd-parity solution for a symmetric potential with $(d\psi/dx)_{x_o} < 0$, then the next-higher-energy odd-parity solution must have $(d\psi/dx)_{x_o} > 0$. In the case of a potential possessing no symmetries, the sign of the value you choose for $(d\psi/dx)_{x_o}$ is irrelevant, as only $|\psi|^2$ has any physical meaning. As to the magnitude of Δx, some quick numerical estimates can serve as a guide. If you are dealing with a potential-well type problem, for example, you can estimate at what values of x your guess for E cuts the walls of the well and then divide the range into a sensibly large number of intervals, say 1000. Carrying the calculation out to a distance of three or four times the cut position should be sufficient to see if ψ is "reasonably" approaching zero as x gets large. If you have doubts as to the accuracy of an energy eigenvalue, try dividing Δx by 2 and repeating the calculation; if the result is the same or closely similar, then Δx was small enough. More accuracy can generally be squeezed out of the solution by decreasing Δx, but this will eventually run up against its own issues of roundoff errors and practicability.

One physical point remains: what of the normalization of ψ, which has nowhere been demanded in the integration scheme? This apparent oversight is actually an advantage in that it allows some flexibility in specifying initial conditions. Consider a potential well with an infinite wall at $x = 0$. Suppose you set $\psi = 0$ and $(d\psi/dx) = 0.01$ at $x = 0$, and experiment with different values of E until an eigenvalue, say E_1, is found. If you then take $(d\psi/dx) = 0.05$, you will find that E_1 is still an eigenvalue. Indeed, it is possible to show that the choice of affects only the amplitude of $\psi(x)$ but not its shape, leaving the energy eigenvalues unaltered (Problem 10.8). Normalization can be taken care of later by multiplying the computed wavefunction by a factor C defined by

$$C \sum_{x_j} |\psi(x_j)|^2 \Delta x = 1.$$

For practical purposes, however, one is usually not interested in normalizing the wavefunction: Energy eigenvalues are the physically measurable quantities.

Example 10.1   Use the numerical integration scheme outlined above to find the first two energy eigenvalues for an electron moving in the linear potential

$$V(x) = \begin{cases} \infty & x \leq 0 \\ \alpha x & x > 0, \end{cases}$$

with $\alpha = 1$ eV/Å.

Table 10.1 Linear potential ground-state numerical coefficient η_1 of $E_1 = \eta_1 \alpha^{2/3} \varepsilon^{1/3}$

Method	η_1
Uncertainty principle	> 1.191
WKB approximation	2.811
Variational	< 2.476
Improved variational	1.560–2.476
Exact	2.338107

This problem was treated with the uncertainty principle in Problem 4.10, by the WKB method in Example 9.1, and by the variational method in Examples 9.7 and 9.9. The energy eigenvalue for quantum state n in this potential was found to be of the form $E_n = \eta_n \alpha^{2/3} \varepsilon^{1/3}$, where η_n is a dimensionless numerical coefficient and where $\varepsilon = \hbar^2/2m$. Table 10.1 lists the values of E_1 yielded by these various methods. The exact value is taken from Abramowitz and Stegun [1].

For the case specified here, $\alpha^{2/3} \varepsilon^{1/3}$ evaluates as

$$\alpha^{2/3} \varepsilon^{1/3} = (1\,\text{eV/Å})^{2/3} \left(\frac{\hbar^2}{2m_e} \right)^{1/3}$$

$$= (1.6022 \times 10^{-9}\,\text{J/m})^{2/3} \left[\frac{(1.0546 \times 10^{-34}\,\text{J s})^2}{2(9.1094 \times 10^{-31}\,\text{kg})} \right]^{1/3}$$

$$= 2.5024 \times 10^{-19}\,\text{J}$$

$$= 1.5619\,\text{eV},$$

that is, $E_n/\eta_n = 1.5619\,\text{eV}$.

Because $V(0) = \infty$, we know that $\psi(0) = 0$. From the discussion in Sect. 3.6, we can infer that the ground and first excited state wavefunctions must possess one and two maxima, respectively, and asymptotically approach zero as $x \to \infty$. Plausible forms for such functions are sketched in Fig. 10.1.

From the numbers given in Table 10.1, we might estimate $\eta \sim 1.8$ for the ground state, equivalent to $E \sim (1.8)(1.5619\,\text{eV}) \sim 2.8\,\text{eV}$. As a sensible first guess for E we can round this off to 3 eV. This proposed energy level would cut the potential both at $x = 0$ Å and $x = 3$ Å. Most of the action for ψ will lie between these limits, and if we want, say, 1000 intervals between these limits, we adopt $x = 0.003$ Å. To be consistent with the infinite wall at $x = 0$, we take $\psi(0) = 0$. $(d\psi/dx)_0$ can be chosen arbitrarily; we take it to be 0.1.

Figure 10.2 shows a screen shot of the first few rows of a Microsoft Excel spreadsheet set up to integrate this problem. The values of E (3 eV) and Δx (0.003 Å) are defined in cells G1 and G2. Columns A through E respectively hold the values of x, $V(x)$, $\psi(x)$, $d\psi/dx$ and $d^2\psi/dx^2$. The values for x, V, ψ, and $d\psi/dx$ at $x = 0$ are entered manually (row 4). Subsequent values for x

advance in steps of Δx; those for $V(x)$ are based on its formulae, and all of the other values for ψ and its derivatives are computed with the algorithm outlined above. For example, $d^2\psi/dx^2$ at $x = 0$ (cell E4) is defined as 0.26246 times (cell B4 − E) [that is (cell B4 − cell G1)] times cell C4; when this formula is copied-and-pasted into the subsequent rows of column E, Excel automatically updates the cell references for $V(x)$ and $\psi(x)$. Similarly, $(d\psi/dx)$ at $x = 0.003$ Å (cell D5) is defined as cell D4 plus cell E4 times Δx, and cell C5, the value of ψ at $x = 0.003$ Å, as cell C4 + (cell D4)Δx + 0.5(cell E4)$(\Delta x)^2$, with like copy-and-paste operations for both columns.

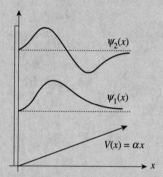

Fig. 10.1 Possible ground and first excited state linear potential wavefunctions

◇	A	B	C	D	E	F	G	H	I	J	K
1	x	V(x)	y	dy/dx	d^2y/dx^2	E =	3				
2						Delta-X	0.003				
3											
4	0	0	0	0.1	0						
5	0.003	0.003	0.0003	0.1	-0.00023598						
6	0.006	0.006	0.0006	0.099999	-0.00047148						
7	0.009	0.009	0.0009	0.099998	-0.00070651						
8	0.012	0.012	0.0012	0.099996	-0.00094106	0.5					
9	0.015	0.015	0.0015	0.099993	-0.00117514						
10	0.018	0.018	0.0018	0.099989	-0.00140873	0.4					
11	0.021	0.021	0.0021	0.099985	-0.00164185						
12	0.024	0.024	0.0024	0.09998	-0.00187448	0.3					
13	0.027	0.027	0.0027	0.099975	-0.00210662						
14	0.03	0.03	0.003	0.099968	-0.00233828	0.2					
15	0.033	0.033	0.0033	0.099961	-0.00256946						
16	0.036	0.036	0.0036	0.099954	-0.00280014	0.1					
17	0.039	0.039	0.0039	0.099945	-0.00303033						
18	0.042	0.042	0.0042	0.099936	-0.00326003	0					
19	0.045	0.045	0.0045	0.099926	-0.00348923						
20	0.048	0.048	0.0048	0.099916	-0.00371794						
21	0.051	0.051	0.0051	0.099905	-0.00394615						
22	0.054	0.054	0.0054	0.099893	-0.00417386						
23	0.057	0.057	0.0057	0.09988	-0.00440108						

Fig. 10.2 Computed wavefunction for the linear potential $V(x) = \alpha x$ ($\alpha = 1$ eV/Å) for $E = 3$ eV. The divergence of the wavefunction indicates that this energy cannot be an eigenvalue

In what follows, the numbers for the first three cycles of the algorithm are worked through in detail to four decimal places; your computer will keep more significant figures than this. Work through these steps with your calculator, referring to Fig. 10.2 as appropriate. For convenience, we denote derivatives with primes and use subscripts to designate x-values, e.g., $\psi''_{0.003}$ denotes the value of $d^2\psi/dx^2$ at $x = 0.003$ Å.

Begin the first cycle. The values for steps (i) and (ii) are described above. For step (iii), (10.9) with $m = 1$ for an electron gives

$$\psi''_{0.000} = 0.26246[V(x) - E]\psi_{0.000} = 0,$$

since $\psi_{0.000} = 0$. This result appears in cell E4. Proceeding to step (iv), we invoke (10.11) to compute the value of ψ at $x = 0.003$ Å:

$$\psi_{0.003} = \psi_{0.000} + \psi'_{0.000}\,\Delta x + \psi''_{0.000}(\Delta x^2/2)$$
$$= 0 + (0.1)(0.003) + (0)(0.003)^2/2 = 3.0000 \times 10^{-4},$$

which appears in cell C5. Step (v) then provides the value of $d\psi/dx$ at $x = 0.003$ Å:

$$\psi'_{0.003} = \psi''_{0.000}\Delta x + \psi'_{0.000} = 0(0.003) + 0.1 = 0.1000,$$

cell D5. Step (vi) sends us back to step (iii) and the start of the second cycle.

Now commence the second cycle of the algorithm at step (iii). From (10.9),

$$\psi''_{0.003} = 0.26246[V(0.003) - E]\psi_{0.003}$$
$$= 0.26246\,[0.003 - 3](3 \times 10^{-4}) = -2.3598 \times 10^{-4}.$$

This result appears in cell E5; note that it depends upon the value of $\psi_{0.003}$ computed in step (iv) of cycle 1. Proceeding to step (iv) of this cycle gives

$$\psi_{0.006} = \psi_{0.003} + \psi'_{0.003}\Delta x + \psi''_{0.003}(\Delta x^2/2)$$
$$= (3 \times 10^{-4}) + (0.1)(0.003) + (-2.3598 \times 10^{-4})(0.003)^2/2$$
$$= 6.0000 \times 10^{-4},$$

where, to an accuracy of four decimal places, the Δx^2 term makes no contribution. This result appears in cell C6. Step (v) now gives

$$\psi'_{0.006} = \psi''_{0.003}\Delta x + \psi'_{0.003}$$
$$= (-2.3598 \times 10^{-4})(0.003) + 0.1 = 9.9999 \times 10^{-2},$$

which appears in cell D6. This brings us to the end of cycle 2. Note that $d\psi/dx$ has decreased slightly from its value in cycle 1.

Cycle 3 proceeds likewise. Looping back to step (iii) fills in cell E6 as

$$\psi''_{0.006} = 0.26246\,[V(0.006) - E]\,\psi_{0.006}$$
$$= 0.26246[0.006 - 3]\,(6 \times 10^{-4}) = -4.7148 \times 10^{-4}.$$

Step (iv) advances us to $x = 0.009$ Å:

$$\psi_{0.009} = \psi_{0.006} + \psi'_{0.006}\Delta x + \psi''_{0.006}(\Delta x^2/2)$$
$$= (6 \times 10^{-4}) + (9.9999 \times 10^{-2})(0.003) + (-4.7148 \times 10^{-4})(0.003)^2/2$$
$$= 8.9999 \times 10^{-4},$$

which appears in cell C7. Step (v) of this cycle fills in cell D7:

$$\psi'_{0.009} = \psi''_{0.006}\,\Delta x + \psi'_{0.006}$$
$$= (-4.7148 \times 10^{-4})(0.003) + 9.9999 \times 10^{-2} = 9.9998 \times 10^{-2}.$$

$d\psi/dx$ has again decreased slightly from its previous value. At this point we again loop back to commence the fourth cycle at step (iii); you should find $\psi''_{0.009} = -7.0651 \times 10^{-4}$ (cell E7), and so on.

The graph in Fig. 10.2 shows that the computed run of ψ for $E = 3$ eV begins to strongly diverge at $x \sim 6$ Å. Increasing the guess for E slightly to 3.1 eV pushes the divergence to a slightly greater value of x. You should find that curve wags from a positive to a negative divergence between $E = 3.6$ and 3.7 eV. Narrowing this down, the wag is found to occur between 3.6510 and 3.6511 eV, corresponding to $\eta \sim 3.6510/1.5619 \sim 2.3375$, about 0.24% low in comparison to the exact value of 2.3381. The sensitivity of the solution to a slight change in E is stunning. Figure 10.3 shows the numerical solution for $E = 3.6510$ eV. Try experimenting with a different initial value of $(d\psi/dx)$; does the eigenvalue change significantly?

On proceeding to E_2, we find that the divergence wags between 6.3841 and 6.3842 eV, corresponding to $\eta_2 \sim 4.0874$. This is about 0.01% low compared to exact value, which is known to be 4.087949. The computed run of ψ for $E = 6.3841$ eV is shown in Fig. 10.4. Note that ψ has two extrema, characteristic of an $n = 2$ wavefunction; it first passes through a positive-valued maximum, turns over, goes through zero, reaches a minimum, and then turns back to the x-axis.

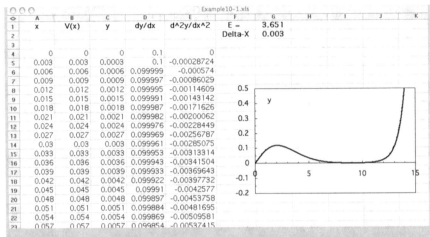

Fig. 10.3 As Fig. 10.2 but with $E = 3.651$ eV

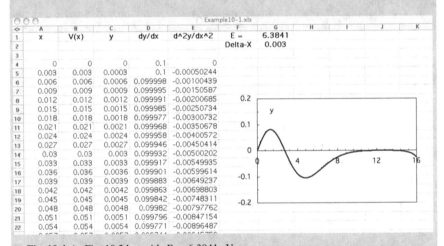

Fig. 10.4 As Fig. 10.2 but with $E = 6.3841$ eV

Example 10.2 The *Pöschl–Teller potential* is given by

$$V(x) = -\frac{\hbar^2}{2m} \frac{\alpha^2 \lambda(\lambda - 1)}{\cosh^2(\alpha x)}, \quad (-\infty \leq x \leq \infty),$$

where α and λ are parameters, both positive. α has dimensions of [length]$^{-1}$, while λ is dimensionless. $\cosh(\alpha x)$ denotes the hyperbolic cosine function of argument αx, which is defined by

$$\cosh(\xi) = \frac{1}{2}(e^\xi + e^{-\xi}).$$

In this example we determine the first two bound states for an electron moving in this potential with $\alpha = 0.3\,\text{Å}^{-1}$ and $\lambda = 5$. Surprisingly, this potential can be solved analytically; the bound states are given by [2]

$$E = -\frac{\hbar^2\alpha^2}{2m}(\lambda - 1 - n)^2 \quad (n \le \lambda - 1; \; n = 0, 1, 2, \dots).$$

This problem differs from Example 10.1 in that we have to deal with the coordinate "x" being both positive and negative. A good place to start is by casting the potential into our system of atomic units and then plotting it. From the discussion in Sect. 10.1, for a particle of mass "m" electron masses, $(2m/\hbar^2) = 0.26246m\,(\text{eV}\,\text{Å}^2)^{-1}$. If α is specified in units of Å^{-1}, we have

$$V(x) = -\frac{\alpha^2\lambda(\lambda - 1)}{(0.26246\,m)\cosh^2(\alpha x)} \quad (\text{eV})$$

Figure 10.5 shows a plot of this function ($m = 1, \alpha = 0.3\,\text{Å}^{-1}, \lambda = 5$). This potential is symmetric about $x = 0$ [where $V = -\hbar^2\alpha^2\lambda(\lambda - 1)/2m = -6.858\,\text{eV}$], and approaches 0 from below as $x \to \pm\infty$ due to the behavior of the cosh function. We can immediately conclude that any bound state must have $-6.858 \le E \le 0\,\text{eV}$. The plot also shows that by $x = +10\,\text{Å}$, $V(x)$ is very close to zero (about -0.07 eV); we take these extremes as our limits of numerical integration. But for its large-x behavior, this potential is reminiscent of the harmonic-oscillator potential of Chap. 5: for the ground state we should expect to find a solution for ψ with a single maximum at $x = 0$.

The spreadsheet developed for Example 10.1 can be used for this example upon changing the definition of $V(x)$ and setting the initial value of x to be $-10\,\text{Å}$. To determine the ground state, it is easiest to work upwards from $E = -6.85$ and seek values of E between which divergent wagging occurs; the first such behavior is found to occur in the vicinity of $E \sim -5.4\,\text{eV}$. Narrowing the estimate yields $E \sim -5.48655\,\text{eV}$, precisely what the exact solution predicts. Figure 10.6 shows a computed ground-state wavefunction for this energy based on choosing $\psi = 0.01$ and $d\psi/dx = 0.001$ at $x = -10\,\text{Å}$ with $\Delta x = 0.002$. As expected, ψ is symmetric about $x = 0$ with one maximum there.

Proceeding to higher guesses for E, another wagging is observed to occur at $E \sim -3.1\,\text{eV}$; refining it gives $E = -3.0862\,\text{eV}$, again in excellent agreement with the exact solution. The corresponding wavefunction is plotted in Fig. 10.7. That this energy corresponds to the first excited-state ($n = 1$) is can be seen by the presence of two maxima in the wavefunction; note that $\psi(0) = 0$ as would be expected for a symmetric potential.

The second excited state for this potential proves to lie at $E \sim -1.373\,\text{eV}$.

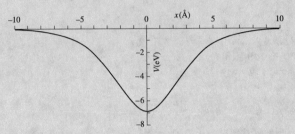

Fig. 10.5 Pöschl–Teller potential for an electron with $\alpha = 0.3\,\text{Å}^{-1}$ and $\lambda = 0.5$

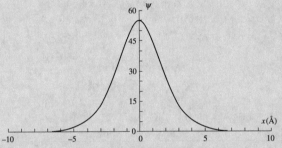

Fig. 10.6 Computed ground-state wavefunction for the potential of Fig. 10.5; $E = -5.48655\,\text{eV}$

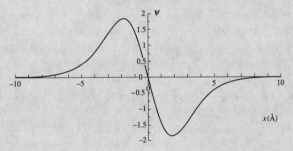

Fig. 10.7 Computed first-excited-state wavefunction for the potential of Fig. 10.5; $E = -3.0862\,\text{eV}$. Note change in vertical scale from Fig. 10.6

A working numerical integration program gives one a powerful, flexible tool for investigating a variety of problems. The method outlined here can be applied as well to the general central-potential radial Eq. (7.2), as only one coordinate is involved; see Problem 10.4.

Problems

NOTE: Results of numerical integrations will vary slightly depending on the program/spreadsheet and operating system used; do not be alarmed if your results differ slightly from those given in Appendix B. All problems in this chapter should be considered to be of intermediate difficulty.

10.1 Write a program or set up a spreadsheet to numerically integrate the one-dimensional Schrödinger equation. Test your program by reproducing the results for the linear potential treated in Example 10.1. What is the energy of an electron in the $n = 3$ state of this potential?

10.2 A potential is given by

$$V(x) = \begin{cases} \infty, & x \le 0 \\ \alpha x^3 & x > 0, \end{cases}$$

where $\alpha = 5\,\text{eV}/\text{Å}^3$. Determine the energy of the ground and first excited states for an electron moving in this potential.

10.3 Consider a potential defined by

$$V(x) = \begin{cases} \infty, & x \le 0 \\ -Axe^{-\alpha x^2} & x > 0. \end{cases}$$

(a) Show (analytically) that the position x_o at which this potential reaches a minimum is $x_o = \sqrt{1/2\alpha}$. Show in addition that $V(x_o) = -Ax_o/\sqrt{e}$.
(b) Consider an electron moving in a potential of this form with $x_o = 1\,\text{Å}$ and $V(x_o) = -10\,\text{eV}$. Determine its ground state energy.

10.4 In atomic units, the radial equation for a central potential can be written as

$$\frac{d^2 R}{dr^2} = \ell(\ell+1)\frac{R}{r^2} - \frac{2}{r}\frac{dR}{dr} - \beta[E - V(r)]R,$$

where $\beta = 0.26242m$.

(a) Write a program or set up a spreadsheet to integrate the radial equation numerically for any value of ℓ.
(b) Show that in atomic units the potential function for the hydrogen atom is given by $V(r) = -14.3993/r$.
(c) Test your program for a few $\ell = 0$ and 1 hydrogen atom states. Use the form of the radial wavefunctions given in Table 7.1 to guide estimates for initial values.
(d) Consider a potential given by $V(r) = \alpha r^2$ with $\alpha = 5\,\text{eV}/\text{Å}^2$. Use your program to deduce the two lowest energy levels for an electron moving in this potential for each of $\ell = 0$ and 1. HINT: Take $R(0) = 1$ and $R(0) = 0$ for $\ell = 0$ and $\ell \ne 0$, respectively.

(e) This potential is the spherical analog of the one-dimensional harmonic oscillator treated in Chap. 5. The radial equation can be solved analytically, and gives energy eigenvalues

$$E = \hbar\sqrt{\frac{2\alpha}{m}}\,(2n + \ell + 3/2), \qquad n = 0, 1, 2, \dots .$$

Compare the energies predicted by this formula to those determined numerically.

10.5 Apply the program developed in Problem 10.4 to the general attractive central potential $\ell = 0$ discussed in Example 9.8. Check your program by applying it to the hydrogen atom, $(K, m, n) = (2, 1, 1)$. Verify the exact values claimed in the example for $(K\mu) = 1$ for $n = 0.5$, 1.0, and 1.5. For what value of the power n is the ground state energy a minimum when $K = 2$?

10.6 In Problem 9.18, the potential

$$V(x) = \begin{cases} \infty & (x < 0) \\ -A\sqrt{x}\,e^{-\alpha x^2} & (0 \le x \le \infty). \end{cases}$$

was investigated by the variational method for $A = 100\,\mathrm{eV}\,\text{Å}^{-1/2}$ and $\alpha = 1.75\,\text{Å}^{-2}$. Solve this potential numerically. How does your variational solution for the ground state compare to the numerical result?

10.7 Set up a program to numerically integrate Schrödinger's equation for the case of a finite rectangular well as treated in Sect. 3.3. Have your program track the wavefunction to $|x| = 2L$. How closely do your numerically-determined energies match the results of Example 3.2?

10.8 It was remarked in Sect. 10.2 that the magnitude of the input value for $(d\psi/dx)_{x_o}$ affects only the amplitude of the resulting ψ, but not its shape or the energy eigenvalues. Prove this. HINT: Assume that your guess for $(d\psi/dx)_{x_o}$ is replaced by a different one, say $k(d\psi/dx)_{x_o}$, where k is a pure number.

10.9 Use your program from Problem 10.4 to investigate the linear radial potential $V(r) = kr$. Set $k = 1\,\mathrm{eV}/\text{Å}$. How do the three lowest energy levels for $\ell = 0$ compare with those of Example 10.1? See Problem 7.8 for a limited analytic treatment of this potential.

10.10 Use your program from Problem 10.4 to investigate the exponential radial potential $V(r) = Ae^{kr}$. Set $A = 5$ eV and $k = 0.2\,\text{Å}^{-1}$. What are the lowest three energy levels for $\ell = 0$ and 1?

10.11 Use your program from Problem 10.4 to determine the lowest three energy levels for $\ell = 0$ and 1 for the Wood–Saxon potential,

$$V(r) = -\frac{A}{1 + e^{(r-b)/d}}.$$

Take $A = 100$ eV, $b = 2$ Å, and $d = 0.03$ Å.

10.12 Problem 7.24 dealt with the case of a mass μ trapped in a cylindrical potential well. Now suppose that you have a more general potential function $V(\rho)$ but with $V = \infty$ for both $z \leq 0$ and $z \geq L$ as in that problem. Show that in this case, Schrödinger's equation for the radial function $R(\rho)$ has the form (atomic units)

$$\frac{d^2 R}{d\rho^2} = -\frac{1}{\rho}\frac{dR}{d\rho} - \left\{ -0.26246\,\mu\,[E - V(\rho)] - \frac{n_z^2 \pi^2}{L^2} - \frac{m^2}{\rho^2} \right\} R,$$

where n_z and m are the same quantum numbers that arose in Problem 7.24. If $V(\rho) = 0$, this is Bessel's equation with m playing the role of the order of the equation. Prepare a program or spreadsheet to numerically integrate this equation for any desired $V(\rho)$. Check your program for the case of an electron trapped in the infinite cylindrical potential of height 50 Å and radius 10 Å of Problem 7.24. Use your program to compute the energy eigenvalues (in eV) for the three lowest states corresponding to $m = 0$, 1, and 2 when $n_z = 1$. How do your results compare with those determined from the analytic solution of Problem 7.24? HINTS: The results given in Appendix B are based on an integration starting at $\rho = 0.002$ (to avoid divergences) with $\Delta\rho = 0.002$ and $(dR/d\rho)_{0.002} = 0.1$. Bessel functions have $R(0) = 1$ for $m = 0$ and $R(0) = 0$ for $m \geq 1$. For $m = 0$, 1, and 2, the three lowest zeros of the Bessel functions are respectively (2.4048, 5.5201, 8.6537), (3.8317, 7.0156, 10.1735) and (5.1356, 8.4172, 11.6198). To incorporate the boundary condition for the edge of the well, you need only find energies that yield $R(10\,\text{Å}) = 0$; your program may go beyond this limit in ρ, but this is of no concern. (Why?)

10.13 Use the program you developed in the preceding problem to determine the energy states for $m = 0$, 1, and 2 (in eV) for an electron trapped in a (cylindrical) potential defined by $V(\rho) = A\rho^2$, $(0 \leq \rho \leq \infty)$ with $V = \infty$ for $z \leq 0$ and $z \geq L$. Take $A = 0.5\,\text{eV}/\text{Å}^2$, $L = 50$ Å, and $n_z = 0$. The results given in Appendix B are based on an integration starting at $\rho = 0.002$ (to avoid divergences) with $\Delta\rho = 0.002$ and $(dR/d\rho)_{0.002} = 1$. Note that since $V(\rho) \neq 0$, the solutions to this problem are *not* Bessel functions.

10.14 The *Lennard–Jones 6-12 potential* is a semi-empirical potential used in the study of intermolecular forces. This has the form

$$V(x) = A\left[\left(\frac{\alpha}{x}\right)^{12} - 2\left(\frac{\alpha}{x}\right)^6\right] \quad (0 \leq x \leq \infty),$$

with $V(x) = \infty$ for $x \leq 0$. Plot this potential for $A = 100$ eV and $\alpha = 2$ Å. For an electron moving in this potential with these values of A and α, how many bound states does it support and what are their energies?

References

1. M. Abramowitz, I.A. Stegun, *Handbook of Mathematical Functions* (Dover, New York, 1965), Table 10.13
2. S. Flügge, *Practical Quantum Mechanics* (Springer-Verlag, Berlin, 1974)

Chapter 11
A Few Results from Time-Dependent Quantum Mechanics: Transition Rates and Probabilities

Summary This brief chapter examines three issues in time-dependent quantum mechanics: (1) How to use the wavefunctions for a system to estimate the rate at which the system will transit between given quantum states; (2) How rules which dictate what transitions are in principle possible can be derived; and (3) How to estimate the probability that a system will be in a given quantum state following a transition. The intent here is not rigorous analyses, but rather to give readers a taste of what can be found in more advanced work.

It was remarked in Chap. 2 that while the primary emphasis in this book is on solutions of the time-independent Schrödinger equation, understanding time-dependent problems is important as it is via interactions *between* stationary states that the characteristics of quantum systems are experimentally deduced. The purpose of this chapter is to give readers a taste of some of the results of time-dependent quantum theory. Specifically, we will look at three very practical, related issues: (i) How frequently we can expect a system to transit between two given quantum states, (ii) Whether or not there are any restrictions on what transitions can in principle occur, and, (iii) If a system currently in state m undergoes a sudden change of some sort (perhaps an alpha decay or the like), what is the probability of it ending up in state n of the altered system? Deriving expressions for predicting such things requires a knowledge of time-dependent solutions to Schrödinger's equation, time-dependent perturbation theory, and radiation theory. Consequently, we will content ourselves with quoting some useful results and illustrating their application with some simple examples. Readers interested in more detailed treatments are urged to consult [1, 2].

In discussing transition phenomena, it is important to bear in mind that the results have meaning only in a statistical sense. A conclusion to the effect that "the number of transitions from state m to state n for some atomic species Z is 1000 per second" means, for practical purposes, that if we start with a large number of atoms of type Z all initially in state m, then, on average, we will find 1000 in state n one second later. A given experiment might reveal 1027 or 945 atoms in state n after one second, but the average over a great many experiments will be 1000. Quantum mechanics provides exact answers to some questions but probabilities to most.

11.1   Transition Frequencies

Suppose that you are given a large number of identical systems, all initially in the same quantum state m; these might be, for example, electrons trapped in infinite rectangular wells. How many of these systems will, on average, have spontaneously transited to a lower-energy state n after one second has elapsed? This quantity, known as the *Einstein coefficient of spontaneous emission*, is designated as A_{mn} and is given approximately by

$$A_{mn} = \frac{16\,\pi^3 q^2}{3\,\varepsilon_o c^3\, h^4}\, (E_m - E_n)^3 \left| \langle \psi_n^* \,|\, x \,|\, \psi_m \rangle \right|^2 \qquad (m > n). \qquad (11.1)$$

(E_m, ψ_m) and (E_n, ψ_n) are the energies and wavefunctions corresponding to states m and n, respectively, and q is the electrical charge of the system constituent (usually an electron) undergoing the transition. This expression results from an analysis of the effects of perturbing radiation incident on a system initially in state m; the charge and position factors arise from the response of the system's electric dipole moment to the perturbing field (see Sects. 39 and 40 of [1]). The units of (11.1) are reciprocal seconds, that is, transitions per second. The reciprocal of A_{mn} thus gives the *lifetime* of state m for transitions to state n. The "x" in the Dirac braket implies a one-dimensional problem; in three-dimensions, one sums the absolute squares of three such integrals, one for each of x, y, and z, integrated over the volume of space appropriate to the system's wavefunction; see Example 11.2. Analogous equations concerning induced emission and absorption can also be derived.

Example 11.1   A particle of mass equal to that of a hydrogen atom and possessing charge e (= 1 electron charge) is moving in the harmonic potential $V(x) = kx^2/2$. Determine the transition frequency between states n and $n-1$. The harmonic oscillator wavefunctions are given in Chap. 5. We have

$$\langle \psi_n^* \,|\, x \,|\, \psi_{n-1} \rangle = A_n A_{n-1} \int_{-\infty}^{\infty} x\, e^{-\alpha^2 x^2}\, H_n(\alpha x)\, H_{n-1}(\alpha x)\, dx,$$

where A_n and A_{n-1} are normalization constants given in (5.39); do not confuse them with the Einstein coefficient $A_{n,n-1}$. On making the usual change of variable $\xi = \alpha x$ with $\alpha = (mk/\hbar^2)^{1/4}$, we have

$$\langle \psi_n^* \,|\, x \,|\, \psi_{n-1} \rangle = \frac{A_n A_{n-1}}{\alpha^2} \int_{-\infty}^{\infty} \xi\, e^{-\xi^2}\, H_n(\xi)\, H_{n-1}(\xi)\, d\xi.$$

This integral was discussed in Example 9.5. In general, Hermite polynomials satisfy

$$\int_{-\infty}^{\infty} \xi\, e^{-\xi^2}\, H_n(\xi)\, H_m(\xi)\, d\xi = \sqrt{\pi}\, [2^{n-1}\, n!\, \delta_m^{n-1} + 2^n\, (n+1)!\, \delta_m^{n+1}].$$

We have $m = n - 1$, so only the first Kronecker delta in this identity survives, leaving

$$\langle \psi_n^* | x | \psi_{n-1} \rangle = \frac{2^{n-1}\, n!\, \sqrt{\pi}\, A_n\, A_{n-1}}{\alpha^2}.$$

On substituting for the normalization factors and α, this reduces to $\langle \psi_n^* | x | \psi_{n-1} \rangle = \sqrt{n}/(\sqrt{2}\alpha)$. Then putting $E_n - E_{n-1} = \hbar\sqrt{k/m}$ gives, after some algebra,

$$A_{n,\,n-1} = \frac{e^2\, k}{6\pi\, \varepsilon_o\, c^3\, m^2}\, n.$$

From Problem 5.8 we can put $k \sim 500$ N/m; setting $n = 2$ and using the mass of a hydrogen atom gives

$$A_{n,\,n-1} \sim 2.05 \times 10^3 \, \text{s}^{-1}.$$

The number of transitions per second is small compared to the frequency of the radiation emitted during a transition, about 10^{15} Hz.

Example 11.2 Determine the transition frequency from the (2, 1, 0) to the (1, 0, 0) state of hydrogen.

The wavefunctions are

$$\psi_{210} = \frac{1}{4\sqrt{2\pi}\, a_o^{5/2}}\, r\, e^{-r/2a_o} \cos\theta$$

and

$$\psi_{100} = \frac{1}{\sqrt{\pi}\, a_o^{3/2}}\, e^{-r/a_o}.$$

To carry out this calculation we require the transformation equations between Cartesian and spherical coordinates: $(x, y, z) = r(\sin\theta \cos\phi, \sin\theta \sin\phi, \cos\theta)$.

For the x-contribution,

$$\langle \psi_{210} \mid x \mid \psi_{100} \rangle = \frac{1}{4\sqrt{2}\,\pi\,a_o^4} \int\limits_0^\infty \int\limits_0^\pi \int\limits_0^{2\pi} r\,e^{-3r/2a_o} \cos\theta\,(r\sin\theta\cos\phi)r^2\,\sin\theta\,d\phi\,d\theta\,dr$$

$$= \frac{1}{4\sqrt{2}\,\pi\,a_o^4} \left\{ \int\limits_0^\infty r^4 e^{-3r/2a_o}\,dr \right\} \left\{ \int\limits_0^\pi \cos\theta\sin^2\theta\,d\theta \right\} \left\{ \int\limits_0^{2\pi} \cos\phi\,d\phi \right\}.$$

These integrals are of the sort encountered in Chap. 7. The integral over ϕ vanishes, rendering $\langle \psi_{210} \mid x \mid \psi_{100} \rangle = 0$. The y-contribution involves an integral of $\sin\phi$ between zero and 2π and is likewise zero. For the z-contribution, we get

$$\langle \psi_{210} \mid z \mid \psi_{100} \rangle = \frac{1}{4\sqrt{2}\,\pi\,a_o^4} \left\{ \int\limits_0^\infty r^4 e^{-3r/2a_o}\,dr \right\} \left\{ \int\limits_0^\pi \cos^2\theta\,\sin\theta\,d\theta \right\} \left\{ \int\limits_0^{2\pi} d\phi \right\}$$

$$= \frac{1}{4\sqrt{2}\,\pi\,a_o^4} \left(\frac{256}{81}\,a_o^5 \right) \left(\frac{2}{3} \right) (2\pi) = \left(\frac{256}{243\sqrt{2}} \right) a_o.$$

From the usual expression for Bohr energy levels, $E_{210} - E_{100} = 3m_e e^4/32\varepsilon_o^2 h^2$. Inserting the expression for the Bohr radius gives (check the algebra!)

$$A_{210,100} = \frac{48}{19{,}683}\,\frac{\pi\,m_e\,e^{10}}{\varepsilon_o^5\,c^3\,h^6}.$$

Substituting the numbers yields a transition rate of 6.27×10^8 per second. The experimental value of this quantity is somewhat less, 4.70×10^8 per second [3]. The difference is attributable to the fact that (11.1) is an approximation; transition rates are also influenced by spin, relativistic, and magnetic effects.

In practice, transition rates are deduced by measuring the energy associated with them: knowing the stationary energy states for some system and hence $E_n - E_m$ (from solving the time-independent Schrödinger equation), a measured rate of energy emission can be transformed into a transition rate.

11.2   Transition Rules

Depending on the properties of the wavefunctions involved, it is clear that the integral in (11.1) above will vanish in some cases. In this event, the corresponding transition is said to be *forbidden*. For any quantum system, it is possible to derive a set of *transition rules* or *selection rules* which indicate in what way the quantum number(s) of the system must change during a transition in order that it not be forbidden.

A convenient example is provided by harmonic-oscillator transitions. The integral appearing in Example 11.1 was of the form

$$\int_{-\infty}^{\infty} \xi\, e^{-\xi^2}\, H_n(\xi)\, H_m(\xi)\, d\xi = \sqrt{\pi}\, [2^{n-1}\, n!\, \delta_m^{n-1} + 2^n\, (n+1)!\, \delta_m^{n+1}].$$

Since this integral is non-zero only when $m = n \pm 1$, it follows that the quantum harmonic-oscillator is restricted to transitions in which n increases or decreases by 1, that is, the system can transit only to immediately adjacent states. The transition rule for this case is then written as $\Delta n = \pm 1$; if a downward transition is possible, so is its upward counterpart. In the case of the infinite one-dimensional well (Problem 11.2), the alternating parity of the wavefunctions restricts permissible transitions to those which involve a change of parity. For the hydrogen atom, the properties of the Laguerre polynomials and spherical harmonics are such that only transitions with $\Delta \ell = \pm 1$ and $\Delta m = \pm 1$ or 0 are permissible; there is no restriction on any change (or not) of the principal quantum number n. A change in ℓ, incidentally, implies a change in the system's angular momentum. If L is to be conserved, where does angular momentum come from or disappear to? The answer is that photons possess intrinsic angular momentum of amount $\hbar$: by absorbing or giving up a photon, an electron can make its transition while L is conserved overall.

The transition frequencies computed via (11.1) correspond to what are known as *electric-dipole transitions*. Other transition mechanisms (quadrupole, octopole, ...) between states are possible but are usually much less frequent than their dipole counterparts. However, it is possible for transitions forbidden to dipole radiation to be permissible by other mechanisms.

Equation (11.1), when evaluated for different combinations of m and n, yields predictions of the *relative intensities* of spectral lines by taking ratios of the results. In the case of multi-dimensional systems, predictions can also be made as to the angular distribution and polarization of the emitted radiation. Predictions along these lines are in accord with the observed characteristics of atomic and molecular spectra. This success of wave mechanics represented a significant improvement in our understanding of spectra that was not available from the Bohr theory.

11.3 The Sudden Approximation

Imagine a quantum harmonic oscillator in some state, say ψ_n. As a result of some process (perhaps a chemical reaction), the force constant k of the system is suddenly reduced to one-half of its original value. What is the probability of finding the "new" system in the "corresponding" state ψ_n or in some other state, ψ_m? "Corresponding" here is meant in the sense of numbering the energy levels of both from the lowest on up. Questions of this sort can be examined with the so-called sudden approximation.

The mathematical statement of this approximation is that if the system was in state n before the change, then the probability of finding it to be in state j after the change is given by

$$P_{n \to j} = \left| \int \psi_j^{new}(x) \, \psi_n^{old}(x) \, dx \right|^2. \tag{11.2}$$

The ψ's are presumed to be normalized and to apply over the same domain. In a three-dimensional problem, dx is replaced by the volume element dV appropriate to the coordinate system in use.

We can give a semi-intuitive justification of (11.2). Short of performing some sort of experiment, we have no idea in which of a number of possible states ψ_j the altered system will find itself after the change. However, we can imagine expressing the new wavefunction as a linear sum over all of the old system wavefunctions as was done in Sect. 9.2. Intuitively, the greater the coefficient multiplying any "former" state ψ_n, the greater will be the probability of finding the system in the "new" n'th state as well. By a development similar to that presented in Sect. 9.2, it is possible to show that the probability of the new system being in state j is given by the square of the absolute value of the expansion coefficient a_j in (9.12). If the ψ's are normalized, the sum over (11.2) above for all possible values of j is unity, in analogy to (9.16).

Example 11.3 The force constant of a harmonic oscillator in the ground state suddenly decreases to one-half of its original value. What is the probability of the system remaining in the ground (lowest energy) state?
We have

$$\psi_0^{old}(x) = \frac{\sqrt{\alpha}}{\pi^{1/4}} e^{-\alpha^2 x^2/2}, \quad \alpha^4 = mk/\hbar^2.$$

Halving the force constant causes α^4 to decrease by a factor of 2:

$$\alpha_{new}^4 = \alpha_{old}^4/2.$$

Setting $\alpha = \alpha_{old}/2^{1/4}$ gives the new wavefunction as

$$\psi_0^{new}(x) = \frac{\sqrt{\alpha}}{\pi^{1/4} 2^{1/8}} e^{-\alpha^2 x^2/2\sqrt{2}}.$$

Equation (11.2) then gives

$$\sqrt{P} = \int_{-\infty}^{\infty} \psi_0^{new}(x)\psi_0^{old}(x)\, dx = \frac{\alpha}{\sqrt{\pi}2^{1/8}} \int_{-\infty}^{\infty} \exp\left[-\left(\frac{1}{2} + \frac{1}{2\sqrt{2}}\right)\alpha^2 x^2\right]dx$$

$$= \frac{2\alpha}{\sqrt{\pi}\,2^{1/8}} \int_{0}^{\infty} \exp\left(-a^2 x^2\right)dx,$$

where

$$a^2 = \frac{(1 + \sqrt{2})}{2\sqrt{2}}\alpha^2.$$

The integral evaluates to $\sqrt{\pi}/2a$, hence

$$P = \frac{4\alpha^2}{\pi\,2^{1/4}}\frac{\pi}{4a^2} = \frac{\alpha^2}{2^{1/4}}\frac{2\sqrt{2}}{(1+\sqrt{2})\alpha^2} = \frac{2^{5/4}}{(1+\sqrt{2})} = 0.985.$$

The probability is overwhelming that the system will remain in the ground state. Note that all of the physical constants canceled out in this calculation: the result must be a pure number.

Summary

The transition frequency A_{mn} (transitions per second) between states m and n for a transiting particle of charge q is given by

$$A_{mn} = \frac{16\pi^3 q^2}{3\varepsilon_o c^3 h^4}(E_m - E_n)^3\left|\langle\psi_n^*|x|\psi_m\rangle\right|^2, \qquad (m > n).$$

Transition is said to be forbidden between states m and n if A_{mn} is zero.

The sudden approximation gives a means of computing the probability $P(n \rightarrow j)$ that a system initially in state n will be in state j after a sudden change in the parameters of the system:

$$P_{n \rightarrow j} = \left|\int \psi_j^{new}(x)\,\psi_n^{old}(x)\,dx\right|^2.$$

Problems

11.1(E) Verify that the units of (11.1) are reciprocal seconds.

11.2(I) Determine the transition frequency for a particle of mass m and charge q between any two states n and j ($n > j$) of the infinite one-dimensional potential well.

11.3(E) Verify by direct calculation that transition between the $(3, 2, 0)$ and $(1, 0, 0)$ states of hydrogen is forbidden.

11.4(I) A hydrogen-like atom with a nuclear charge of $+Ze$ is initially in the $(1, 0, 0)$ state. As a result of an inverse β-decay, $Z \rightarrow Z - 1$. Determine the probability that the new species remains in the $(1, 0, 0)$ state. Evaluate your answer numerically for $Z = 2$.

11.5(I) A particle of mass m is in state $\psi_n(x)$ of an infinite rectangular well of width L ($0 \leq x \leq L$). The right wall of the well suddenly expands to $x = KL$ with $K > 1$, that is, the well now occupies $0 \leq x \leq KL$. Determine the probability that the particle will be in any state $\psi_k(x)$ after the expansion. (Take the original wavefunction to be zero between $x = L$ and $x = KL$.) Following the expansion, in what state is the particle most likely to be found?

References

1. L. Pauling, E.B. Wilson, *Introduction to Quantum Mechanics with Applications to Chemistry* (Dover Publications, New York, 1985), Sect. 40
2. A.P. French, E.F. Taylor, *An Introduction to Quantum Physics* (W. W. Norton, New York, 1978), see Chaps. 8 and 14
3. H.L. Anderson (ed.), *A Physicists Desk Reference: The Second Edition of Physics Vade Mecum* (American Institute of Physics, New York, 1989), p. 99

Appendix A
Miscellaneous Derivations

A.1 Heisenberg's Uncertainty Principle

The proof of the uncertainty principle given here is adopted from Sect. 12 of Leonard Schiff, *Quantum Mechanics*: 3rd edition (McGraw-Hill, New York, 1968). This proof assumes one-dimensional wavefunctions, utilizes the time-dependent Schrödinger equation of Chap. 2, assumes that wavefunctions vanish asymptotically (strictly, $\psi \rightarrow 0$ at the limits of its domain), and requires an integral identity from functional analysis.

We begin by reiterating the definitions of Δx and Δp from Sect. 4.3:

$$\Delta x = \sqrt{\langle x^2 \rangle - \langle x \rangle^2} \tag{A.1}$$

and

$$\Delta p = \sqrt{\langle p^2 \rangle - \langle p \rangle^2} = \sqrt{\langle p^2 \rangle}. \tag{A.2}$$

In writing Δp we have made use of the proof in Sect. 4.2 that $\langle p \rangle = 0$ for any well-behaved wavefunction.

It proves easier to work with the squares of these quantities,

$$\Delta x^2 = \langle x^2 \rangle - \langle x \rangle^2 \tag{A.3}$$

and

$$\Delta p^2 = \langle p^2 \rangle. \tag{A.4}$$

We begin with the Δx part of the Uncertainty Principle. It is helpful to introduce a new operator α defined by

$$\alpha = x - \langle x \rangle. \tag{A.5}$$

© The Editor(s) (if applicable) and The Author(s), under exclusive license to Springer Nature Switzerland AG 2022
B. C. Reed, *Quantum Mechanics*, https://doi.org/10.1007/978-3-031-14020-4

Now consider the integral

$$X = \int (\alpha\psi)(\alpha^*\psi^*)\,dx, \tag{A.6}$$

where an asterisk denotes complex conjugation and where the limits of integration are those of the domain of ψ. Invoking the definition of α gives

$$X = \int (x\psi - \langle x\rangle\,\psi)\left(x^*\psi^* - \langle x\rangle\,\psi^*\right)dx$$
$$= \int xx^*\psi\psi^*dx - \langle x\rangle\int x\psi\psi^*dx - \langle x\rangle\int x^*\psi\psi^*dx + \langle x\rangle^2\int\psi\psi^*dx.$$

Since $x^* = x$, the first integral here is just $\langle x^2\rangle$, and the middle two integrals are both equal to $\langle x\rangle$. If ψ is normalized, the last integral will be equal to unity. Hence

$$X = \langle x^2\rangle - 2\langle x\rangle^2 + \langle x\rangle^2 = \langle x^2\rangle - \langle x\rangle^2 = \Delta x^2. \tag{A.7}$$

We will return to this result later.

Now we consider Δp^2. In analogy to (A.6), consider the quantity

$$P = \int (p\psi)(p^*\psi^*)\,dx = \int \left(-\iota\hbar\frac{d\psi}{dx}\right)\left(+\iota\hbar\frac{d\psi^*}{dx}\right)dx$$
$$= \hbar^2\int (\psi)'(\psi^*)'dx, \tag{A.8}$$

where p is the momentum operator and a prime denotes differentiation with respect to x. Now consider the second derivative

$$\frac{d^2}{dx^2}(\psi^*\psi) = 2(\psi')(\psi^*)' + (\psi^*)''(\psi) + (\psi^*)(\psi''). \tag{A.9}$$

But for a factor of 2, the first term on the right side of this expression is the integrand appearing in (A.8). Hence we can write (A.8) as

$$P = \frac{\hbar^2}{2}\int (\psi^*\psi)''dx - \frac{\hbar^2}{2}\int (\psi^*)''(\psi)dx - \frac{\hbar^2}{2}\int (\psi^*)(\psi'')dx. \tag{A.10}$$

The first integral in this expression vanishes as follows:

$$\int (\psi^*\psi)''\,dx = \int \frac{d}{dx}\left[\frac{d}{dx}(\psi^*\psi)\right]dx = \int d\left[\frac{d}{dx}(\psi^*\psi)\right]$$
$$= \left[(\psi^*)'(\psi) + (\psi^*)(\psi')\right]_{\text{limits}} = 0.$$

What remains of (A.10) is then

$$P = -\frac{\hbar^2}{2} \int (\psi^*)''(\psi)dx - \frac{\hbar^2}{2} \int (\psi^*)(\psi'')dx. \qquad \text{(A.11)}$$

Now, Δp^2 is given by (A.4) as

$$\Delta p^2 = \int (\psi^*) (p^2\psi) \, dx = -\hbar^2 \int (\psi^*) (\psi'') \, dx. \qquad \text{(A.12)}$$

Compare (A.11) and (A.12). The second integral in (A.11) is identical to (A.12) but for a factor of 2. Also, the first integral in (A.11) is just the complex conjugate of the second one. If we make the assumption (proof below) that conjugation leaves (A.12) unchanged, that is, that we can write

$$\Delta p^2 = -\hbar^2 \int (\psi^*)'' (\psi) \, dx, \qquad \text{(A.13)}$$

then (A.11) reduces to

$$P \equiv \int (p\psi)(p^*\psi^*)dx \equiv \Delta p^2. \qquad \text{(A.14)}$$

From (A.6), (A.7), and (A.14) we now have

$$\Delta x^2 \Delta p^2 = XP = \left\{ \int (\alpha\psi)(\alpha^*\psi^*)dx \right\} \left\{ \int (p\psi)(p^*\psi^*)dx \right\}. \qquad \text{(A.15)}$$

The reason for going through this manipulation is that (A.15) is of the form of a theorem from functional analysis known as Schwarz's inequality. This theorem states that for two functions, $f(x)$ and $g(x)$, both possibly complex and which run over the same domain, they satisfy the inequality

$$\left\{ \int f^*(x)f(x)dx \right\} \left\{ \int g^*(x)g(x)dx \right\} \geq \left| \left\{ \int f^*(x)g(x) \, dx \right\} \right|^2, \qquad \text{(A.16)}$$

where the limits of integration are those of the domain of $f(x)$ and $g(x)$. A proof of this inequality can be found in any good text on mathematical physics. Applying this to (A.15) gives

$$\Delta x^2 \Delta p^2 \geq \left| \int (\alpha^*\psi^*)(p\,\psi)dx \right|^2. \qquad \text{(A.17)}$$

Since α is a real operator, we can write $\alpha^* = \alpha$. Evaluating the integral gives

$$
\begin{aligned}
\int (\alpha * \psi*)(p\psi)\,dx &= \int \left[(x\psi^*) - \langle x \rangle \, \psi^*\right](p\psi)\,dx \\
&= \left\{ \int \psi^* x \,(p\psi)\,dx - \langle x \rangle \int \psi^*\,(p\psi)\,dx \right\} \\
&= \left\{ \langle xp \rangle - \langle x \rangle \langle p \rangle \right\}.
\end{aligned}
\tag{A.18}
$$

Because $\langle p \rangle = 0$, the second term within the curly brackets vanishes.
Now consider the integral for $\langle xp \rangle$ in more detail:

$$
\langle xp \rangle = \int \psi^* x \,(p\psi)\,dx = -\iota\hbar \int \psi^* x \left(\frac{d\psi}{dx}\right)\,dx = -\iota\hbar \int \psi^* x\,d\psi.
$$

Integrating by parts gives

$$
\langle xp \rangle = -\iota\hbar \left\{ [\psi^* x\psi]_{\lim\,its} - \int \psi\,d(\psi^* x) \right\}.
$$

If ψ vanishes at its limits, then the first term within the curly brackets will be zero.
Expanding out the integral gives

$$
\langle xp \rangle = \iota\hbar \left\{ \int \psi \, \psi^*\,dx + \int \psi x\,d\psi^* \right\}.
\tag{A.19}
$$

The first integral on the right side of this expression will be equal to unity if ψ is normalized. In the second integral, convert $d\psi^*$ back to a derivative: $d\psi^* = (d\psi^*/dx)dx$, and then put the derivative in terms of the momentum operator $d\psi^* = -(1/\iota\hbar)(p\psi^*)dx$. This gives

$$
\langle xp \rangle = \iota\hbar \left\{ 1 - \frac{1}{\iota\hbar} \int \psi x(p\psi^*)dx \right\} = \iota\hbar - \langle xp \rangle,
$$

where we have invoked the assumption that the order of ψ and ψ^* in an expectation-value integral is irrelevant. Hence we have

$$
\langle xp \rangle = \frac{\iota\hbar}{2}.
\tag{A.20}
$$

Equations (A.17), (A.18) and (A.20) reduce to

$$
\Delta x \Delta p \geq \frac{\hbar}{2},
$$

the formal statement of the uncertainty principle.

One step remains to complete the proof: that it is legitimate to write [see (A.12) and (A.13)]

$$\int (\psi^*)'' \, (\psi) \, dx = \int (\psi^*) \, (\psi'') \, dx. \tag{A.21}$$

The proof begins with the normalization condition, which we write in Dirac notation:

$$\langle \psi^* | \psi \rangle = 1. \tag{A.22}$$

Now, even if ψ should be time-dependent, this condition must always be satisfied, that is, the time-derivative of (A.22) must vanish:

$$\frac{\partial}{\partial t} \langle \psi^* | \psi \rangle = \left\langle \frac{\partial \psi^*}{\partial t} \Big| \psi \right\rangle + \left\langle \psi^* \Big| \frac{\partial \psi}{\partial t} \right\rangle = 0. \tag{A.23}$$

To deal with the time derivatives, invoke the time-dependent Schrödinger equation and its complex conjugate (2.28)

$$\left(\frac{\partial \psi}{\partial t} \right) = \frac{1}{\iota \hbar} \left[-\frac{\hbar^2}{2m} \frac{\partial^2 \psi}{\partial x^2} + V \psi \right] \tag{A.24}$$

and

$$\left(\frac{\partial \psi^*}{\partial t} \right) = -\frac{1}{\iota \hbar} \left[-\frac{\hbar^2}{2m} \frac{\partial^2 \psi^*}{\partial x^2} + V \psi^* \right]. \tag{A.25}$$

Substituting these into (A.23), canceling a factor of $(\iota \hbar)$, and expanding out the Dirac brackets gives

$$\frac{\partial}{\partial t} \langle \psi^* | \psi \rangle = \frac{\hbar^2}{2m} \langle (\psi^*)'' | \psi \rangle - \langle V \psi^* | \psi \rangle - \frac{\hbar^2}{2m} \langle \psi^* | \psi'' \rangle + \langle \psi^* | V \psi \rangle = 0. \tag{A.26}$$

Since the potential V is purely multiplicative, the two terms involving V cancel each other. What remains of (A.26) then gives

$$\langle (\psi^*)'' | \psi \rangle = \langle \psi^* | \psi'' \rangle, \tag{A.27}$$

exactly as presumed in (A.21).

A.2 Normalization of Hermite Polynomials

We saw in Chap. 5 that the solution to Schrödinger's equation for the harmonic oscillator potential is

$$\psi_n(x) = A_n H_n(\xi) e^{-\xi^2/2} = A_n H_n(\alpha x) e^{-\alpha^2 x^2/2} \tag{A.28}$$

where A_n is a normalization constant,

$$A_n = \sqrt{\frac{\alpha}{\sqrt{\pi} 2^n n!}}, \tag{A.29}$$

where α is given by

$$\alpha = \left(mk/\hbar^2\right)^{1/4}, \tag{A.30}$$

and where $H_n(\xi)$ is a polynomial in even or odd powers of ξ:

$$H_n(\xi) = \sum_{\substack{j=0,2,4,\dots \\ \text{or } j=1,3,5,\dots}}^{n} a_j \xi^j, \tag{A.31}$$

The expansion coefficients a_j satisfy the recursion relation given in (5.25):

$$\frac{a_{k+2}}{a_k} = \frac{(2k+1-\lambda)}{(k+1)(k+2)} = \frac{2(k-n)}{(k+1)(k+2)}. \tag{A.32}$$

In using n as the upper limit of the index of summation in (A.31), we have in mind an oscillator in the n'th possible quantum state, that is, of energy $E_n = (n+1/2)\hbar\omega$, which gives λ in (A.32) as $\lambda = 2n+1$ from (5.34).

The purpose of this section is to show how the normalization constant in (A.29) arises. This proof is rather lengthy, and is most easily treated in three steps:

(i) Equations (A.31) and (A.32) are first transformed into an equivalent form where the index of summation proceeds in unit steps, as opposed to steps of two.
(ii) The result of step (i) is used to transform $H_n(\xi)$ into a more compact form known as a generating function.
(iii) The generating function of step (ii) is then used to simplify the normalization integral for $\psi_n(\xi)$.

A bonus of this approach is that the generating function can be used to prove some useful identities and recurrence relationships for Hermite polynomials.

To begin step (i), you should be able to convince yourself that an equivalent way of writing (A.31) is

$$H_n(\xi) = \sum_{\substack{j=0,2,4,\dots \\ \text{or } j=1,3,5,\dots}}^{n} a_j \xi^j = \sum_{k=0,1,2,\dots}^{[n/2]} b_k \xi^{n-2k}, \tag{A.33}$$

where $[n/2]$ designates the greatest integer less than or equal to $n/2$. If you are in doubt about this, choose a value of n and expand both sums; you will see that the powers of ξ match one-for-one. The "b" summation here has coefficients that are numbered consecutively: $b_0, b_1, b_2, \ldots b_n$; the value of this is that we can account for either an even or odd-parity solution with one expression once the value of n has been chosen. Notice that the sum with the b coefficients runs from the highest power of ξ down to the lowest, whereas the a sum goes from lowest to highest.

We need to know how the a and b coefficients in (A.33) are related. Matching terms in the same power of ξ shows that $b_0 = a_n$, $b_1 = a_{n-2}$, $b_2 = a_{n-4}$, and so forth; that is, in general, $b_j = a_{n-2j}$. This means that two successive b coefficients are related as

$$\frac{b_j}{b_{j-1}} = \frac{a_{n-2j}}{a_{n-2j+2}} \equiv \frac{a_m}{a_{m+2}}, \tag{A.34}$$

where we have set $m = n - 2j$. From (A.32) this becomes

$$\frac{b_j}{b_{j-1}} \equiv \frac{a_m}{a_{m+2}} = \frac{(m+1)\,(m+2)}{2\,(m-n)}, \tag{A.35}$$

or, with $m = n - 2j$,

$$\frac{b_j}{b_{j-1}} = -\frac{(n-2j+1)\,(n-2j+2)}{4j}, \tag{A.36}$$

which we cast as

$$b_j = (-1)\left[\frac{(n-2j+1)\,(n-2j+2)}{4j}\right] b_{j-1}. \tag{A.37}$$

For reasons that will become clear later, it is handy to develop an expression for b_j that involves only the indices j, n, and the first coefficient in the b-summation, b_0. To do this, abbreviate the large square bracket appearing in (A.37) as $f(j)$. Strictly, the square bracket is a function of n as well, but n is fixed for a given quantum state; our concern here is with j. We can then write (A.37) as

$$b_j = (-1)f(j)\,b_{j-1}. \tag{A.38}$$

By applying (A.37) recursively on replacing j with $j-1$, we can express b_{j-1} in terms of b_{j-2} in the form $b_{j-1} = (-1)f(j-1)\,b_{j-2}$. This puts (A.38) into the form

$$b_j = (-1)^2 f(j)\,f(j-1)\,b_{j-2}. \tag{A.39}$$

By continuing this logic we eventually arrive at b_0:

$$b_j = (-1)^j f(j)\,f(j-1)\,f(j-2)\ldots f(j=2)f(j=1)\,b_0. \tag{A.40}$$

This expression contains a product of j factors of the function f, each with a different argument. Using (A.37), we can write this out as

$$b_j = \frac{(-1)^j}{2^{2j}} \left[\frac{(n-2j+1)(n-2j+2)}{j} \right] \left[\frac{(n-2j+3)(n-2j+4)}{j-1} \right] \cdots$$
$$\cdots \left[\frac{(n-3)(n-2)}{2} \right] \left[\frac{(n-1)(n)}{1} \right] b_0. \tag{A.41}$$

The factor of 2^{2j} in the denominator here comes from the product of j factors of 4 in the denominator of (A.37); we write $4j$ in this way as it proves handy in going from (A.48) to (A.49) below. From the pattern of terms in (A.41), it is evident that the product of the denominators of the square brackets is just $j!$, while the numerators give a product of terms $n(n-1)(n-2) \ldots (n-2j+1)$:

$$b_j = \frac{(-1)^j}{2^{2j} j!} \left[n(n-1)(n-2) \ldots (n-2j+1) \right] b_0. \tag{A.42}$$

This expression, when back-substituted into (A.33) [changing the index k in (A.33) to j as we do so in order that we can use (A.42)] allows us to write the n-th order Hermite polynomial as

$$H_n(\xi) = b_0 \sum_{j=0}^{[n/2]} \frac{(-1)^j}{2^{2j} j!} \left[n(n-1)(n-2) \ldots (n-2j+1) \right] \xi^{n-2j}. \tag{A.43}$$

At this point, we could back-substitute (A.43) into (A.28), multiply by $e^{-\xi^2/2}$, square, and integrate over $-\infty \le \xi \le \infty$ to effect the desired normalization of $\psi_n(\xi)$. Evidently, the leading coefficient b_0 would then emerge as a function of n. However, because of the messy product of factors within the square bracket in (A.43), the proof of (A.29) does not go in this way. At this point, a manipulation is made that looks curious, although it is entirely legitimate. This is to arbitrarily set $b_0 = 2^n$ and write

$$H_n(\xi) = 2^n \sum_{j=0}^{[n/2]} \frac{(-1)^j}{2^{2j} j!} \left[n(n-1)(n-2) \ldots (n-2j+1) \right] \xi^{n-2j}. \tag{A.44}$$

The value of this choice will become clear presently. This choice for b_0 does not upset the eventual normalization of the $\psi_n(\xi)$. This is because it is the A_n of (A.29) that are determined by normalization, so this change in the definition of $H_n(\xi)$ is simply absorbed into (A.28) and hence into the A_n. Equation (A.44) is in fact the formal mathematical definition of a Hermite polynomial of order n.

With (A.44) in hand, we can now begin step (ii) of the normalization procedure. The first sub-step here involves showing how (A.44) can be considerably compacted. This involves utilizing the fact that the j'th derivative of a power can in general be written as

$$\frac{d^J}{dx^J}\left(\xi^N\right) = \left[\prod_{i=0}^{J-1}(N-i)\right]\xi^{N-J}. \tag{A.45}$$

For our purposes, apply this with $J = 2j$ and $N = n$, where j and n are the indices used in (A.44):

$$\frac{d^{2j}}{dx^{2j}}\left(\xi^n\right) = \left[\prod_{i=0}^{2j-1}(n-i)\right]\xi^{n-2j} = \left[n(n-1)(n-2)\ldots(n-2j+1)\right]\xi^{n-2j}. \tag{A.46}$$

Remarkably, this result is exactly what appears within the sum in (A.44); it was in anticipation of this that we obtained (A.42) for b_j in the first place. We can then write a general Hermite polynomial as

$$H_n(\xi) = 2^n \sum_{j=0}^{[n/2]} \frac{(-1)^j}{2^{2j}j!}\frac{d^{2j}}{d\xi^{2j}}\left(\xi^n\right), \tag{A.47}$$

which gives yet another definition of the Hermite polynomials.

We now make a subtle but crucial modification to this result. As it is written, the upper limit of j is $[n/2]$. We could just as well make this upper limit equal to infinity, because no derivatives will survive beyond $j = [n/2]$. *Be sure to understand this manipulation.* This gives

$$H_n(\xi) = 2^n \sum_{j=0}^{\infty} \frac{(-1)^j}{2^{2j}j!}\frac{d^{2j}}{d\xi^{2j}}\left(\xi^n\right). \tag{A.48}$$

This expression can be further compacted by incorporating the factors of 2^n and 2^{2j} into (respectively) the factor of ξ^n of which derivative is being taken and into the $d\xi^{2j}$ in the denominator of the derivative:

$$H_n(\xi) = \sum_{j=0}^{\infty} \frac{(-1)^j}{j!}\frac{d^{2j}}{d(2\xi)^{2j}}\left[(2\xi)^n\right]. \tag{A.49}$$

Be sure to understand the notation here: the derivative is being taken with respect to 2ξ. It was to achieve this manipulation that b_0 was set to 2^n in (A.44).

We now perform an even more curious-looking manipulation. Multiply both sides of (A.49) by $t^n/n!$, where t is a dummy variable ($t \neq 0$), and sum over all possible values of n from zero to infinity:

$$\sum_{n=0}^{\infty} H_n(\xi)\frac{t^n}{n!} = \sum_{n=0}^{\infty}\left\{\sum_{j=0}^{\infty}\frac{(-1)^j}{j!}\frac{d^{2j}}{d(2\xi)^{2j}}\left[(2\xi)^n\right]\right\}\frac{t^n}{n!}. \tag{A.50}$$

The factor of t^n on the right side of (A.50) can be brought inside the curly brackets and combined with the factor of $(2\xi)^n$ to give $(2\xi t)^n$. We can then factor the order of the summations without losing any terms:

$$\sum_{n=0}^{\infty} H_n\left(\xi\right) \frac{t^n}{n!} = \left\{ \sum_{j=0}^{\infty} \frac{(-1)^j}{j!} \frac{d^{2j}}{d\left(2\xi\right)^{2j}} \right\} \sum_{n=0}^{\infty} \frac{(2\xi t)^n}{n!}. \tag{A.51}$$

The second summation on the right side here is of exactly the form of that for an exponential series:

$$e^z = \sum_{n=0}^{\infty} \frac{z^n}{n!}, \tag{A.52}$$

with $z = 2\xi t$. This means that we can compact (A.51) to

$$\sum_{n=0}^{\infty} H_n\left(\xi\right) \frac{t^n}{n!} = \sum_{j=0}^{\infty} \frac{(-1)^j}{j!} \frac{d^{2j}}{d\left(2\xi\right)^{2j}} \left(e^{2\xi t} \right). \tag{A.53}$$

The derivative of $e^{2\xi t}$ with respect to 2ξ is $te^{2\xi t}$; taking the $2j$'th derivative will give $(t^{2j})e^{2\xi t}$, and (A.53) reduces to

$$\sum_{n=0}^{\infty} H_n\left(\xi\right) \frac{t^n}{n!} = \sum_{j=0}^{\infty} \frac{(-1)^j}{j!} t^{2j} e^{2\xi t}. \tag{A.54}$$

It was in order to get this exponential formulation that the upper index of summation was switched to infinity in going from (A.47) to (A.48); otherwise, we could not have invoked (A.52).

The factor of $e^{2\xi t}$ on right side of (A.54) can be brought out in front of the sum as it does not involve the summation index j. What remains inside the right-hand sum can then be further simplified using the exponential-series notation of (A.52):

$$\sum_{j=0}^{\infty} \frac{(-1)^j}{j!} t^{2j} e^{2\xi t} = e^{2\xi t} \sum_{j=0}^{\infty} \frac{(-1)^j}{j!} \left(t^2\right)^j = e^{2\xi t} \sum_{j=0}^{\infty} \frac{\left(-t^2\right)^j}{j!} = e^{2\xi t - t^2}, \tag{A.55}$$

that is,

$$\sum_{n=0}^{\infty} H_n\left(\xi\right) \frac{t^n}{n!} = e^{2\xi t - t^2}. \tag{A.56}$$

The function of the right side of (A.56) is known as the *generating function* for Hermite polynomials.

We are now ready to begin step (iii) of the normalization. If the wavefunctions $\psi_n(x)$ are to be properly orthonormalized, then we know from Chap. 5 and the orthogonally theorem of Sect. 4.6 that any two of them, say $\psi_n(x)$ and $\psi_m(x)$, must satisfy

$$A_n A_m \int_{-\infty}^{\infty} H_n(x) H_m(x) e^{-\alpha^2 x^2} dx = \delta_n^m, \qquad (A.57)$$

where δ_n^m is the Kronecker delta symbol introduced in Sect. 4.6. A proof of this expression will actually arise independently in what follows. Since our various manipulations throughout this section have been in terms of $\xi = \alpha x$, it is useful to put this integral in terms of ξ:

$$A_n A_m \int_{-\infty}^{\infty} H_n(x) H_m(x) e^{-\alpha^2 x^2} dx = \frac{A_n A_m}{\alpha} \int_{-\infty}^{\infty} H_n(\xi) H_m(\xi) e^{-\xi^2} d\xi = \delta_n^m, \quad (A.58)$$

which follows from $d\xi = \alpha \, dx$; α is given by (A.30). Now, (A.56) has the $H_n(\xi)$ polynomials within a sum from which there is no way to easily extract them for use in (A.58). We therefore evaluate (A.58) indirectly. This is done by writing the left side of (A.56) twice, once each for $H_n(\xi)$ and $H_m(\xi)$, and taking the product of the two expressions. In doing this we use the dummy variable s in place of t in the $H_m(\xi)$ version of (A.56) to help keep things straight:

$$\sum_{n=0}^{\infty} \sum_{m=0}^{\infty} \frac{t^n}{n!} \frac{s^m}{m!} H_n H_m = e^{2\xi t - t^2} e^{2\xi s - s^2}, \qquad (A.59)$$

where we suppress the dependence of the H's on ξ for brevity. Multiplying through this expression by $e^{-\xi^2}$ and integrating over ξ gives

$$\sum_{n=0}^{\infty} \sum_{m=0}^{\infty} \frac{t^n}{n!} \frac{s^m}{m!} \int_{-\infty}^{\infty} H_n H_m e^{-\xi^2} d\xi = \int_{-\infty}^{\infty} e^{2\xi t - t^2} e^{2\xi s - s^2} e^{-\xi^2} d\xi. \qquad (A.60)$$

The integral on the right side of this expression can be cast as

$$\int_{-\infty}^{\infty} e^{2\xi t - t^2} e^{2\xi s - s^2} e^{-\xi^2} d\xi = e^{2st} \int_{-\infty}^{\infty} e^{-(\xi - s - t)^2} d\xi. \qquad (A.61)$$

This integral has an exact solution. Setting $w = \xi - s - t$ and $d\xi = dw$ (s and t are arbitrary, but presumed to remain fixed once chosen) gives

$$e^{2st} \int_{-\infty}^{\infty} e^{-(\xi - s - t)^2} d\xi = e^{2st} \int_{-\infty}^{\infty} e^{-w^2} dw = \sqrt{\pi} e^{2st}. \qquad (A.62)$$

The e^{2st} term here can be put into a summation notation via (A.52) as

$$e^{2st} = \sum_{n=0}^{\infty} \frac{2^n t^n s^n}{n!}.$$
(A.63)

The selection of n as an index of summation here is inspired by its appearance in (A.60). We could equally well have chosen m; the same conclusions would eventually result. Gathering (A.61)–(A.63) into (A.60) gives

$$\sum_{n=0}^{\infty} \sum_{m=0}^{\infty} \frac{t^n}{n!} \frac{s^m}{m!} \int_{-\infty}^{\infty} H_n H_m e^{-\xi^2} d\xi = \sqrt{\pi} \sum_{n=0}^{\infty} \frac{2^n t^n s^n}{n!}.$$
(A.64)

Rearrange this expression by bringing the sum on the right side within that on the left side:

$$\sum_{n=0}^{\infty} \frac{t^n}{n!} \left\{ \sum_{m=0}^{\infty} \frac{s^m}{m!} \int_{-\infty}^{\infty} H_n H_m e^{-\xi^2} d\xi - \sqrt{\pi} 2^n s^n \right\} = 0.$$
(A.65)

Complicated as it looks, this expression can be regarded as a polynomial in powers of t. Since t and n (and s and m for that matter) are arbitrary, the only way this expression can be satisfied for all possible choices of t and n is if the expression within the curly brackets is always zero:

$$\sum_{m=0}^{\infty} \frac{s^m}{m!} \int_{-\infty}^{\infty} H_n H_m e^{-\xi^2} d\xi = \sqrt{\pi} 2^n s^n.$$
(A.66)

Expand the left side of this expression:

$$\frac{s^0}{0!} \int_{-\infty}^{\infty} H_n H_0 e^{-\xi^2} d\xi + \frac{s^1}{1!} \int_{-\infty}^{\infty} H_n H_0 e^{-\xi^2} d\xi$$

$$+ \cdots + \frac{s^n}{n!} \int_{-\infty}^{\infty} H_n H_n e^{-\xi^2} d\xi + \cdots = \sqrt{\pi} 2^n s^n.$$
(A.67)

Matching powers of s shows that we must have, for all terms where $m \neq n$,

$$\int_{-\infty}^{\infty} H_n H_m e^{-\xi^2} d\xi = 0 \quad (m \neq n),$$
(A.68)

and, from the term where $m = n$,

$$\int_{-\infty}^{\infty} H_n^2 e^{-\xi^2} d\xi = \sqrt{\pi} 2^n n!. \tag{A.69}$$

Equation (A.68) reproduces what we would have expected from the orthogonality relation of (A.57). Equation (A.69) gets us to the normalization of the harmonic-oscillator wavefunctions. Combining it with (A.58) gives

$$\frac{A_n^2}{\alpha} \left(\sqrt{\pi} 2^n n! \right) = 1 \Rightarrow A_n = \sqrt{\frac{\alpha}{\sqrt{\pi} 2^n n!}}, \tag{A.70}$$

as claimed in (A.29). *This completes the harmonic oscillator wavefunctions normalization.*

It was remarked at the beginning of this section that the generating function appearing in (A.56) can be used to prove some useful identities and recursion relationships for the Hermite polynomials. We now do this. The first thing we will prove is an identity given in the Summary in Chap. 5:

$$H_k(\xi) = (-1)^k e^{\xi^2} \frac{\partial^k}{\partial \xi^k} \left[e^{-\xi^2} \right].$$

Begin with the left side of (A.56), defining it as $G(n, \xi, t)$:

$$G(n, \xi, t) = \sum_{n=0}^{\infty} H_n \frac{t^n}{n!} = H_0 + H_1 \frac{t}{1!} + H_2 \frac{t^2}{2!} + \cdots H_j \frac{t^j}{j!} + \cdots. \tag{A.71}$$

Now imagine taking the partial derivative of G with respect to t. The first term of the series (H_0) will disappear; all of the remaining terms will get multiplied by their original power of t as those powers of t themselves decrease by one. The lowest-order term surviving after the first derivative will be just H_1, multiplied by no numbers or factors of t. Upon taking the second derivative, the H_1 term will vanish and the lowest-order surviving term will simply be H_2, again with no multiplicative numbers or factors of t. Extending this logic to taking k derivatives, we will be left with an expression of the form

$$\frac{\partial^k G}{\partial t^k} = H_k(\xi) + (\text{factor}) H_{k+1}(\xi) t + (\text{factor}) H_{k+2}(\xi) t^2 + \cdots, \tag{A.72}$$

where "factor" denotes purely numerical factors, different for each term.

Evaluating (A.72) at $t = 0$ then gives

$$\left(\frac{\partial^k G}{\partial t^k}\right)_{t=0} = H_k\left(\xi\right). \tag{A.73}$$

Now apply the same logic to the right side of (A.56), completing the square on the exponential similarly to what we did with (A.61):

$$G = e^{\xi^2} e^{-(t-\xi)^2}. \tag{A.74}$$

We want the k'th partial derivative of G with respect to t:

$$\frac{\partial^k G}{\partial t^k} = e^{\xi^2} \frac{\partial^k}{\partial t^k}\left[e^{-(t-\xi)^2}\right]. \tag{A.75}$$

Put $w = t - \xi$. Since we are dealing with a partial derivative, it is legitimate to write $\partial/\partial w = \partial/\partial t$, that is, to write

$$\frac{\partial^k G}{\partial t^k} = e^{\xi^2} \frac{\partial^k}{\partial w^k}\left[e^{-w^2}\right]. \tag{A.76}$$

Now evaluate this expression at $t = 0$. In this case $w = -\xi$ and we have

$$\left(\frac{\partial^k G}{\partial t^k}\right)_{t=0} = e^{\xi^2} \frac{\partial^k}{\partial (-\xi)^k}\left[e^{-\xi^2}\right] = (-1)^k e^{\xi^2} \frac{\partial^k}{\partial \xi^k}\left[e^{-\xi^2}\right]. \tag{A.77}$$

Combining this result with (A.73) gives a revised form of the generating function:

$$H_k\left(\xi\right) = (-1)^k e^{\xi^2} \frac{\partial^k}{\partial \xi^k}\left[e^{-\xi^2}\right], \tag{A.78}$$

justifying the expression given in the Summary in Chap. 5.

We now prove two recursion relations for the Hermite polynomials. Begin again with (A.56), suppressing the ξ-dependence of the H_n for brevity:

$$\sum_{n=0}^{\infty} H_n \frac{t^n}{n!} = e^{2\xi t - t^2}.$$

As above, we begin by taking the partial derivative of this with respect to t:

$$\sum_{n=0}^{\infty} H_n \frac{n\, t^{n-1}}{n!} = (-2t + 2\xi)\, e^{2\xi t - t^2}. \tag{A.79}$$

Circular as it seems, replace the exponential on the right side of (A.79) with (A.56):

$$\sum_{n=0}^{\infty} H_n \frac{n\, t^{n-1}}{n!} = (-2t + 2\xi) \sum_{n=0}^{\infty} H_n \frac{t^n}{n!}. \tag{A.80}$$

Now factor out the sum on the right side of this expression, carrying the factor of t in the $-2t$ term within the sum to make the power of t there be $n + 1$:

$$\sum_{n=0}^{\infty} H_n \frac{n\, t^{n-1}}{n!} = -2 \sum_{n=0}^{\infty} H_n \frac{t^{n+1}}{n!} + 2\xi \sum_{n=0}^{\infty} H_n \frac{t^n}{n!}. \tag{A.81}$$

Now, the sum of the left side of (A.81) could just as well start at $n = 1$ since the $n = 0$ term vanishes; further, we can put $n/n! = 1/(n - 1)!$. With this in mind, perform a change of index $j = n - 1$ similar to what we did in Sect. 5.3:

$$\sum_{n=0}^{\infty} H_n \frac{n t^{n-1}}{n!} = \sum_{n=1}^{\infty} H_n \frac{t^{n-1}}{(n - 1)!} = \sum_{j=0}^{\infty} H_{j+1} \frac{t^j}{j!} = \sum_{n=0}^{\infty} H_{n+1} \frac{t^n}{n!}. \tag{A.82}$$

Substitute (A.82) back into (A.81):

$$\sum_{n=0}^{\infty} H_{n+1} \frac{t^n}{n!} = -2 \sum_{n=0}^{\infty} H_n \frac{t^{n+1}}{n!} + 2\xi \sum_{n=0}^{\infty} H_n \frac{t^n}{n!}. \tag{A.83}$$

Look at the middle term in this expression. We can bring its power of t down to n by the following manipulation:

$$\sum_{n=0}^{\infty} H_n \frac{t^{n+1}}{n!} = \sum_{j=1}^{\infty} H_{j-1} \frac{t^j}{(j - 1)!} = \sum_{j=1}^{\infty} H_{j-1} \left(\frac{j}{j!} \right) t^j = \sum_{j=0}^{\infty} H_{j-1} \left(\frac{j}{j!} \right) t^j. \tag{A.84}$$

The first step in (A.84) is a change of index: $j = n + 1$. The second step is a restatement of $(j - 1)!$, and the last step (changing the starting value of j from one to zero) is allowable as this would introduce no extra terms because of the j in the numerator introduced in the second step. Resetting the index of summation to n and incorporating this result into (A.83) gives

$$\sum_{n=0}^{\infty} H_{n+1} \frac{t^n}{n!} = -2 \sum_{n=0}^{\infty} H_{n-1} \left(\frac{n}{n!} \right) t^n + 2\xi \sum_{n=0}^{\infty} H_n \frac{t^n}{n!}, \tag{A.85}$$

or, on rearranging,

$$\sum_{n=0}^{\infty} \left[H_{n+1} - 2\xi H_n + 2n H_{n-1} \right] \frac{t^n}{n!} = 0. \tag{A.86}$$

In analogy to (A.67) for powers of s, (A.86) is a polynomial in t; the only way it can hold for all values of t is if the term in square brackets is zero. This gives us a recursion relation between three successive Hermite polynomials,

$$H_{n+1} - 2\xi H_n + 2n H_{n-1} = 0, \tag{A.87}$$

as claimed in Sect. 5.4.

Our last proof involves a recursion relation between H_{n-1} and the derivative of H_n. Start again with (A.56),

$$\sum_{n=0}^{\infty} H_n \frac{t^n}{n!} = e^{2\xi t - t^2}.$$

Now take a partial derivative with respect to ξ:

$$\sum_{n=0}^{\infty} \left(\frac{dH_n}{d\xi}\right) \frac{t^n}{n!} = (2t) \, e^{2\xi t - t^2}. \tag{A.88}$$

Again invoke (A.56) for the exponential:

$$\sum_{n=0}^{\infty} \left(\frac{dH_n}{d\xi}\right) \frac{t^n}{n!} = (2t) \sum_{n=0}^{\infty} H_n \frac{t^n}{n!}. \tag{A.89}$$

On the right side of this, bring the factor of t that is outside the sum within the sum:

$$\sum_{n=0}^{\infty} \left(\frac{dH_n}{d\xi}\right) \frac{t^n}{n!} = 2 \sum_{n=0}^{\infty} H_n \frac{t^{n+1}}{n!}. \tag{A.90}$$

To get both sides of (A.90) to the same power of t, do a change of index $j = n + 1$ on the right-side sum:

$$2 \sum_{n=0}^{\infty} H_n \frac{t^{n+1}}{n!} = 2 \sum_{j=1}^{\infty} H_{j-1} \frac{t^j}{(j-1)!} = 2 \sum_{j=0}^{\infty} H_{j-1} \frac{j}{j!} t^j = 2 \sum_{n=0}^{\infty} H_{n-1} \frac{n}{n!} t^n. \tag{A.91}$$

This result, when back-substituted into (A.89), gives

$$\sum_{n=0}^{\infty} \left(\frac{dH_n}{d\xi}\right) \frac{t^n}{n!} = 2 \sum_{n=0}^{\infty} H_{n-1} \frac{n}{n!} t^n, \tag{A.92}$$

that is,

$$\sum_{n=0}^{\infty} \left(\frac{dH_n}{d\xi} - 2n H_{n-1}\right) \frac{t^n}{n!} = 0. \tag{A.93}$$

Again we have a polynomial in t which must vanish for every value of t:

$$\frac{dH_n}{d\xi} = 2n\,H_{n-1}. \qquad (A.94)$$

This result is useful in that in allows one to compute the value of the derivative of a Hermite polynomial at some value of ξ in terms of the next-lowest order polynomial evaluated at the value of ξ concerned.

A.3 Explicit Series Form for Associated Legendre Functions

From Chap. 6, the azimuthal solutions for the angular part of Schrödinger's equation for a central potential are

$$\Theta_{\ell,m}(\theta) = \sqrt{\frac{2\ell+1}{2}\frac{(\ell-m)!}{(\ell+m)!}}\,P_{\ell,m}(\cos\theta) \qquad (|m| \le \ell), \qquad (A.95)$$

where $P_{\ell,m}(\cos\theta)$ denotes an Associated Legendre function,

$$P_{\ell,m}(x) = \frac{1}{2^{\ell}\,\ell!}(1-x^2)^{m/2}\frac{d^{\ell+m}}{dx^{\ell+m}}(x^2-1)^{\ell}, \qquad (A.96)$$

where we have $x = \cos\theta$. This expression is valid for m positive, negative, or zero.

For computational purposes, it can be convenient to express $\Theta(\theta)$ directly as a series in terms of, say, $\cos\theta$. This can be achieved as follows.

The binomial theorem states that the sum of any two quantities raised to a power can be expressed as

$$(a+b)^p = \sum_{k=0}^{p}\binom{p}{k}a^{p-k}b^k, \qquad (A.97)$$

where the binomial coefficients $\binom{p}{k}$ are given by

$$\binom{p}{k} = \frac{p!}{k!(p-k)!}. \qquad (A.98)$$

Applying this to $(x^2-1)^{\ell}$ gives

$$(x^2-1)^{\ell} = \sum_{k=0}^{\ell}(-1)^k\binom{\ell}{k}(x^{2\ell-2k}). \qquad (A.99)$$

This expression is a polynomial in powers of x ranging from 0 to 2ℓ in steps of two, which, according to (A.96), we need to differentiate $(\ell + m)$ times with respect to x. That is, we desire

$$\frac{d^{\ell+m}}{dx^{\ell+m}}(x^2 - 1)^\ell = \sum_{k=0}^{\ell}(-1)^k \binom{\ell}{k} \frac{d^{\ell+m}}{dx^{\ell+m}}(x^{2\ell-2k}). \tag{A.100}$$

Since m runs from $-\ell$ to $+\ell$, the order of the derivative may be any integer from 0 to 2ℓ; consequently, any number of terms in the series, from all of them down to only the highest-order one, may potentially survive the differentiation. But, only those whose power $2(\ell - k)$ exceeds the highest order of the derivative, $(\ell + m)$, will survive the differentiation. Hence, we need only differentiate so long as $2(\ell - k) \geq (\ell + m)$, that is, so long as $k \leq (\ell - m)/2$. Since the derivative of a power can in general be written as

$$\frac{d^J}{dx^J}(x^N) = \left[\prod_{i=0}^{J-1}(N - i)\right]x^{N-J}, \tag{A.101}$$

we find, with $N = 2\ell - 2k$ and $J = \ell + m$,

$$\frac{d^{\ell+m}}{dx^{\ell+m}}(x^2 - 1)^\ell = \sum_{k=0}^{\left[\frac{\ell-m}{2}\right]}\left[(-1)^k \binom{\ell}{k} \prod_{i=o}^{\ell+m-1}(2\ell - 2k - i)\right]x^{\ell-m-2k}, \tag{A.102}$$

where $\left[\frac{\ell-m}{2}\right]$ designates the greatest integer less than or equal to $(\ell - m)/2$.

The term within the large square bracket can be simplified. The product is first compacted by writing it in a binomial-coefficient form:

$$\prod_{i=0}^{\ell+m-1}(2\ell - 2k - i) = [2\ell - 2k]\,[2\ell - 2k - 1]\,[2\ell - 2k - 2]$$
$$\dots[2\ell - 2k - (\ell + m - 1)]$$
$$= \frac{(2\ell - 2k)!}{[2\ell - 2k - (\ell + m - 1) - 1]!} = \frac{(2\ell - 2k)!}{[2\ell - 2k - (\ell + m)]!}. \tag{A.103}$$

Now, the definition of the binomial coefficients above can be rearranged to give

$$\binom{p}{k} = \frac{p!}{k!(p - k)!} \Rightarrow \frac{p!}{(p - k)!} = k!\binom{p}{k}, \tag{A.104}$$

which, when applied to (A.103) yields

$$\prod_{i=o}^{\ell+m-1}(2\ell - 2k - i) = \frac{(2\ell - 2k)!}{[2\ell - 2k - (\ell + m)]!} = (\ell + m)!\binom{2\ell - 2k}{\ell + m}, \tag{A.105}$$

leading to

$$\frac{d^{\ell+m}}{dx^{\ell+m}}(x^2-1)^\ell = \sum_{k=0}^{\left[\frac{\ell-m}{2}\right]}\left[(-1)^k\binom{\ell}{k}(\ell+m)!\binom{2\ell-2k}{\ell+m}x^{\ell-m-2k}\right]. \quad \text{(A.106)}$$

Combining this result with the definition of $\Theta(\theta)$ gives

$$\begin{aligned}
\Theta_{\ell,m}(\theta) &= \sqrt{\frac{2\ell+1}{2}\frac{(\ell-m)!}{(\ell+m)!}}\,P_{\ell,m}(\cos\theta) \\
&= \frac{1}{2^\ell\,\ell!}(1-x^2)^{m/2}\sqrt{\frac{2\ell+1}{2}\frac{(\ell-m)!}{(\ell+m)!}}\frac{d^{\ell+m}}{dx^{\ell+m}}(x^2-1)^\ell \\
&= \frac{1}{2^\ell\,\ell!}(1-x^2)^{m/2}\sqrt{\frac{2\ell+1}{2}\frac{(\ell-m)!}{(\ell+m)!}} \\
&\quad \sum_{k=0}^{\left[\frac{\ell-m}{2}\right]}\left[(-1)^k\binom{\ell}{k}(\ell+m)!\binom{2\ell-2k}{\ell+m}x^{\ell-m-2k}\right]
\end{aligned} \qquad \text{(A.107)}$$

or, with $x=\cos\theta$,

$$\begin{aligned}
\Theta(\theta) &= \frac{1}{2^\ell\ell!}\sqrt{\frac{2\ell+1}{2}(\ell+m)!\,(\ell-m)!}(\sin^m\theta) \\
&\quad \sum_{k=0}^{\left[\frac{\ell-m}{2}\right]}\left[(-1)^k\binom{\ell}{k}\binom{2\ell-2k}{\ell+m}(\cos^{\ell-m-2k}\theta)\right],
\end{aligned} \qquad \text{(A.108)}$$

the desired explicit form of $\Theta(\theta)$ as written in (6.90) and (6.91).

A.4 Proof That $Y_{\ell,-m}=(-1)^m Y_{\ell,m}^*$

This proof is adapted from that given in M. E. Rose, *Elementary Theory of Angular Momentum*, John Wiley & Sons, Inc., New York, 1957, Appendix III.

To prove this relationship, it is easiest to begin with the Associated Legendre functions $P_{\ell,m}$ of Sect. 6.5:

$$\Theta_{\ell,m}(\theta) = \sqrt{\frac{2\ell+1}{2}\frac{(\ell-m)!}{(\ell+m)!}}\,P_{\ell,m}(\cos\theta) \qquad (|m|\le\ell), \qquad \text{(A.109)}$$

where

$$P_{\ell,m} = \frac{1}{2^\ell \, \ell \,!}(1 - x^2)^{m/2} \frac{d^{\ell+m}}{dx^{\ell+m}}(x^2 - 1)^\ell. \tag{A.110}$$

If we replace m by $-m$, this becomes

$$P_{\ell,-m} = \frac{1}{2^\ell \, \ell \,!}(1 - x^2)^{-m/2} \frac{d^{\ell-m}}{dx^{\ell-m}}(x^2 - 1)^\ell, \tag{A.111}$$

where we have suppressed the x-dependence of $P_{\ell,m}$ for brevity.

Two results from calculus are central to this proof. The first is that the jth derivative of x^n can be written as

$$\frac{d^j}{dx^j}(x^n) = \left[\prod_{i=0}^{j-1}(n - i)\right] x^{n-j} = \frac{n!}{(n-j)!}x^{n-j}. \tag{A.112}$$

The second is a similar result, a way of writing the jth derivative of a product of two functions of x, $A(x)$ and $B(x)$:

$$\frac{d^j}{dx^j}(AB) = \sum_{n=0}^{j} \binom{j}{n} \left(\frac{d^{j-n}A}{dx^{j-n}}\right) \left(\frac{d^n B}{dx^n}\right), \tag{A.113}$$

where $\binom{j}{n}$ denotes a binomial coefficient. This is known as Leibniz's rule for differentiating a product.

Now write (A.111) as

$$P_{\ell,-m} = \frac{1}{2^\ell \, \ell \,!}(1 - x^2)^{-m/2} \frac{d^{\ell-m}}{dx^{\ell-m}}[(x + 1)^\ell (x - 1)^\ell],$$

and apply Liebniz's rule, taking $A = (x + 1)^\ell$ and $B = (x - 1)^\ell$. The result is

$$P_{\ell,-m} = \frac{1}{2^\ell \ell!}(1 - x^2)^{-m/2} \sum_{n=0}^{\ell-m} \frac{(\ell - m)!}{n! \, (\ell - m - n)!} \left[\frac{d^{\ell-m-n}(x + 1)^\ell}{dx^{\ell-m-n}}\right] \left[\frac{d^n(x - 1)^\ell}{dx^n}\right]. \tag{A.114}$$

Apply (A.112) to each of the derivatives appearing here. This yields

$$P_{\ell,-m} = \frac{(1 - x^2)^{-m/2}(\ell - m)!}{2^\ell \, \ell \,!} \sum_{n=0}^{\ell-m} \frac{\ell! \, \ell!(x + 1)^{m+n}(x - 1)^{\ell-n}}{n! \, (\ell - m - n)! \, (m + n)! \, (\ell - n)!}. \tag{A.115}$$

Now we make a curious manipulation of the product $(x + 1)^{m+n}(x - 1)^{\ell-n}$. First, break up the first term into $(x + 1)^m(x + 1)^n$. Then, multiply by a factor of $(x - 1)^m$ while dividing by the same factor, that is, write

$$(x+1)^{m+n}(x-1)^{\ell-n} = (x+1)^m(x+1)^n(x-1)^{\ell-n}\left[\frac{(x-1)^m}{(x-1)^m}\right]$$

$$= (x^2-1)^m(x+1)^n(x-1)^{\ell-n-m}.$$

The factor of $(x^2-1)^m$ is now extracted from within the sum in (A.115), and combined with that of $(1-x^2)^{-m/2}$ appearing outside the sum to give

$$P_{\ell,-m} = \frac{(-1)^m(1-x^2)^{m/2}(\ell-m)!}{2^\ell \, \ell!} \sum_{n=0}^{\ell-m} \frac{\ell! \, \ell!(x+1)^n(x-1)^{\ell-n-m}}{n! \, (\ell-m-n)! \, (m+n)! \, (\ell-n)!}.$$
(A.116)

The reason for this manipulation is that by concocting a factor of $(1-x^2)^{m/2}$ out front, we make the prefactor in (A.116) look like that which appears in the expression for $P_{\ell,m}$ (A.110).

Now, both multiply and divide (A.116) by a factor of $(\ell+m)!$, placing the multiplicative factor inside the sum:

$$P_{\ell,-m} = \frac{(-1)^m(1-x^2)^{m/2}(\ell-m)!}{2^\ell \, \ell! \, (\ell+m)!} \sum_{n=0}^{\ell-m} \frac{(\ell+m)! \, \ell! \, \ell!(x+1)^n(x-1)^{\ell-n-m}}{n! \, (\ell-m-n)! \, (m+n)! \, (\ell-n)!}.$$
(A.117)

Now examine the sum appearing in this expression in more detail, writing it as

$$\sum_{n=0}^{\ell-m} \frac{(\ell+m)!}{(m+n)! \, (\ell-n)!} \left[\frac{\ell! \, (x+1)^n}{n!}\right]\left[\frac{\ell! \, (x-1)^{\ell-n-m}}{(\ell-m-n)!}\right].$$
(A.118)

Look back to (A.112). With it, we can write the square bracketed terms in (A.118) as

$$\frac{\ell! \, (x+1)^n}{n!} = \frac{d^{\ell-n}(x+1)^\ell}{dx^{\ell-n}}$$

and

$$\frac{\ell! \, (x-1)^{\ell-n-m}}{(\ell-m-n)!} = \frac{d^{m+n}(x-1)^\ell}{dx^{m+n}}.$$

That is, the sum in (A.118) can be written as

$$\sum_{n=0}^{\ell-m} \frac{(\ell+m)!}{(m+n)! \, (\ell-n)!} \left[\frac{d^{\ell-n}(x+1)^\ell}{dx^{\ell-n}}\right]\left[\frac{d^{m+n}(x-1)^\ell}{dx^{m+n}}\right].$$
(A.119)

In this expression, make a change in the index of summation to a new index s defined as $s = n + m$. This takes the lower and upper limits of summation to $s = m$ and $s = \ell$, respectively:

$$\sum_{s=m}^{\ell} \frac{(\ell+m)!}{s!\,(\ell+m-s)!} \left[\frac{d^{\ell+m-s}(x+1)^{\ell}}{dx^{\ell+m-s}} \right] \left[\frac{d^{s}(x-1)^{\ell}}{dx^{s}} \right]. \tag{A.120}$$

Look carefully at the derivatives which appear here. As the index s advances from $s = m$ up to $s = \ell$, the order of the derivative in the first square bracket decreases from ℓ down to m, while, at the same time, that in the second square bracket increases from m up to ℓ. If either term were to be differentiated more than ℓ times, the result would be zero. (Why?) Consequently, for the first square bracket, if the index s were to start at a lower value than m, all derivatives of $(x+1)^{\ell}$ would be zero until s reached m; similarly, if the upper index of s were increased beyond ℓ, all derivates of $(x-1)^{\ell}$ would be zero for $s > \ell$. Hence, we can safely both reduce the lower limit of s to zero while increasing its upper limit to $\ell + m$ without introducing any more terms in the sum than already appear:

$$\sum_{s=0}^{\ell+m} \frac{(\ell+m)!}{s!\,(\ell+m-s)!} \left[\frac{d^{\ell+m-s}(x+1)^{\ell}}{dx^{\ell+m-s}} \right] \left[\frac{d^{s}(x-1)^{\ell}}{dx^{s}} \right]. \tag{A.121}$$

The value of this curious manipulation will become clear momentarily.

Now, circuitous as it may seem, apply Leibniz's rule to the product $(x + 1)^{\ell}(x - 1)^{\ell}$:

$$\frac{d^{\ell+m}}{dx^{\ell+m}} \left[(x+1)^{\ell}(x-1)^{\ell} \right] = \sum_{s=0}^{\ell+m} \binom{\ell+m}{s} \left[\frac{d^{\ell+m-s}(x+1)^{\ell}}{dx^{\ell+m-s}} \right] \left[\frac{d^{s}(x-1)^{\ell}}{dx^{s}} \right].$$

The sum here is precisely that appearing in (A.121); this is why the limits on the index s were modified between (A.120) and (A.121). Hence we can write (A.121) as

$$\sum_{s=0}^{\ell+m} \frac{(\ell+m)!}{s!\,(\ell+m-s)!} \left[\frac{d^{\ell+m-s}(x+1)^{\ell}}{dx^{\ell+m-s}} \right] \left[\frac{d^{s}(x-1)^{\ell}}{dx^{s}} \right]$$

$$= \frac{d^{\ell+m}}{dx^{\ell+m}} \left[(x+1)^{\ell}(x-1)^{\ell} \right].$$

This expression is an alternate version of the sum in (A.117). Back-substituting this into that expression and writing $(x + 1)^{\ell}(x - 1)^{\ell} = (x^2 - 1)^{\ell}$ gives

$$\begin{aligned} P_{\ell,-m} &= \frac{(-1)^{m}(1-x^2)^{m/2}(\ell-m)!}{2^{\ell}\,\ell!\,(\ell+m)!} \frac{d^{\ell+m}}{dx^{\ell+m}}(x^2-1)^{\ell} \\ &= (-1)^{m}\frac{(\ell-m)!}{(\ell+m)!}P_{\ell,m}. \end{aligned} \tag{A.122}$$

Now, look back to (A.109) and replace m with $-m$,

$$\Theta_{\ell,-m}(\theta) = \sqrt{\frac{2\ell+1}{2}\frac{(\ell+m)!}{(\ell-m)!}}\, P_{\ell,-m},$$

or, with (A.122),

$$\Theta_{\ell,-m}(\theta) = (-1)^m \sqrt{\frac{2\ell+1}{2}\frac{(\ell+m)!}{(\ell-m)!}}\left(\frac{(\ell-m)!}{(\ell+m)!}\right) P_{\ell,m}$$

$$= (-1)^m \sqrt{\frac{2\ell+1}{2}\frac{(\ell-m)!}{(\ell+m)!}}\, P_{\ell,m}.$$

Comparing this to (A.109) shows that we have

$$\Theta_{\ell,-m}(\theta) = (-1)^m \Theta_{\ell,m}(\theta).$$

Now, the overall spherical harmonic is defined as

$$Y_{\ell,m}(\theta,\phi) = \Theta_{\ell,m}(\theta)\,\Phi_m(\phi).$$

Since the azimuthal part of this satisfies (Sect. 6.5)

$$\Phi_{-m}(\phi) = \Phi_m^*(\phi),$$

then

$$Y_{\ell,-m} = (-1)^m Y_{\ell,m}^*,$$

as claimed.

A.5   Radial Nodes in Hydrogen Wavefunctions

From (7.63), (7.76), and the fact that the series for $F(\rho)$ terminates at an upper index of the $n-\ell-1$, the hydrogenic radial wavefunctions are

$$R_{n\ell}(r) = \frac{1}{r}(r/a_o)^{\ell+1}\exp(-r/na_o)\sum_{k=0}^{n-\ell-1} b_k\,(r/a_o)^k$$

$$= \left(\sum_{k=0}^{n-\ell-1}\frac{b_k}{a_o^{k+\ell+1}}\,r^{k+\ell}\right)\exp(-r/na_o).$$

When computing radial probability distributions $P(r)$, we have

$$P\,(r) = 4\pi\, r^2\, R^2_{n\,\ell}\,(r) = 4\,\pi\, r^2\, \left(\sum_{k=0}^{n-\ell-1} \frac{b_k}{a_o^{k+\ell+1}}\, r^{k+\ell}\right)^2 \exp\left(-2r/na_o\right).$$

The factor of r^2 outside the large bracket can be carried inside the bracket if we write it simply as r, since the large bracket is itself squared. Combining this factor of r with the factor of $r^{k+\ell}$ within the sum gives

$$P(r) = 4\pi \left(\sum_{k=0}^{n-\ell-1} \frac{b_k}{a_o^{k+\ell+1}}\, r^{k+\ell+1}\right)^2 \exp(-2r/na_o). \qquad (A.123)$$

Label the sum in (A.123) as $\Re(r)$; this represents a polynomial in r whose lowest and highest orders are, respectively, $(\ell + 1)$ (when $k = 0$) and n (when $k = n - \ell - 1$), that is,

$$\begin{aligned}
\Re(r) &= \left(\sum_{k=0}^{n-\ell-1} \frac{b_k}{a_o^{k+\ell+1}}\, r^{k+\ell+1}\right) \\
&= \left(\frac{b_o}{a_o^{\ell+1}}\, r^{\ell+1} + \frac{b_1}{a_o^{\ell+2}}\, r^{\ell+2} + \frac{b_2}{a_o^{\ell+3}}\, r^{\ell+3} + \cdots + \frac{b_{n-\ell-1}}{a_o^n}\, r^n\right),
\end{aligned}$$

or

$$\Re(r) = r^{\ell+1} \left(\frac{b_o}{a_o^{\ell+1}} + \frac{b_1}{a_o^{\ell+2}}\, r + \frac{b_2}{a_o^{\ell+3}}\, r^2 + \cdots + \frac{b_{n-\ell-1}}{a_o^n}\, r^{n-\ell-1}\right).$$

Fundamentally, $\Re(r)$ is a polynomial of order $(n - \ell - 1)$. Since any polynomial of order p has p zeros, $P(r)$ must have $(n - \ell - 1)$ nodes, as asserted in Chap. 7.

The proof that $r_{mp} = n^2 a_o$ when $\ell = n - 1$ proceeds similarly. In this case the upper limit of k in (A.123) becomes zero, and we have

$$P_{\ell=n-1}(r) = 4\pi \left(\sum_{k=0}^{0} \frac{b_k}{a_o^{k+n}}\, r^{k+n}\right)^2 \exp(-2r/na_o) = 4\pi \left(\frac{b_o}{a_o^n}\right)^2 r^{2n} \exp(-2r/na_o).$$

Differentiating gives

$$\frac{dP_{\ell=n-1}(r)}{dr} = 4\pi \left(\frac{b_o}{a_o^n}\right)^2 \left\{2nr^{2n-1} - \frac{2}{na_o}\, r^{2n}\right\} \exp(-2r/na_o).$$

Setting the derivative to zero to maximize $P_{\ell=n-1}(r)$ gives $r_{mp} = n^2 a_o$ as claimed.

Appendix B
Answers to Selected Odd-Numbered Problems

1.1 1.28 Å, 1.14 Å, 1.14 Å, 1.56 Å, 1.24 Å

1.3 $T = 300$: $5.48 \times 10^{14}\,\mathrm{m}^{-3}$; $T = 2.7$: $400\,\mathrm{m}^{-3}$

1.5 $v_n = e^2/2\varepsilon_o h n \approx c/(137\,n)$. Relativistic effects negligible.

1.7 $F_C = (\pi m_e^2 e^6/4\,\varepsilon_o^3 h^4\,n^4) = (8.233 \times 10^{-8}/n^4)$ N

1.9 $F_{elec}/F_{grav} \sim 2.3 \times 10^{39}$. Electricity dominates!

1.13 $\lambda_{He} - \lambda_H \sim -1.72$ Å

1.15 For transitions down to orbit n, $\lambda_{lowest} = R_H^{-1}\,(n^2)$;
$\lambda_{highest} = R_H^{-1}\left[\frac{n^2(n+1)^2}{2n+1}\right]$

1.17 $E_n = n^2 h^2/4\pi^2 m L^2$

1.19 0.39 Å. Use QM

1.21 $\lambda = \frac{h}{\sqrt{3mkT}}$; ~ 1.46 Å

1.23 $E_{photon}/E_{recoil} = 2mc\lambda/h \sim 7.6 \times 10^8$

2.3 One possibility: $-\frac{\hbar^2}{2m}\left(\frac{d\psi}{dx}\right)^2 + V\psi^2 = E\psi^2$

2.5 $A = 2k^{3/2}$

B. C. Reed, *Quantum Mechanics*, https://doi.org/10.1007/978-3-031-14020-4

2.7 $V(x) = E + \frac{\hbar^2}{2m}\left(4k^2x^2 - 2k\right)$

3.3 $v = nh/2mL$, $v > c$ for $n \sim 82$; $E \sim 253$ keV

3.5   0.0271

3.9   Four

3.11 $E_1 = 1.187$ eV, $E_2 = 4.667$ eV

3.13 $\xi_{true} = 0.9149$, $\xi_{approx} = 1.0398$

3.17 $T \sim 1.89 \times 10^{-70}$

3.21 $\ln P \approx -\frac{4A\sqrt{2mV_o}}{3\hbar}(1 - E/V_o)^{3/2}$; $P \sim 2.0 \times 10^{-4}$

3.23 $T \sim 10^{-10^{36}}$

3.25   Yes

4.1 $1.678\,\text{m}$, $2.816\,\text{m}^2$, $2.825\,\text{m}^2$, $0.095\,\text{m}$

4.3 $\sqrt{2\alpha}$, $1/2\alpha$, $-\hbar^2\alpha^2/2m$

4.5 $\Delta p \geq 1.76 \times 10^{-40}$ kg(m/s) $\Delta v = 80$ sec^{-1}

4.7 $A^2 = \frac{2a^3}{\pi}$; $V(x) = E + \frac{\hbar^2}{m}\frac{(x^2-a^2)}{(a^2+x^2)^2}$; $\Delta x\,\Delta p = \hbar$

4.9 $\Delta x\Delta p$
$$= \left[\Gamma^{-3/2}\left(\tfrac{n+1}{q}\right)\sqrt{(1 + nq - q)\Gamma\left(\tfrac{n-1}{q}\right)\left\{\Gamma\left(\tfrac{n+1}{q}\right)\Gamma\left(\tfrac{n+3}{q}\right) - \Gamma^2\left(\tfrac{n+2}{q}\right)\right\}}\right]\frac{\hbar}{2}$$
If $q > 0$, the only restriction is $n > 1$

4.11 $\Delta m \geq 9.54 \times 10^{-55}$ kg

4.17 $E > 9534$ MeV

4.19 $\langle E \rangle = (61/12)E_1$

4.23 $\beta^2 = 12mk/\hbar^2$;   not valid

5.1 [Force] $\equiv c^4/G \sim 1.21 \times 10^{44}$ N; independent of h

5.7 $P_{out} \sim 0.1116$

5.13 $\alpha = K_\alpha m^{1/(p+2)} A^{1/(p+2)} \hbar^{-2/(p+2)}$; $\varepsilon = K_\varepsilon m^{p/(p+2)} A^{-2/(p+2)} \hbar^{-2p/(p+2)}$.
Fails for $p = -2$. Taking $K_\alpha = 2^{1/(p+2)}$ and $K_\varepsilon = 2^{p/(p+2)}$ reduces
Schrödinger's equation to $d^2\psi/d\xi^2 + (\lambda - |\xi|^p)\,\psi = 0$.

6.1 Illustrative values: $E_{111} = 64.7$, $E_{113} = 148.1$, $E_{323} = 499.1$, all in eV

6.3 (b) and (e) are separable

6.9 $L^2 = -\hbar^2 \left\{ (y^2 + z^2)\frac{\partial}{\partial x^2} + (x^2 + z^2)\frac{\partial}{\partial y^2} + (x^2 + y^2)\frac{\partial}{\partial z^2} \right.$
$\left. -2x\frac{\partial}{\partial x} - 2y\frac{\partial}{\partial y} - 2z\frac{\partial}{\partial z} - 2xy\frac{\partial^2}{\partial x \partial y} - 2zx\frac{\partial^2}{\partial x \partial z} - 2yz\frac{\partial^2}{\partial y \partial z} \right\}$

$$L^2 = -\hbar^2 \left\{ z^2 \frac{\partial^2}{\partial \rho^2} + \rho^2 \frac{\partial^2}{\partial z^2} + \frac{(z^2+\rho^2)}{\rho^2} \frac{\partial^2}{\partial \phi^2} - 2z\frac{\partial}{\partial z} + \frac{(z^2-\rho^2)}{\rho} \frac{\partial}{\partial \rho} - 2\rho z \frac{\partial^2}{\partial z \partial \rho} \right\}$$

7.3 $P_{out} = \frac{\xi}{(\xi - \eta \cot \xi + \eta \xi \csc^2 \xi)}$; 0.1807

7.5 $V_0 = 36.6$ MeV

7.7 $V_0 > 940$ MeV

7.9 5.764×10^{-14} m

7.11 $\langle r \rangle = 5a_o$, $\langle 1/r \rangle = 1/4a_o$, $P(3.9a_o, 4.1a_o) = 0.039$

7.13 $r(50\%) = 1.337a_o$

8.5 1.1×10^{-6} eV

8.7 4.8768 mass units

9.1 $\phi(x) = x^2 = -\frac{2L^2}{\pi^3} \sum_{n=1}^{\infty} \frac{[(-1)^n (n^2\pi^2 - 2) + 2]}{n^3} \sin\left(\frac{n\pi x}{L}\right)$

9.3 $E_n \sim (3\alpha hn/8\sqrt{2m})^{2/3}$

9.5 $n \gg 2.42$

9.7 $\quad \frac{2\sqrt{2mV_o}}{\alpha} \int\limits_{0}^{\ln(1+\beta)} \sqrt{1+\beta-e^y}\,dy \sim nh;\ E=\beta V_0,\ y=\alpha x,\ \alpha \sim 0.01639\ \text{Å}$

9.9 $\quad E \sim \frac{n^2h^2}{8mL^2} + \alpha L^2\left(\frac{1}{3} - \frac{1}{2\pi^2 n^2}\right)$

9.11 $\quad E \sim E_n^0 + \beta a^2\left(\frac{1}{3} - \frac{1}{2\pi^2 n^2}\right);\ \psi \sim \psi_n^0(x) - \frac{64\beta ma^{7/2}}{\sqrt{2}\pi^{5/2}\hbar^2} \sum\limits_{k\neq n} \frac{(-1)^{k-n}\,nk}{(k+n)^3(k-n)^3}\, \frac{\sin(k\pi r/a)}{r}$

9.13 $\quad E \sim \hbar\omega/2 + 3\beta/4\alpha^4;\ \psi \sim A_0 e^{-\xi^2/2}\left\{1 - \frac{\beta}{\alpha^4\hbar\omega}\left[\frac{3}{8}H_2(\xi) + \frac{1}{64}H_4(\xi)\right]\right\}$

9.15 $\quad \beta < \pi m_e^2 e^6/120\varepsilon_o^3 h^4 = 1.713\ \text{eV/Å}$

9.17 $\quad \lambda = 1.519 \times 10^{-6}\ \text{m}$

9.19 $\quad E = \varepsilon\beta - \frac{A\sqrt{2\beta}}{\sqrt{\alpha+2\beta}}(\varepsilon = \hbar^2/2m);\ \beta \sim 0.428542\text{Å}^{-2};\ E \sim -2.193\,\text{eV}$

9.21 $\quad E \leq -m_e e^4/32\pi^2\varepsilon_o^2\hbar^2$ (hydrogen ground-state energy)

9.23 $\quad \beta_{\min} = \left[\frac{n\alpha(n+2)!}{2^{n+2}(\hbar^2/2m)}\right]^{1/(n+2)};\ E \leq \frac{1}{4}\left(\frac{\hbar^2}{2m}\right)^{n/(n+2)}\left(1+\frac{2}{n}\right)\{\alpha n(n+2)!\}^{2/(n+2)}$

10.1 $\quad E_3 \sim 8.6217\ \text{eV}$

10.3 $\quad E_1 \sim -3.800\ \text{eV}$

10.5 $\quad W \sim -0.962$ at $n \sim 0.85$ for $K = 2$

10.9 $\quad \sim 3.65, 6.38, 8.62$ eV for $\ell = 0$; identical to linear case

10.11 $\quad \ell = 0 : E \sim -79.041, -51.192, -25.448$ eV; $\ell = 1 :$
$\qquad E \sim -65.991, -38.346, -14.004$ eV

10.13 $\quad m = (0, 1, 2) : (2.7594, 5.5193, 8.2805)$ eV

11.5 $\quad \sqrt{P_{j\to n}} = \begin{cases} \frac{1}{\pi\sqrt{K}}\left\{\frac{\sin\left[(j-n/K)\pi\right]}{(j-n/K)} - \frac{\sin\left[(j+n/K)\pi\right]}{(j+n/K)}\right\} & (n \neq jK) \\[2ex] \frac{2}{\sqrt{K}}\left\{\frac{1}{2} - \frac{K\sin(2n\pi/K)}{4n\pi}\right\} & (n = jK) \end{cases}$

Appendix C
Integrals and Trigonometric Identities

The gamma function $\Gamma(n)$ appearing in (**38**), can be expressed in terms of the factorial function via $\Gamma(n+1) = n!$, and can be extended to non-integer arguments. Note that $\Gamma(1/2) = \sqrt{\pi}$. Integral (**29**) is a generalization of (**26**)–(**28**); similarly (**37**) and (**38**) generalize (**30**)–(**36**).

(**1**) $\quad \int \frac{dx}{a^2+x^2} = \frac{1}{a}\tan^{-1}\left(\frac{x}{a}\right)$

(**2**) $\quad \int \frac{dx}{(a^2+x^2)^n} = \frac{1}{2a^2(n-1)}\left[\frac{x}{(a^2+x^2)^{n-1}} + (2n-3)\int \frac{dx}{(a^2+x^2)^{n-1}}\right]$

(**3**) $\quad \int \frac{x^2 dx}{(a^2+x^2)} = x - a^2\int \frac{dx}{a^2+x^2}$

(**4**) $\quad \int \frac{x^2 dx}{(a^2+x^2)^{n+1}} = -\frac{x}{2n(a^2+x^2)^n} + \frac{1}{2n}\int \frac{dx}{(a^2+x^2)^n}$

(**5**) $\quad \int \sqrt{a+bx}\,dx = \frac{2}{3b}\sqrt{(a+bx)^3}$

(**6**) $\quad \int \sqrt{x^2 \pm a^2}\,dx = \frac{1}{2}\left[x\sqrt{x^2 \pm a^2} \pm a^2\ln\left(x + \sqrt{x^2 \pm a^2}\right)\right]$

(**7**) $\quad \int \sqrt{a^2 - x^2}\,dx = \frac{1}{2}\left[x\sqrt{a^2 - x^2} + a^2\text{Sin}^{-1}\left(\frac{x}{|a|}\right)\right]$

(**8**) $\quad \int \sin^2(ax)dx = \frac{x}{2} - \frac{1}{4a}\sin(2ax)$

(**9**) $\quad \int \cos^2(ax)dx = \frac{x}{2} + \frac{1}{4a}\sin(2ax)$

(**10**) $\quad \int \sin(mx)\sin(nx)dx = \frac{\sin[(m-n)x]}{2(m-n)} - \frac{\sin[(m+n)x]}{2(m+n)} \quad (m^2 \neq n^2)$

© The Editor(s) (if applicable) and The Author(s), under exclusive license to Springer Nature Switzerland AG 2022
B. C. Reed, *Quantum Mechanics*, https://doi.org/10.1007/978-3-031-14020-4

(11) $\int \cos(mx)\cos(nx)dx = \frac{\sin[(m-n)x]}{2(m-n)} + \frac{\sin[(m+n)x]}{2(m+n)}$ $(m^2 \neq n^2)$

(12) $\int \sin(mx)\cos(nx)dx = -\frac{\cos[(m-n)x]}{2(m-n)} - \frac{\cos[(m+n)x]}{2(m+n)}$ $(m^2 \neq n^2)$

(13) $\int \sin(ax)\cos(ax)dx = \frac{1}{2a}\sin^2(ax)$

(14) $\int \sin(ax)\cos^m(ax)dx = -\frac{\cos^{m+1}(ax)}{(m+1)a}$

(15) $\int \sin^m(ax)\cos(ax)dx = \frac{\sin^{m+1}(ax)}{(m+1)a}$

(16) $\int x\sin(ax)dx = \frac{1}{a^2}\sin(ax) - \frac{x}{a}\cos(ax)$

(17) $\int x\cos(ax)dx = \frac{1}{a^2}\cos(ax) + \frac{x}{a}\sin(ax)$

(18) $\int x^2\sin(ax)dx = \frac{2x}{a^2}\sin(ax) - \left(\frac{a^2x^2-2}{a^3}\right)\cos(ax)$

(19) $\int x^2\cos(ax)dx = \frac{2x\cos(ax)}{a^2} + \left(\frac{a^2x^2-2}{a^3}\right)\sin(ax)$

(20) $\int x\sin^2(ax)dx = \frac{x^2}{4} - \frac{x\sin(2ax)}{4a} - \frac{\cos(2ax)}{8a^2}$

(21) $\int x\cos^2(ax)dx = \frac{x^2}{4} + \frac{x\sin(2ax)}{4a} + \frac{\cos(2ax)}{8a^2}$

(22) $\int x^2\sin^2(ax)dx = \frac{x^3}{6} - \left(\frac{x^2}{4a} - \frac{1}{8a^3}\right)\sin(2ax) - \frac{x\cos(2ax)}{4a^2}$

(23) $\int x^2\cos^2(ax)dx = \frac{x^3}{6} + \left(\frac{x^2}{4a} - \frac{1}{8a^3}\right)\sin(2ax) + \frac{x\cos(2ax)}{4a^2}$

(24) $\int xe^{ax}dx = \frac{e^{ax}}{a^2}(ax - 1)$

(25) $\int x^m e^{ax}dx = \frac{x^m e^{ax}}{a} - \frac{m}{a}\int x^{m-1}e^{ax}dx$

(26) $\int\limits_0^\infty xe^{-ax}dx = \frac{1}{a^2}$

(27) $\int\limits_0^\infty x^2 e^{-ax}dx = \frac{2}{a^3}$

(28) $\int\limits_0^\infty x^3 e^{-ax} dx = \frac{6}{a^4}$

(29) $\int\limits_0^\infty x^n e^{-ax} dx = \frac{\Gamma(n+1)}{a^{n+1}}$ $(a > 0; n > -1)$

(30) $\int\limits_0^\infty e^{-ax^2} dx = \frac{\sqrt{\pi}}{2\sqrt{a}}$

(31) $\int\limits_0^\infty x e^{-ax^2} dx = \frac{1}{2a}$

(32) $\int\limits_0^\infty x^2 e^{-ax^2} dx = \frac{\sqrt{\pi}}{4a^{3/2}}$

(33) $\int\limits_0^\infty x^3 e^{-ax^2} dx = \frac{1}{2a^2}$

(34) $\int\limits_0^\infty x^4 e^{-ax^2} dx = \frac{3\sqrt{\pi}}{8a^{5/2}}$

(35) $\int\limits_0^\infty x^5 e^{-ax^2} dx = \frac{1}{a^3}$

(36) $\int\limits_0^\infty x^{2n} e^{-ax^2} dx = \frac{1 \cdot 3 \cdot 5 \cdots (2n-1)}{2^{n+1} a^n} \sqrt{\frac{\pi}{a}}$

(37) $\int\limits_0^\infty x^{2n+1} e^{-ax^2} dx = \frac{n!}{2a^{n+1}}$ $(a > 0)$

(38) $\int\limits_0^\infty x^n e^{-ax^p} dx = \frac{\Gamma(k)}{pa^k}$ $[n > -1, p > 0, a > 0, k = (n+1)/p]$

(39) $\sin \alpha \sin \beta = \frac{1}{2} \cos (\alpha - \beta) - \frac{1}{2} \cos (\alpha + \beta)$

(40) $\cos \alpha \cos \beta = \frac{1}{2} \cos (\alpha - \beta) + \frac{1}{2} \cos (\alpha + \beta)$

(41) $\sin \alpha \cos \beta = \frac{1}{2} \sin (\alpha + \beta) + \frac{1}{2} \sin (\alpha - \beta)$

(42) $\cos \alpha \sin \beta = \frac{1}{2} \sin (\alpha + \beta) - \frac{1}{2} \sin (\alpha - \beta)$

Appendix D
Physical Constants

The National Institute of Standards and Technology (NIST) maintains an extensive listing of physical constants at physics.nist.gov/constants. The values here are taken from that list, truncated to four decimal places.

Quantity	Symbol	Value	Unit
Speed of light in vacuum	c	2.9979×10^8	m s^{-1}
Planck's constant	h	6.6261×10^{-34}	J-s
Planck's constant	$\hbar$	1.0546×10^{-34}	J-s
Gravitational constant	G	6.6743×10^{-11}	$\text{m}^3 \text{ kg}^{-1} \text{ s}^{-2}$
Elementary charge	e	1.6022×10^{-19}	C
Electron mass	m_e	9.1094×10^{-31}	kg
Proton mass	m_p	1.6726×10^{-27}	kg
Neutron mass	m_n	1.6749×10^{-27}	kg
Atomic mass unit	u	1.6605×10^{-27}	kg
Bohr radius	a_o	5.2918×10^{-11}	m
Rydberg constant	R_∞	1.0974×10^7	m^{-1}
Avogadro's number	N_A	6.0221×10^{23}	mol^{-1}
Boltzmann constant	k	1.3806×10^{-23}	J K^{-1}
Permeability constant	μ_o	$4\pi \times 10^{-7}$	$\text{T m amp}^{-1} = \text{kg m C}^{-2}$
Permittivity constant	ε_o	8.8542×10^{-12}	$\text{C}^2 \text{ N}^{-1} \text{ m}^{-2} = \text{C}^2 \text{ s}^2 \text{ kg}^{-1} \text{ m}^{-3}$

© The Editor(s) (if applicable) and The Author(s), under exclusive license to Springer Nature Switzerland AG 2022
B. C. Reed, *Quantum Mechanics*, https://doi.org/10.1007/978-3-031-14020-4

Index

© The Editor(s) (if applicable) and The Author(s), under exclusive license to Springer Nature Switzerland AG 2022
B. C. Reed, *Quantum Mechanics*, https://doi.org/10.1007/978-3-031-14020-4

Printed in the United States
by Baker & Taylor Publisher Services